# Lecture Notes in Physics

W0231925

Springer-Verlag Berlin Heidelberg GmbH

# The Editorial Policy for Proceedings

The series Lecture Notes in Physics reports new developments in physical research and teaching – quickly, informally, and at a high level. The proceedings to be considered for publication in this series should be limited to only a few areas of research, and these should be closely related to each other. The contributions should be of a high standard and should avoid lengthy redraftings of papers already published or about to be published elsewhere. As a whole, the proceedings should aim for a balanced presentation of the theme of the conference including a description of the techniques used and enough motivation for a broad readership. It should not be assumed that the published proceedings must reflect the conference in its entirety. (A listing or abstracts of papers presented at the meeting but not included in the proceedings could be added as an appendix.)

When applying for publication in the series Lecture Notes in Physics the volume's editor(s) should submit sufficient material to enable the series editors and their referees to make a fairly accurate evaluation (e.g. a complete list of speakers and titles of papers to be presented and abstracts). If, based on this information, the proceedings are (tentatively) accepted, the volume's editor(s), whose name(s) will appear on the title pages, should select the papers suitable for publication and have them refereed (as for a journal) when appropriate. As a rule discussions will not be accepted. The series editors and Springer-Verlag will normally not interfere with the detailed editing except in fairly obvious cases or on technical matters.

Final acceptance is expressed by the series editor in charge, in consultation with Springer-Verlag only after receiving the complete manuscript. It might help to send a copy of the authors' manuscripts in advance to the editor in charge to discuss possible revisions with him. As a general rule, the series editor will confirm his tentative acceptance if the final manuscript corresponds to the original concept discussed, if the quality of the contribution meets the requirements of the series, and if the final size of the manuscript does not greatly exceed the number of pages originally agreed upon. The manuscript should be forwarded to Springer-Verlag shortly after the meeting. In cases of extreme delay (more than six months after the conference) the series editors will check once more the timeliness of the papers. Therefore, the volume's editor(s) should establish strict deadlines, or collect the articles during the conference and have them revised on the spot. If a delay is unavoidable, one should encourage the authors to update their contributions if appropriate. The editors of proceedings are strongly advised to inform contributors about these points at an early stage.

The final manuscript should contain a table of contents and an informative introduction accessible also to readers not particularly familiar with the topic of the conference. The contributions should be in English. The volume's editor(s) should check the contributions for the correct use of language. At Springer-Verlag only the prefaces will be checked by a copy-editor for language and style. Grave linguistic or technical shortcomings may lead to the rejection of contributions by the series editors. A conference report should not exceed a total of 500 pages. Keeping the size within this bound should be achieved by a stricter selection of articles and not by imposing an upper limit to the length of the individual papers. Editors receive jointly 30 complimentary copies of their book. They are entitled to purchase further copies of their book at a reduced rate. As a rule no reprints of individual contributions can be supplied. No royalty is paid on Lecture Notes in Physics volumes. Commitment to publish is made by letter of interest rather than by signing a formal contract. Springer-Verlag secures the copyright for each volume.

# The Production Process

The books are hardbound, and the publisher will select quality paper appropriate to the needs of the author(s). Publication time is about ten weeks. More than twenty years of experience guarantee authors the best possible service. To reach the goal of rapid publication at a low price the technique of photographic reproduction from a camera-ready manuscript was chosen. This process shifts the main responsibility for the technical quality considerably from the publisher to the authors. We therefore urge all authors and editors of proceedings to observe very carefully the essentials for the preparation of camera-ready manuscripts, which we will supply on request. This applies especially to the quality of figures and halftones submitted for publication. In addition, it might be useful to look at some of the volumes already published. As a special service, we offer free of charge LATEX and TEX macro packages to format the text according to Springer-Verlag's quality requirements. We strongly recommend that you make use of this offer, since the result will be a book of considerably improved technical quality. To avoid mistakes and time-consuming correspondence during the production period the conference editors should request special instructions from the publisher well before the beginning of the conference. Manuscripts not meeting the technical standard of the series will have to be returned for improvement.

For further information please contact Springer-Verlag, Physics Editorial Department II, Tiergartenstrasse 17, D-69121 Heidelberg, Germany

Annie Steinchen (Ed.)

# Dynamics
# of Multiphase Flows
# Across Interfaces

Springer

Editor

Annie Steinchen
Faculté des Sciences de St Jérôme
Laboratoire de Thermodynamique
Bd Escadrille Normandie Niemen
F-13397 Marseille Cédex 20, France

Cataloging-in-Publication Data applied for.

Die Deutsche Bibliothek - CIP-Einheitsaufnahme

**Dynamics of multiphase flows across interfaces** / Annie
Steinchen (ed.). - Berlin ; Heidelberg ; New York ; Barcelona ;
Budapest ; Hong Kong ; London ; Milan ; Paris ; Santa Clara ;
Singapore ; Tokyo : Springer, 1996
   (Lecture notes in physics ; Vol. 467)
   ISBN 978-3-662-14082-6
NE: Steinchen, Annie [Hrsg.]; GT

ISBN 978-3-662-14082-6          ISBN 978-3-540-49620-5 (eBook)
DOI 10.1007/978-3-540-49620-5

© Springer-Verlag Berlin Heidelberg 1996
Originally published by Springer-Verlag Berlin Heidelberg New York in 1996
Softcover reprint of the hardcover 1st edition 1996

Typesetting: Camera-ready by the authors
Cover design: Springer-Verlag Design & Production
SPIN: 10520036       55/3142-543210 - Printed on acid-free paper

# Preface

This book is the result of two meetings of the European network "Human Capital and Mobility" devoted to "Dynamics of Multiphase Flows across Interfaces" The first meeting (Part I) was held in Bellevue (near Paris) on October 14th-15th, 1994, and was organized by R. Prud'homme, head of the subgroup "Fundamental equations for surfaces" The second meeting (Part II), organized by the scientific editor of this volume, included all the themes studied by the network and was held at Aussois on January 6th-8th, 1995.

The participants were scientists of seven countries of the European Union: Belgium, France, Italy, Germany, Great Britain, Nothern Ireland, and Spain.

The role of interfaces in the exchanges between phases has received growing recognition in the last thirty years. However, several questions of matter on the fundamental point of view theoretical and experimental.remain open. The aim of the network is to clarify the issues under four themes, chosen as subjects of investigation:

1) fundamental constitutive equations for surfaces,
2) surface instabilities,
3) boiling, evaporation, condensation,
4) surfaces in external fields.

One of the most difficult problems in actual multiphase systems is the attempt to reconcile the treatment of ideal surfaces of zero thickness with the reality of the physical surface as region of space containing matter and energy. Due to the overlapping of the length scales describing the interfacial region, tools of both *microscopic and macroscopic* approaches (statistical mechanics, rational mechanics, thermodynamics and hydrodynamics) are needed. The differences between macroscopic and microscopic approaches cannot be ignored. Combining both approaches avoids severe contradictions and provides new insights into what enables experimental observations of important technological interest to be modelled successfully.

The current studies of static and dynamic fluid interfaces involve various approaches. At present a combination of various methods and tools is required to describe the complexity of the chemical and physical behaviour of fluid surfaces. The participating laboratories offer experience in different complementary fields. The common denominator for all the contributions is the simultaneous use of concepts from surface chemistry and physics and from hydrodynamics, where external force fields can be introduced. Theoretical and experimental work is equally represented and even united in many of the papers. My hope is that this volume should be a reference document for physicists, physico-chemists, and chemical engineers interested by the role of interfaces in the transfer processes in multiphase systems.

I wish here to acknowledge the European Union for having supported our research programme. Several contributors are also indebted to their national space agencies as well as to the European Space Agency for sponsoring the part of the project concerned with phase transfer in the absence of gravity.

Marseille, December 1995                                          A. Steinchen

# CONTENTS

## PART I

### Basic Constitutive Relations for Surfaces

## PART II

### 1 - Fundamental Equations

### 2 - Surface Instabilities

# LIST OF PARTICIPANTS TO THE BELLEVUE MEETING

Gatignol Renée
Modélisation en Mécanique URA 229
Université de Paris 6 case 162
4, Place Jussieu 75252 Paris Cédex 05

Gouin Henri
Modélisation en Mécanique
Fac.des Sciences St Jérôme
IUSTI Bd Escadrille Normandie Niemen
13397 Marseille Cédex 20-France

Jamet Nicolas
Modélisation en Mécanique URA 229
Université de Paris 6 case 162
4, Place Jussieu 75252 Paris Cédex 05

Lebon Georgy
Mécanique-Thermod- irréversible
Université de Liège-B5 Sart Tilman
4000 Liège - Belgium

Liggieri Libero
CNR - IFCAM
6, via de Marini
16149  Genova -Italy

Padday John
Nether Crutches
Jordans Bucks HP 9 2TA - United Kingdom

Prud'homme Roger
Modélisation en Mécanique URA 229
Université de Paris 6 case 162
4, Place Jussieu 75252 Paris Cédex 05

Sanfeld Albert
LISA Université Paris 7, Tour 44 ,3ème étage
2, Place Jussieu
75251 Paris Cédex 05 - France

Seppecher Pierre
Analyse non linéaire appliquée
Université de Toulon
Av. de l'Université, 83130 Lagarde - France

Steinchen Annie
Fac.des Sciences St Jérôme
IUSTI Bd Escadrille Normandie Niemen
13397 Marseille Cédex 20-France

Velarde Manuel Garcia
Instituto Pluridisciplinar
Iniversidad de Complutense
Paseo Juan XXIII , 28040 Madrid - Spain

# LIST OF PARTICIPANTS TO THE AUSSOIS
# WORKSHOP

Arlabos Patricia
Fac.des Sciences St Jérôme
IUSTI Bd Escadrille Normandie Niemen
13397 Marseille Cédex 20-France

Bois André
Laboratoire de Physique des liquides
Université de Provence
1, Place Victor Hugo
13331 Marseille - France

Casses Patrick
CNRS Labo. d' Aérothermique
4 ter Route des Gardes
92190 Meudon-France

Cerisier Pierre
Fac.des Sciences St Jérôme
IUSTI Bd Escadrille Normandie Niemen
13397 Marseille Cédex 20-France

Chacha Mama
Fac.des Sciences St Jérôme
IUSTI Bd Escadrille Normandie Niemen
13397 Marseille Cédex 20-France

Cuadros Francisco
Dept. de Fisica
Universidad de Extremadura
Av. de Elvas
06071 Badajoz - Spain

Di Marco P.
Dept. di Energetica
Universita degli Studi di Pisa
Via Diotisalvi, 2
56126 Pisa - Italia

Earnshaw John
Deptmt of Pure and Applied Physics
Queen's University Belfast
Belfast BT7 1NN - Northern Ireland

Grassi Walter
Dept. di Energetica
Universita degli Studi di Pisa
Via Diotisalvi, 2
56126 Pisa - Italia

Lebon Georgy
Mécanique-Thermod- irréversible
Université de Liège-B5 Sart Tilman
4000 Liège - Belgium

Liggieri Libero
CNR - IFCAM
6, via de Marini
16149  Genova -Italy

Lin Michel
Fac.des Sciences St Jérôme
IUSTI Bd Escadrille Normandie Niemen
13397 Marseille Cédex 20-France

Padday John
Nether Crutches
Jordans Bucks HP 9 2TA - United
Kingdom

Passerone Alberto
CNR-ICFAM
6,Via de Marini
16149 Genova -Italy

Pétré Georges.
ULB Chimie Phys. E.P. C.P 165
50, Av F.D. Roosevelt
1050 Bruxelles -Belgium

Picker George
LATTUM
Technische Universität München
21 Arcisstrasse 8000 München 2 -
Deutschland

Prud'homme Roger
Modélisation en Mécanique URA 229
Université de Paris 6 case 162
4, Place Jussieu 75252 Paris Cédex 05

Sanfeld Albert
LISA Université Paris 7, Tour 44 ,3ème étage
2, Place Jussieu
75251 Paris Cédex 05 - France

Sefiane Khellil
Thermodynamique Université d'Aix-Marseille 3
Faculté des Sciences de St Jérôme
Bd Escadrille Normandie Niemen
13397 Marseille Cédex 20-France

Simon Blaise
Fac.des Sciences St Jérôme
IUSTI Bd Escadrille Normandie Niemen
13397 Marseille Cédex 20-France

Steinchen Annie
Fac.des Sciences St Jérôme
IUSTI Bd Escadrille Normandie Niemen
13397 Marseille Cédex 20-France

Straub Johannes
LATTUM
Technische Universität München
21 Arcisstrasse 8000 München 2 - Deutschland

Winter Johannes
LATTUM
Technische Universität München
21 Arcisstrasse 8000 München 2 - Deutschland

# BALANCE EQUATIONS
# FOR FLUID CURVILINEAR MEDIA

Roger Prud'homme

Laboratoire de Modélisation en Mécanique, Université Pierre et Marie Curie &
C.N.R.S.,
Case 162, 4, place Jussieu, F75252 Paris Cedex 05, France

## Basic equations for fluid curvilinear media

Examples of fluid lines are numerous. Common line between capillary surfaces, some fluid flows along moving and deformable pipes or nozzles, free continuous fluid jets, fluid bridges between two attached drops can sometimes be considered as curvilinear fluid media [1, 2, 3].

If we assume that these media verify 3D continuous hypothesis at a small scale, which we will call microscopic (but not molecular) scale, 1D balance equations have to be deduced from 3D balance equations.

The 3D balance equations must be written in an appropriate form permitting easily the transformation into 1D relation. To do this, we have to distinguish firstly the direction $(/\,/)$ which is tangent to the line from the plane $(\perp)$ which is normal to this line at a current point.

The classical balance equation for the property $F$ is :

$$\frac{d_v F}{dt} + \Phi_{VF} = P_F \tag{1}$$

where $d_v F\,/\,dt$ is the time derivative of the quantity $F$ which is contained in the control volume $(D)$ moving with the local velocity $\vec{V}$, $\Phi_{VF}$ is the flux or $F$ through the moving surface $(\partial D)$ of the volume $(D)$ and $P_F$ is the source term.

Using local variables, or excess local quantities, we write :

$$\frac{d_v}{dt} \int_D \rho_F \, d\vartheta + \int_{\partial D} \vec{J}_{VF} \cdot \vec{n} dS = \int_D p_F \, d\vartheta \tag{2}$$

where $\rho_F$ is the $F$ density (or excess density) per unit volume, $\vec{J}_{VF}$ the local flux density corresponding to the velocity $\vec{V}$ and $p_F$ the production of $F$ per unit volume. The choice between absolute and excess quantities to use in the equations

depends on the problem to be solved where some conditions must often be verified to ensure for instance the convergence of integrals of asymptotic analysis.
The classical local form is deduced from (2) :

$$\frac{d_v\rho_F}{dt} + \rho_F \vec{\nabla}.\vec{V} + \vec{\nabla}.\vec{J}_{VF} = p_F \tag{3}$$

If we choose the velocity $\vec{V}$ and the derivation operators in an appropriate manner, equation (3) can be written in the form :

$$\frac{d_v\rho_F}{dt} + \rho_F \vec{\nabla}_{//}.\vec{V} + \vec{\nabla}_{\perp}.\vec{J}_{VF} + \vec{\nabla}_{//}.\vec{J}_{VF} = p_F \tag{4}$$

To obtain 1D balance equations along the physical fluid lines, it is useful to use particular curvilinear coordinates. It is generally possible to choose orthogonal curvilinear coordinates in such a way that the mean line is a particular line of the system given by fixed values of coordinates $x_1$ and $x_2$. Evidently, such a coordinate system is a moving one [4, 5].
We remind hereafter some properties of the transformation between cartesian orthogonal coordinates and such a system. We introduce $\vec{x} = \{x, y, z\}$ and $t$ respectively for the position and the time in the cartesian system and $\vec{\xi} = \{x_1, x_2, x_3\}$ and $t$ for the position and time in the curvilinear system and we have :

$$\vec{x} = \vec{x}(\vec{\xi}, t) \tag{5}$$

At a given value of $t$ we can define :

$$\vec{h}_i = \frac{\partial \vec{x}}{\partial x_i} \quad , \quad |h_i| = h_i \quad \text{and} \quad \vec{h}_i = h_i \, \vec{e}_i \tag{6}$$

where $\{\vec{e}_1, \vec{e}_2, \vec{e}_3\}$ form a direct orthogonal basis.
We can also define $\chi_i$ :

$$d\chi_i = h_i \, dx_i \tag{7}$$

So, for a curve $(C_i)$ obtained by fixing the values of

$x_j$ and $x_k (j \neq k \neq i = 1,2,3)$, the curvature vector $\vec{C}_i$ is the following :

$$\vec{C}_i = \frac{\partial \vec{e}_i}{\partial \chi_i} = \frac{1}{h_i} \frac{\partial \vec{e}_i}{\partial x_i} = -\frac{1}{h_i} \left( \frac{h_{i,j}}{h_j} \vec{e}_j + \frac{h_{i,k}}{h_k} \vec{e}_k \right) \tag{8}$$

and for a surface $(S_i)$, obtained for a given value of $\chi_i$ the mean curvature $1/R_i$ is the following :

$$1/R_i = \frac{1}{h_i} \left( \frac{h_{j,i}}{h_j} + \frac{h_{k,i}}{h_k} \right) . \tag{9}$$

For a given velocity field $\vec{V}(\vec{x},t)$, written in curvilinear coordinates as $\vec{V}(\vec{\xi},t)$, we can define : the line stretch of curve $(C_3)$ as :

$$\frac{1}{d\chi_3} \frac{d(d\chi_3)}{dt} = \frac{1}{h_3} \left( V_{3,3} + V_1 \frac{h_{3,1}}{h_1} + V_2 \frac{h_{3,2}}{h_2} \right) , \tag{10}$$

and the same for $(C_1)$ and $(C_2)$, the surface stretch of surface $(S_3)$ as :

$$\frac{1}{dS_3} \frac{d(dS_3)}{dt} = \frac{1}{d\chi_1} \frac{d(d\chi_1)}{dt} + \frac{1}{d\chi_2} \frac{d(d\chi_2)}{dt} \tag{11}$$

and the volume expansion :

$$\frac{1}{d\vartheta} \frac{d(d\vartheta)}{dt} = \sum_{i=1}^{3} \frac{1}{d\chi_i} \frac{d(d\chi_i)}{dt} = \vec{\nabla}.\vec{V} \tag{12}$$

If we suppose that :

- the velocity $\vec{V}$ and in particular its component $\vec{V}_\perp$ normal to the line, varies very slightly through the thickness of this line that is that along the surface $S_3$ : $V_{1,1} = V_{2,2} \cong 0$,

- the curvature of $(S_3)$ is negligible : $1/R_3 << 1/L$, where $L$ is a reference hydrodynamic length,

- the curves $(C_1)$ and $(C_2)$ have also small curvatures $\vec{e}_2.\vec{C}_1$ and $\vec{e}_1.\vec{C}_2$, then, we can put :

$$\rho_F^\ell = \int_{S_3} \rho_F dS \,, \quad \vec{J}_{VF}^\ell = \int_{S_3} \vec{J}_{VF} dS \,, \quad \rho_F^\ell = \int_{S_3} p_F dS \,. \tag{13}$$

and we obtain the line balance equation of the property $F$ :

$$\frac{d_v \rho_F^\ell}{dt} + \rho_F^\ell \, \vec{\nabla}_{//}.\vec{V} + \varphi_{VF}^\ell + \vec{\nabla}_{//}.\vec{J}_{VF}^\ell = p_F^\ell \tag{14}$$

with :

$$\begin{cases} \varphi_{VF}^\ell = \int_S \vec{\nabla}_\perp.\vec{J}_{VF}\, dS = \int_C \vec{J}_{VF\perp}.\vec{n}dc \quad \text{(peripheric term, see Fig.1)} \\[2mm] \vec{\nabla}_{//}.\vec{J}_{VF}^\ell = \dfrac{\partial \vec{J}_{VF//}^\ell}{\partial s} + \vec{J}_{VF\perp}^\ell.\vec{C}_3 \,, \quad ds = d\chi_3 = h_3 dx_3 \\[2mm] \vec{\nabla}_{//}.\vec{V} = \dfrac{\partial V_{//}}{\partial s} + \vec{V}_\perp.\vec{C}_3 \,, \quad \text{stretch given by eq. (10).} \end{cases}$$
$$\tag{15}$$

## Comparison with the case of a fluid interface

We can remark here that the eq. (14) has been obtained in the same manner as an interface balance equation. In this last case, we start from an equation analogous to (4) with the difference that the direction ($\perp$) is normal to the interface and that (//) is tangential to the interface. We obtain by integration along the direction $x_3$ normal to the interface :

$$\frac{d_v \rho_F^S}{dt} + \rho_F^S \, \vec{\nabla}_{//}.\vec{V} + [J_{VF\perp}] + \vec{\nabla}_{//}.\vec{J}_{VF}^S = p_F^S \,. \tag{16}$$

(see Fig. 1), $\vec{\nabla}_{//}.\vec{V}$ being the stretch given by eq. (11).

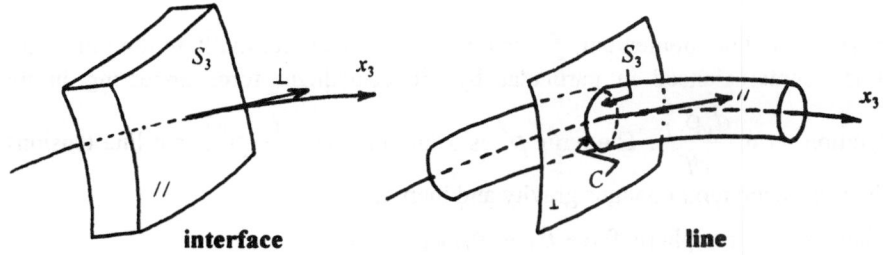

Fig. 1

Tangential and normal directions

The hypotheses are similar : one can construct a mixed velocity vector $\vec{V}$ which presents very small variations at the crossing of the interface whose thickness does not vary locally (for a line, it does not vary crossing the line in the surface $S_3$) and the procedure is analogous :

- writing the (excess) 3D balance equations in a system of varying curvilinear coordinates,

- integrating along the $x_3$ coordinate (on the surface $S_3$ for a line).

## Application to a pipe flow

We suppose that the pipe has its proper deformation and that the fluid has a uniform velocity in a cross section $(S_3)$ except through a thin boundary layer near the wall. The effect of this layer is characterized by a friction force $p_f$ per unit length (Fig. 2).

We obtain the mass balance equation :

$$\frac{d_v \rho^{\ell}}{dt} + \rho^{\ell}\, \vec{\nabla}_{//}.\vec{V} = 0 \quad , \quad \rho^{\ell}_{M} = \rho^{\ell} \tag{17}$$

and the momentum balance equation

$$\frac{d_v\vec{\rho}_I^\ell}{dt} + \vec{\rho}_I^\ell \,\vec{\nabla}_{//}.\vec{V} + \vec{p}_\perp + p_{//}\,\vec{e}_3 + \vec{\nabla}_{//}p^\ell = \rho^\ell\left(\vec{g} - \vec{\gamma}_e - \vec{\gamma}_c\right) \qquad (18)$$

where $\vec{\rho}_I^\ell$ is the line momentum, $\vec{p}_\perp$ is a line force acting normally to the pipe. This force is counterbalanced in particular by the centrifugal force appearing in the acceleration term $\dfrac{d_v\vec{\rho}^{\,\ell}_{\,I}}{dt}$. The term $p^\ell$ is a line pressure $\left(\left(-p^\ell\right)\right.$ is a line tension$\left.\right)$ and the right-hand term contains gravity and inertial forces.

We have for the peripheric force $\vec{p}_\perp + p_{//}\,\vec{e}_3 = \vec{\varphi}\,^\ell_{VI}$.

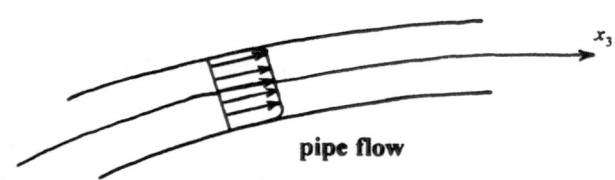

<pre>
                                                    x₃
                                                  →
                         pipe flow
</pre>

Fig. 2

Velocity profile

## Application to Euler's first law common line

J.C. Slattery [5] writes the impulsion balance equation for the common line between three capillary surfaces. In this case, the peripheric force results from the actions of the capillary surfaces.

If we neglect the mass transfer at the common line and if we neglect the effect of inertial forces on the common line, the momentum balance on the common line may be expressed in terms of surface tension (Fig. 3) :

$$\gamma^{(AB)}\vec{v}^{(AB)} + \gamma^{(AS)}\vec{v}^{(AS)} + \gamma^{(BS)}\vec{v}^{(BS)} = 0, \qquad (19)$$

where $\vec{v}^{(AB)}$, $\vec{v}^{(AS)}$ and $\vec{v}^{(BS)}$ are the unit vectors that are normal to the common line and that are both tangent to and directed into the *A-B*, *A-S* and *B-S* dividing surfaces respectively.

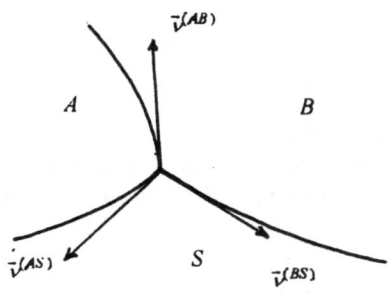

Fig. 3

Common line

## The problem of constitutive relations

We have given a general procedure for obtaining balance equations for fluid curvilinear media. A more detailed analysis have to be made in each case to establish the governing equations for mass, species, momentum and energy in the adequate form. In addition, it is necessary, before solving the system, to close it with constitutive relations. Some of them can be obtained from 3D constitutive relations by simple integration. But the problem is not always easy to solve for several reasons. Two of them are the following :
    - 3D constitutive relations are often themselves unknown,
    - when we know such 3D constitutive relations, they are generally nonlinear and thus difficult to treat for 3D-2D or 3D-1D mathematical transition.
    This paper has been partly written in relation with a research presently developed in the "Direction Scientifique de l'Energétique" of ONERA.

## References

[1] Pétré, G., Sanfeld, A. (Eds) (1991) : "Capillarity today", *Lecture Notes in Physics, 386.*
[2] Aris, R. (1962) : "Vectors, tensors and the basic equations of fluid mechanics", *Prentice Hall Publ., NJ.*
[3] Germain, P. (1986) : "Mécanique", *Ecole Polytechnique, Ellipses.*
[4] Prud'homme, R. (1988) : "Fluides hétérogènes et réactifs : écoulements et transferts", *Lecture Notes in Physics, 304.*
[5] Slattery, J.C. (1990) : "Interfacial transport phenomena", *Springer-Verlag.*

# The second gradient theory applied to interfaces: Models of Continuum Mechanics for fluid interfaces

Henri GOUIN

Laboratoire de Mécanique, Case 322
Université Aix-Marseille III, Faculté des Sciences et Techniques
Avenue Escadrille Normandie-Niemen, 13397 Marseille Cedex 20, France

**Abstract.** A thermodynamical model of continuous media based on second gradient theory is able to study the motions in fluid interfaces. We present a review of some results obtained with the internal capillarity theory for the boundary conditions of such media, the problem of microscopic drops and bubbles and the waves in the vicinity of the critical point of a pure fluid.

**Keywords.** Interfaces, second gradient theory, boundary conditions, drops, bubbles, waves.

## 1. Introduction

The second gradient theory, conceptually more straightforward than Laplace's theory can be used to build a theory of capillarity [18, 8, 13, 17]. Such a theory is able to take into account systems in which fluid interfaces are present. The *internal capillarity* is one of the simplest cases [2, 4].

A mathematical limit analysis associated with the thickness of the interface when the size of the layer goes toward zero and the behaviour of the layer between fluid phases yields the model of material surfaces [1, 4, 15, 7].

Such a theory is able to calculate the superficial tension as well in the case of thin interfaces as thick interfaces. It is possible to obtain the radius of nucleation of microscopic drops and bubbles and to develop a macroscopic theory as Laplace's theory [5]. The stability of interfaces is investigated with differential or partial derivative equations.

The static model in continuum mechanics of second gradient theory is extended to dynamics. The equation of motion is able to introduce a stress tensor. In the fluid case, the theory does not lead to an isotropic stress tensor. Contact forces are in fact of a different nature than the ones associated with the Cauchy stress tensor. Classical conditions with the tetrahedron construction due to Cauchy are not efficient to study the non-linear behaviour of fluids endowed with *internal capillarity*. We deduce contact forces concentrated on edges representing boundaries of Cauchy surfaces of separation [10]. For example, such conditions are necessary to study the stability of thin films in contact with curved solid walls [3,14].

In interfacial layers, the theory of internal capillarity, different from the one of classical continuum media fits with molecular models issued from the mean field theory [11,12]. The theory interprets dynamically phenomena in the vicinity of the critical point [9] (in this case, the size of the interface grows to be a macroscopic one), and authorises a modelisation -at least qualitative- of the dynamic change of phases between bulks in fluids or fluid mixtures.

## 2 The equations of motion and the thermodynamic of fluids endowed with internal capillarity

The equation of motion is the one classically given in the literature with additive terms of second gradient type [8, 13].

In the present approach, the only difference with the classical case is due to an internal mass energy that is a function of volumic mass, entropy as well as gradient of volumic mass [4].

$$\varepsilon = f(\rho, s, \text{grad } \rho)$$

In a bulk where the gradients are null, the internal mass energy becomes:

$$\varepsilon = f(\rho, s, 0) = \alpha(\rho, s)$$

Practically, we consider the following expression:

$$\varepsilon = \alpha + Q/(2\rho) \quad \text{with } Q = C \, (\text{grad } \rho)^2$$

The simplest model able to take into account the volumic mass and its gradient uses an unique additive quantity, the constant C of internal capillarity. In macroscopic units its value is very low and is taken into account only in the thickness of the interfacial layer. The equation of motion is

$$\rho \, \Gamma + \text{grad } P + \rho \, \text{grad } (\Omega - C \, \Delta\rho) - \text{div } \sigma_V = 0$$

where $P$ denotes the thermodynamical pressure (as van der Waals or other equations of state with the same behaviour), $\Omega$ is the extraneous force potential, $\sigma_V$ is the stress tensor due to viscosity, $\Gamma$ is the acceleration vector. For an isothermal and non-viscous flow, the equation of motion is:

$$\Gamma + \text{grad}(\mu - C \, \Delta\rho + \Omega) = 0$$

where $\mu$ denotes the chemical potential of the fluid.

If we take into account dissipative effects as Navier-Stokes' -(it would be of course coherent in second gradient theory to add to the viscosity terms accounting for the influence of higher order derivatives of the velocity field)- equation of energy introduces a supplementary vector $\sigma_v V$. With the terms of heat flux vector and radiant heating (q et r), the equation of energy yields [13]:

$$\partial e/\partial t + \text{div } [(e+P)V] - \text{div } W - \text{div } (\sigma_v V) + \text{div } q - r = \rho \, \partial\Omega/\partial t$$

where $e = \rho(1/2 \, V^2 + \varepsilon + \Omega)$ is the total volumic energy. This brings us to add the additional vector $W = C \, (d\rho/dt) \, \text{grad } \rho$. It is important to note that this term has the dimension of a heat flux vector and occurs even in the conservative case.

The equation of entropy is the same than in the classical case,

$$\rho \, T \, ds/dt + \text{div } q - r - \Psi = 0$$

with $\Psi$ is the function of dissipation of stresses due to viscosity and T the Kelvin temperature.

The second law of thermodynamics is expressed mechanically as $\Psi \geq 0$ and implies the Planck inequality:

$$\rho \, T \, ds/dt + \text{div} \, q - r \geq O$$

This inequality is the same than for classical fluids. Supposing that the Fourier law is expressed in a very general manner with:

$$q \cdot \text{grad} \, T \leq 0,$$

the Clausius-Duhem inequality is deduced directly [4]:

$$\rho \, (ds/dt) + \text{div}(q/T) - r/T \geq 0.$$

## 3 Boundary conditions for fluids endowed with internal capillarity

The principle of virtual power applied for media endowed with internal capillarity is able to obtain the boundary conditions. The conditions are different from those of elastic media. It is not possible to obtain these conditions by the way of the tetrahedron construction by Cauchy. The conditions take into account not only the normal vector but also the curvature of the surface. We obtain also contact forces concentrated on the edges of the boundary.

On the boundary $(S_t)$ of the continuous medium, we obtain [3, 10]:

$$\rho \, u \, V + [\, P - C/2 \, (\text{grad} \, \rho)^2 - C \, \rho \Delta \rho + C \, (\text{grad} \, \rho)(\text{grad} \, \rho)^t - 2A/R_m] \, n - \text{grad}_{tg} \, A = P$$

$$- u \, e_1 - [\, P - 2A/R_m] \, n^t \, V - n^t \, \partial(AV)/\partial x \, n + \text{div} \, (AV) + C \, \partial \rho/\partial t \, \text{grad} \, \rho = \kappa$$

$$- A \, n^t \, V = \lambda \quad \text{et} \quad A = S$$

with $A = C\rho \, d\rho/dn$, n the normal vector to $(S_t)$, $R_m$ the mean radius of curvature, u the velocity of the fluid through the surface $(S_t)$, $e_1$ the free total energy of the fluid, P, $\kappa$, $\lambda$ et S are terms balancing the capillary fluid forces through the surface.

On the edges $(\Gamma_t)$ of $(S_t)$ we obtain the additional conditions:

$$A \, n' = R \quad \text{et} \quad A \, n' \, V = \mu$$

where n' represents the unit vector $t \times n$ where t is the unit vector tangent to $(\Gamma_t)$ and R and $\mu$ are terms balancing the capillary fluid forces on the edges.

We notice that such a theory -equivalent to the one issue from molecular mean field theory [11, 12] - is convenient for representing the motion or the equilibrium of fluid interfaces. This is the objective of the three next paragraphs.

## 4 Calculus of the superficial tension for planar or curved interfaces

The surfaces of equal density are convenient to parametrise interfaces. We assume they are parallel surfaces and so determine a system of orthogonal coordinates. With classical notations, subscript 3 is relevant of the normal direction to the surface of equal density following the increasing densities. Neglecting the body forces, the normal component of the equation of motion yields:

$$(1/h_3)(\, \partial P/\partial x_3) = C \, \rho \, (1/h_3) \, (\partial \Delta \rho/\partial x_3)$$

In the macroscopic case, the interface is not the one of bubble or drop of molecular size. Developing a mathematical limit analysis where we assume the parameter associated with the thickness of the interface goes toward zero, the integration of the previous equation on the third coordinate line normal to the interface yields:

$$P - P_v = C \rho \, \Delta\rho - C \int_{x_{3,v}}^{x_{3,1}} \Delta\rho \, (\partial\rho/\partial x_3) \, dx_3$$

$R_m$ denotes the radius of mean curvature to the surfaces of equal density, we obtain:

$$P - P_v = C [ \rho \, \Delta\rho - 1/2 \, (grad \, \rho)^2 ] + 2C/R_m \int_{x_{3,v}}^{x_{3,1}} (grad \, \rho)^2 \, h_3 \, dx_3$$

Noting $dn = h_3 \, dx_3$, this equation may be written

$$P_1 - P_v = 2 \, \sigma \, /R_m \qquad \text{with} \qquad \sigma = C \int_{n_v}^{n_1} (grad \, \rho)^2 \, dn,$$

This expresses Laplace's equation for dynamic of interfaces. $\sigma$ is interpreted as the fluid surface tension [4] (the same form than for the static of planar interfaces).

## 5 Microscopic drops or bubbles

In 1948, Tolman [16] established, using a Gibbs-like approach to Laplace theory, a relation between the radius and the surface tension of bubbles in equilibrium with their liquid phase. Experimental investigations concerning bubbles of quasi-molecular dimension were performed [6]. By means of the *internal capillary* model, we are able to evaluate the difference between the free energy of the homogeneous fluid of density $\rho_0$ in a domain $D$ and the free energy of a small bubble of volume V contained in the same domain and with the same total mass. The energies of nucleation in the two cases of Laplace theory and the theory of internal capillarity (Second gradient theory) are the following [5]:

$$W = 4 \, \pi \, R \, \sigma + 4/3 \, R^3 \, [ \, \psi(\rho_v) - \psi(\rho_1) + \mu(\rho_1)( \rho_1 - \rho_v )] \quad \text{(Laplace)}$$

$$W = \int_{R^3} [ \, \psi(\rho) - \psi(\rho_1) + \mu(\rho_1)( \rho_1 - \rho ) + C/2 \, (grad \, \rho)^2 \, ] \, dv \quad \text{(Second gradient theory)}$$

where $\psi$ denotes the free volumic energy of the homogeneous fluid.

In second gradient theory, the equilibrium of a bubble surrounded by its liquid phase with density $\rho_1$ is represented by a spherical symmetric profile of density associated with the differential equation:

$$C \, d^2\rho/dr^2 + 2 \, C/r \, d\rho/dr = \mu(\rho) - \mu(\rho_1)$$

We can define the surface tension and the radius of a bubble by identifying the nucleation energies and the pressure differences computed in both theories. We propose a radius R and surface tension σ for microscopic bubbles in the case where classically the Laplace theory is no more efficient

$$R = [\, 2C \int_0^\infty r^2 \rho'^2 \, dr]^{1/3} \, [\, \psi(\rho_l) - \psi(\rho_v) + \mu(\rho_l)(\rho_v - \rho_l)\,]^{-1/3}$$

$$\sigma = [\, C/4 \int_0^\infty r^2 \rho'^2 \, dr]]^{1/3} \, [\, \psi(\rho_l) - \psi(\rho_v) + \mu(\rho_l)(\rho_v - \rho_l)\,]^{2/3}$$

These expression are used with non convex equations of state (like van der Waals). Some numerical experiments were performed. We could observe that for bubbles whose the radius is close to the critical one, the predicted variations of the surface tension fit with the experimental observations [6].

## 6 Material waves in the vicinity of the critical point

One studies the one dimensional problem of isothermal waves with velocity V and volumic mass ρ which are function only of one variable ζ:

$$V = V(X - q\,t) \text{ and } \rho = \rho(X - q\,t)$$

where q denotes the celerity of the wave in a reference space and $\zeta = X - q\,t$.

For a one-dimensional problem, the equation of motion for a non-viscous isothermal capillary fluid yields :

$$C \frac{d^2\rho}{dx^2} = \mu - \mu_o + \frac{q^2}{2\rho^2} + \Omega$$

where $\mu_o$ is a constant of integration. In the vicinity of the critical point, the chemical potential of a one-component fluid has the following expansion:

$$\mu = \mu^c - \mu^c_{o1} \, \Delta T - \mu^c_{11} \, \Delta\rho \, \Delta T + \frac{1}{6} \mu^c_{3o} \, (\Delta\rho)^3$$

where $\Delta\rho = \rho - \rho_c$ and $\Delta T = T_c - T$ denote the distance of the density ρ and the Kelvin temperature T to their values at the critical point. The coefficients are the values of partial derivatives of μ at the critical point. In the unidimensional isothermal case, neglecting the body forces, the equation of waves yields:

$$C \frac{d^2\rho}{dx^2} = - A \, \Delta\rho \, \Delta T + B \, (\Delta\rho)^3 + \frac{q^2}{2\rho^2} + a_o$$

with $A = \mu^c_{11}$ , $B = \frac{1}{6} \mu^c_{3o}$ and $a_o$ is constant at given temperature.

Due to the fact we are close from the critical point, $c = \dfrac{q}{\rho_c}$ is the *celerity of the wave* in the physical space.

The second gradient theory is able to foresee the existence of solitons in the vicinity of the critical point of a fluid.

Two kinds of material waves were investigated [9]:
- moving liquid-vapour interfaces below the critical temperature with a superficial tension more important than the one at equilibrium
- moving solitary waves do not existing at equilibrium.

Above the critical temperature, the fluid behaves like an elastic medium and the waves are supersonic with respect to the isothermal sound celerity.

Below the critical temperature, the fluid behaves like non rigid medium and the wave velocities can take any value.

For wave speed, it is important now to compare, favourably or unfavourably, the model with experimental results involving different materials.

It will be also important to introduce the viscous terms for studying motions in the vicinity of the critical point.

# References

[1] M. Barrere and R. Prud'homme, Equations fondamentales de l'aérothermochimie, Masson, 1973.

[2] P. Casal, Cahier du Groupe Français de Rhéologie, CNRS VI, n°3, 1961, p.31-37 and C.R. Acad. Sc, t. 256, 1963, p. 3820-3822.

[3] P. Casal, C.R. Acad. Sc, t. 274, A,1972, p.1571-1574.

[4] P. Casal and H. Gouin, Journal de Mécanique Théorique et Appliquée, vol. 7, 1988, p. 689-718.

[5] F. Dell'Isola, H. Gouin, P. Seppecher, C.R. Acad. Sc, t. 320, IIb, 1995, p. 211-216.

[6] L.R. Fisher and J.N. Israelachvili, Chem. Phys. Letters, vol. 76, n°2, 1980, p. 325-328.

[7] R. Gatignol and P. Seppecher, Journal de Mécanique Théorique et Appliquée, spécial, 1986, p. 225-247.

[8] P. Germain, Journal de mécanique, vol. 12, N°2, 1973, p. 235-274.

[9] H. Gouin, C.R. Acad. Sc, t. 317, II, 1993, p. 1263-1268.

[10] H. Gouin and F. Cubisol , C.R. Acad. Sc, 1995, to appear.

[11] Y. Rocard, Thermodynamique, Masson, 1967.

[12] R. Rowlinson and B. Widom, Molecular theory of Capillarity, Clarendon Press, 1984.

[13] J. Serrin ed., New perspectives in thermodynamics, Springer, 1986.

[14] P. Seppecher, European J. Mechanics/B Fluids, t. 12, 1993, p. 69-84.

[15] A. Steichen, R. Defay, A. Sanfeld, Journal de physique, t. 68, 1971, p. 835-840, 1240-1244, 1323-1328.

[16] R.C. Tolman, J. Chem. Phys., 16, 1948, p. 758-774.

[17] C.Truesdell and W. Noll, The Non-linear Field Theories of Mechanics, Encyclopedia of Physics, III/3, Springer, 1965.

[18] J. D. van der Waals, Archives Néerlandaises, vol. 28, 1894-1895, p. 121-209.

# ASYMPTOTIC MODELLING
# OF FLUID-FLUID INTERFACES

Renée GATIGNOL

Laboratoire de Modélisation en Mécanique
Université Pierre et Marie Curie & CNRS
Case 162, 4 Place Jussieu, F75252 PARIS Cedex 05, France

**Abstract:** The interfacial region between two fluids is seen as a transition layer through which the gradients of physical parameters are large. This is the "microscopic" point of view. A "macroscopic" description is then introduced where the two fluids are separated by a surface with no thickness, called "interface". The asymptotic modelling consists of considering that the transition layer has a thickness $\delta$ that is very small compared to the radius of curvature $R$ of the isodensity surfaces located inside the layer. In the asymptotic limit $\delta/R \rightarrow 0$, the layer is seen as a discontinuity surface: an "interface". The usual interfacial balance laws are re-established and discussed.

Due to the discrepancy between the normal component of the material velocity and the displacement velocity of the interface, an extra term appears in the irreversible production of the interfacial entropy. This new term links the difference between these two normal components to a deviation from the equilibrium given by the Laplace equation.

## 1. Introduction

In this paper, we wish to discuss the local jump conditions of a two fluid system (fluid 1 and fluid 2 ) separated by a moving surface of arbitrary shape. Furthermore we assume that this surface possesses material properties like mass, energy, entropy densities.

For the description of the region separating the two fluids, we can adopt two points of view. First a "microscopic" view in which we have a transition layer through which the gradients of the physical quantities are large. Second a "macroscopic" view in which the fluids are separated by a surface, with no thickness, called "interface". Our purpose is to give a way of obtaining the jump conditions for the interface from the local 3D-dimensionnal description of the transition layer.

This way may be achieved via an asymptotic approach. It consists of considering that the transition layer has a thickness $\delta$ very small compared to the radius of curvature $R$ of the isodensity surfaces located inside the layer. In the limit $\delta/R \rightarrow 0$, the layer appears as a discontinuity surface: interface.

A very large amont of litterature treats the jump conditions through a fluid-fluid interface. In a number of classic works such as [1], the interface is seen as a geometric surface, namely a surface of discontinuities without any interfacial mass density or other surface material properties. Interfaces without interfacial mass density but with an internal energy per unit of area related to the phenomenon of surface tension have been also studied for many years. The first papers introducing this type of jump conditions are reviewed in the paper by Delhaye [2] and we only mention the very important works of Scriven [3].

We know that for an interface with no interfacial mass, one cannot define the material velocity of the interface [4]. In their work [5], Bedeaux, Albano and Mazur discuss this for an interface between two immiscible fluids (fluid 1 and fluid 2 ). By denoting $\vec{V}_s$ the material velocity of the interface, they choose the tangential component $\vec{V}_{s//}$ of $\vec{V}_s$ as being $\left(\vec{V}_{1//} + \vec{V}_{2//}\right)/2$ where $\vec{V}_{1//}$ and $\vec{V}_{2//}$ are the tangential components of the velocities $\vec{V}_1$ and $\vec{V}_2$ of the fluids 1 and 2 near the interface. In order to justify this choice, the authors introduce a finite value of the surface mass density $\rho_a$ and they use the linear phenomenological laws between fluxes and thermodynamic forces occuring in the surface entropy production rate. However, the limit process used to obtain the result is not clear. If it is carried out carefully, one is left again with $\vec{V}_{s//}$ unspecified, as was noticed by Prosperetti [4].

For an interface between two fluids with mass transfer, we face the same problem for a good definition of the material velocity. To be precise, at every moment in time and at each point of the interface we defined two velocities: the material velocity $\vec{V}_s$ and the velocity $\vec{W}$ of the geometric surface which coincides at each time with the material interface (in fact only the normal component of $\vec{W}$ is meaningful); we denote by $\vec{V}_{s\perp}$ the normal material velocity and by $\vec{W}_\perp$ the normal displacement velocity of the interface.

In their works, Barrère and Prud'homme [6] and Delhaye [5], consider an interface with a surface mass density and they introduce the velocities $\vec{V}_s$ and $\vec{W}$. However, they suppose that the normal velocities $\vec{V}_{s\perp}$ and $\vec{W}_\perp$ are equal. According to this assumption the particles belonging to the interface cannot leave the interface and hence only move along the interface.

In the previous mentionned works, the material properties on the interface are introduced by analogy with the tridimensional continuous medium. The balance laws for the surface quantities are given in the same form as the analogous balance laws for the bulk volume quantities. However, the exchanges between the interface and the surroundings fluids are allowed.

Another approach consists of considering a thin transition layer between the two fluids. A parameter associated with a physical quantity in the layer matches asymptotically the corresponding values in the bulk phases. Only the order of magnitude of the thickness of the layer is well defined. In many papers [7, 8, 9, 10, 11, 12, 13, 14], the question referring to the formulation of the convenient jump conditions is discussed with this point of view. Two types of quantities are defined on the interface: the "true" quantities by integration over the thickness of the layer [8], and the "excess" quantities [9]. It is easy to define the physical meaning of the "true" quantities but there is some difficulty in assigning a finite thickness to the layer. It is also easy to introduce the "excess" quantities but there is some difficulty in locating the position of the interface inside the transition layer.

In this paper, we adopt the "true" quantities point of view and, following Ishii [8], we define the interface quantities by integrating the corresponding bulk parameters characterizing the medium in the transition layer over the thickness of the layer. So we obtain expressions for the mass per unit of area, the mean velocity $\vec{V}_s$ of the particles located in the layer, the surface internal energy, and so on. In the same way, we define the source terms on the interface, the surface stress tensor and the energy current. We can derive balance laws where exchanges of mass, momentum and energy between the surface and the bulk fluids are allowed.

In Section 2 we introduce the surface quantities and we derive the general form for the balance laws. In Section 3 we pay attention to the detailed balance laws for the mass, momentum and energy. In Section 4 the entropy inequality is derived and the entropy production term is discussed. In the present approach the normal components of $\vec{V}_s$ and $\vec{W}$ are different, so we obtain an extra term in the interface entropy production. Finally, in Section 5 we consider an interface without mass density as the limiting case of an interface with the mass density $\rho_a$ tending to zero. It seems that in this limit case the distinction between the normal components of $\vec{V}_s$ and $\vec{W}$ is irrelevant.

## 2. Interfacial quantities and balance laws

Here the interfacial region between the two fluids is seen as a thin transition layer through which the gradients of the various physical parameters are large. We denote by 1 and 2 the two fluids and by $S$ a geometric surface inside the interfacial layer. The normal unit vector to $S$ pointing from fluid 1 to fluid 2 is denoted by $\vec{\xi}$. Any physical quantity $\psi$ is defined everywhere, has a large gradient inside the layer and is such that it tends asymptotically to the corresponding values $\psi_1$ and $\psi_2$ in the bulk fluids. The thickness of the interfacial layer is assumed to be finite and the boundaries of the layer are at the distances $\xi_1$ and $\xi_2$ from the geometric surface $S$ (Fig. 1). On the boundaries we have: $\psi = \psi_1$ or $\psi = \psi_2$.

We define the interfacial mass density by the following integral through the layer along the normal direction:

$$\rho_a = \int_{\xi_1}^{\xi_2} \rho \, d\xi \tag{1}$$

where $\rho$ is the mass density per unit volume. Then we introduce the mean velocity $\vec{V}_s$ of the fluid particles located inside the transition layer by the relation (2) where $\vec{V}$ is the velocity of the continuum medium inside the layer:

$$\rho_a \vec{V}_s = \int_{\xi_1}^{\xi_2} \rho \vec{V} \, d\xi \tag{2}$$

In the same way, following Ishii [8], we write for any physical quantity $\psi$ :

$$\psi_a = \rho_a \psi_s = \int_{\xi_1}^{\xi_2} \rho \psi \, d\xi \tag{3}$$

We notice that $\psi_s$ is a quantity per unit of interfacial mass and $\psi_a$ is a quantity per unit of area on the surface. We emphasize that $\vec{V}_s$ , which appears as a barycentric velocity, may be seen as the material velocity of the particle belonging to the interface at the time and point considered. This velocity $\vec{V}_s$ is generally different from the velocity $\vec{W}$ of the surface $S$.

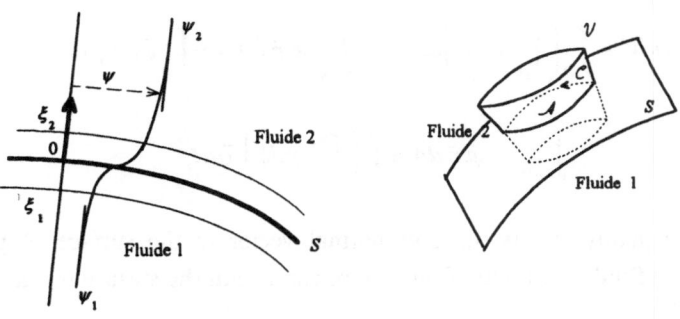

Figure 1                                Figure 2

The transition layer                     The controle volume $\mathcal{V}$

We now proceed to write the balance laws in a thin volume which contains the transition layer. The interface $S$ moves with the velocity $\vec{W}$. The velocity $\vec{W}$ is not necessarily normal to $S$, and we are allowed to define later a convenient tangential component. For the moment, only the normal velocity $\vec{W}_\perp = (\vec{W}.\vec{\xi})\vec{\xi}$ is well defined. Let $\mathcal{A}$ be a piece of $S$ limited by the curve $\mathcal{C}$. We denote by $\vec{\upsilon}$ the unit vector tangent to $S$ and by $\vec{\tau}$ the tangential vector to $\mathcal{C}$ such that $\vec{\xi} = \vec{\upsilon} \wedge \vec{\tau}$. We consider a volume $\mathcal{V}$ cutting off the surface $S$ along the curve $\mathcal{C}$. The lateral surface of $\mathcal{V}$ is made of normals to $S$. Finally the surfaces $\mathcal{A}_1$ and $\mathcal{A}_2$ are at the distances $\xi_1$ and $\xi_2$ from $S$ (Fig. 2).

We write down the classical integral balance law in $\mathcal{V}$ for the quantity $\rho\psi$ defined per unit of volume. The vector $\vec{J}$ is the current density associated to the quantity $\rho\psi$ ,such that $\vec{J}.\vec{n}$ is the rate of the quantity lost across the limited surface $\partial \mathcal{V}$ of $\mathcal{V}$ per unit of area ($\vec{n}$ is the outward unit normal vector to $\partial \mathcal{V}$), and the quantity $\rho \phi$ corresponds to the production rate per unit volume. We have [1]:

$$\frac{\delta}{\delta t} \int_{\mathcal{V}} \rho \psi \, dV + \sum_{k=1}^{2} \int_{A_k} \left\{ \rho \psi \left( \vec{V} - \vec{W} \right) + \vec{J} \right\} . \vec{n} \, dA$$

$$+ \int_{(\partial \mathcal{V})_{lat.}} \left\{ \rho \psi \left( \vec{V} - \vec{W} \right) + \vec{J} \right\} . \vec{n} \, dA = \int_{\mathcal{V}} \rho \phi \, dV \quad (4)$$

In the first term the time derivative $\delta / \delta t$ is associated with the velocity $\vec{W}$.

Now, we suppose that the transition layer is very thin, namely we have the condition: (layer thickness) / (mean curvature radius of $S$) very small compared to 1. So we can consider that the volume $\mathcal{V}$ is a small cylindrical volume, and we can write:

$$\int_{\mathcal{V}} \varphi \, dV \approx \int_{A} \left( \int_{\xi_1}^{\xi_2} \varphi \, d\xi \right) dA \; , \qquad \int_{A_k} \vec{\varphi} . \vec{n} \, dA \approx - \int_{A} \vec{\varphi}_k . \vec{n}_k \, dA$$

$$\int_{(\partial \mathcal{V})_{lat.}} \vec{\varphi} . \vec{n} \, dA \approx \int_{C} \left( \int_{\xi_1}^{\xi_2} \vec{\varphi} \, d\xi \right) . \vec{v} \, dl$$

In the second equality, $\vec{n}_k$ is the unit normal vector to the surface $S$ pointing outward from the fluid $k$. Finally, from (4) we can obtain the surface balance law in a integral form:

$$\frac{\delta}{\delta t} \int_{A} \left( \int_{\xi_1}^{\xi_2} \rho \psi \, d\xi \right) dA = \sum_{k=1}^{2} \int_{A} \left\{ \rho_k \psi_k \left( \vec{V}_k - \vec{W} \right) + \vec{J}_k \right\} . \vec{n}_k \, dA$$

$$- \int_{C} \left( \int_{\xi_1}^{\xi_2} \left\{ \rho \psi \left( \vec{V} - \vec{W} \right) + \vec{J} \right\} d\xi \right) . \vec{v} \, dl + \int_{A} \left( \int_{\xi_1}^{\xi_2} \rho \phi \, d\xi \right) dA \quad (5)$$

Now, we want to introduce the interfacial quantities by expressions similar to the expression (3). However even though the velocity $\vec{W}$ appears in the line integral, it is only in the form of the product $\vec{W} . \vec{v}$. Like Ishii [8], we adopt for the tangential component of $\vec{W}$ the following choice (Fig. 3):

$$\vec{W} = \vec{W}_{//} + \vec{W}_{\perp} = \vec{V}_{s//} + \vec{W}_{\perp} \quad (6)$$

So, the velocity $\vec{W}$ includes two velocities: the material tangential velocity $\vec{W}_{//} = \vec{V}_{s//}$ and the normal geometric velocity $\vec{W}_{\perp}$ of $S$. With this choice for $\vec{W}$ the interfacial current density which appears in the third term of the equation (5) may be written in the following form:

19

$$\vec{J}_a = \int_{\xi_1}^{\xi_2} \left\{ \rho \psi \left( \vec{V} - \vec{V}_s \right) + \vec{J} \right\} d\xi \tag{7}$$

In $\vec{J}_a$, there are two contributions: the first is related to the fluctuations of the fluid about the mean velocity $\vec{V}_s$ and the second corresponds to the mean value of $\vec{J}$ in the layer.

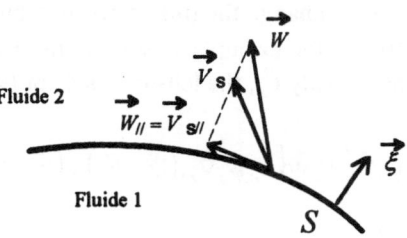

Figure 3

The different velocities on the interface

Now, with the definition (7) for $\vec{J}_a$ the balance law is written in the integral form:

$$\frac{\delta}{\delta t} \int_A \rho_a \psi_s \, dA = \sum_{k=1}^{2} \int_A \left\{ \rho_k \psi_k \left( \vec{V}_k - \vec{W} \right) + \vec{J}_k \right\} . \vec{n}_k \, dA - \int_C \vec{J}_a . \vec{v} d\ell + \int_A \rho_a \phi_s \, dA \tag{8}$$

We obtain the local balance law by classical techniques using the surface divergence theorem and the surface transport theorem. These two theorems are written with intrinsic notation for some vector $\vec{B}$ which is not necessarily tangent to $S$, and for some scalar quantity $\varphi_a$ defined on $S$:

$$\frac{\delta}{\delta t} \int_C \vec{B} . \vec{v} d\ell = \int_A \vec{\nabla}_s . \vec{B}_{//} \, dA \, , \qquad \frac{\delta}{\delta t} \int_C \varphi_a \, dA = \int_A \left( \frac{\delta_s}{\delta t} \varphi_a + \varphi_a \vec{\nabla}_s . \vec{W} \right) dA$$

The notations are given the appendix. These two theorems are shown, for example, in the book by Barrère and Prud'homme [6] or in the paper by Prosperetti [4)]. We shall only note that $\vec{\nabla}_s . \vec{B}_{//}$ is the surface divergence of $\vec{B}$ and $\delta_s / \delta t$ is the convective time derivative associated with the velocity $\vec{W}$ . In this last derivative, the tangential material velocity along the surface $S$ are taken into account. Finally, the local balance equation is written in the following form:

$$\frac{\delta_s}{\delta t} \left( \rho_a \psi_s \right) + \rho_a \psi_s \vec{\nabla}_s . \vec{W} = \sum_{k=1}^{2} \left\{ \rho_k \psi_k \left( \vec{V}_k - \vec{W} \right) + \vec{J}_k \right\} . \vec{n}_k - \vec{\nabla}_s . \vec{J}_{a//} + \rho_a \phi_s \tag{9}$$

Here we may note three things: First, on the right-hand side we only have the tangential vector $\vec{J}_{a//}$ , defined by: $\vec{J}_{a//} = \left(1 - \vec{\xi}\vec{\xi}\right).\vec{J}_a$ . According to the definition (7), the tangential density $\vec{J}_{a//}$ corresponds to the tangential fluctuations of the fluid particles in the layer and to the mean value of $\vec{J}_{//}$ . Second, the time derivative on the left-hand side of (9) is a convective time derivative with the geometric velocity $\vec{W} = \vec{V}_{s//} + \vec{W}_{\perp}$ of $S$. Last, if we change the time derivative on the left-hand side, we must introduce an extra term: for example if we use the material time derivative $D_s / Dt$ associated with the velocity $\vec{V}_s$, the left-hand side of (9) is replaced by

$$\frac{D_s}{Dt}\left(\rho_a \psi_s\right) + \rho_a \psi_s \vec{\nabla}_s.\vec{V}_s + \left\{\rho_a \psi_s \vec{\nabla}_s.\left(\vec{W} - \vec{V}_s\right) + \left(\vec{W} - \vec{V}_s\right).\vec{\nabla}\left(\rho_a \psi_s\right)\right\}$$

In many approaches, the interfacial balance laws are postulated in the integral or in the local form, with a material time derivative in the left-hand side similar to those of (8) or (9). However, these balance laws are only valid if $\vec{W}$ and $\vec{V}_s$ are identical. This hypothesis is made in Delhaye's works [2] and also in [6]. In the work of Ishii [8] these two velocities are assumed different and an equation similar to (9) is obtained. A mathematic approach is given in [15]. Of course, in the case of two immiscible fluids, the two velovities $\vec{W}$ and $\vec{V}_s$ are identical as we shall see in the next section, and we have no trouble with the time derivatives.

## 3. Interfacial balance of mass, momentum and energy

### 3.1 Interfacial balance of mass

For the interfacial balance of mass, we take: $\psi = 1$ , $\vec{J} = \vec{0}$ and $\phi = 0$. So using the definitions (1), (2) and (7) we have $\psi_a = \rho_a$ , $\vec{J}_a = \vec{0}$ and $\phi_a = 0$ . We notice that $\vec{J}_a = \vec{0}$ according to the definition (2) of $\vec{V}_s$. Finally the balance law for mass is:

$$\frac{\delta_s}{\delta t}\rho_a + \rho_a \vec{\nabla}.\vec{W} = \sum_{k=1}^{2}\left\{\rho_k\left(\vec{V}_k - \vec{W}\right).\vec{n}_k\right\} \tag{10}$$

As usual, we can write this equation in a condensed form: $\dot{m}_a = \dot{m}_1 + \dot{m}_2$ where $\dot{m}_k$ is the loss of fluid $k$ per unit area of $S$ and per unit time: i.e. $\dot{m}_k = \rho_k\left(\vec{V}_k - \vec{W}\right).\vec{n}_k$

Using this balance law for the mass, we give the surface balance law (9) a new form:

$$\rho_a \frac{\delta_s}{\delta t}\psi_s = \sum_{k=1}^{2}\left\{\dot{m}_k\left(\psi_k - \psi_s\right) + \vec{J}_k.\vec{n}_k\right\} - \vec{\nabla}_s.\vec{J}_{a//} + \rho_a \phi_s \tag{11}$$

## 3.2 Momentum interfacial balance law

Now we give $\psi$, $\vec{J}$ and $\phi$, the following quantities respectively: the velocity $\vec{V}$, the opposite of the volume stress tensor $\mathbf{S}$ of the continuum medium located inside the layer, and the vector $\vec{f}$ which represents the external force per unit mass. So we have for $\psi_s$, $\vec{J}_a$ and $\phi_s$ the quantities $\vec{V}_s$, $-\mathbf{S}_a$ and $\vec{f}_s$. Hence, from the definition (7) and the velocity $\vec{V}_s$, the expression for $\mathbf{S}_a$ is:

$$\mathbf{S}_a = \int_{\xi_1}^{\xi_2} \left\{ \mathbf{S} - \rho(\vec{V} - \vec{V}_s)(\vec{V} - \vec{V}_s) \right\} d\xi \tag{12}$$

As in the general case, in $\mathbf{S}_a$ we have two parts: the first is due to the fluctuations of the fluid particles in the transition layer and the second corresponds to the mean value of $\mathbf{S}$. With these definitions and from (11) we obtained the following balance equation for the momentum:

$$\rho_a \frac{\delta_s}{\delta t} \vec{V}_s = \sum_{k=1}^{2} \left\{ \dot{m}_k (\vec{V}_k - \vec{V}_s) - \mathbf{S}_k . \vec{n}_k \right\} + \vec{\nabla}_s . \left\{ (1 - \vec{\xi}\vec{\xi}) . \mathbf{S}_a \right\} + \rho_a \vec{f}_s \tag{13}$$

According to its definition, the stress tensor $\mathbf{S}_a$ is symmetric. It is possible, by physical arguments [10] and by using the balance law for the angular momentum to get to the conclusion that the tensor $\mathbf{S}_a$ has the following transversality property: $(\mathbf{S}_a . \vec{\xi}) \wedge \vec{\xi} = \vec{0}$. In other words $(\mathbf{S}_a . \vec{\xi})$ is parallel to $\vec{\xi}$, so that $\vec{\xi}$ is an eigenvector of the tensor $\mathbf{S}_a$. Also, we can say that $\mathbf{S}_a$ transforms each vector $\vec{v}$ tangent to $S$ to a vector tangent to $S$. This property for $\mathbf{S}_a$ is often assumed [6, 8]. In the paper by Bedeaux, Albano and Mazur [5], as well in [11] this property was given as a result of the balance equations for the momentum and for the angular momentum in the case $\rho_a = 0$. Here we shall assume that this property is true. So, from now on we have $(\mathbf{S}_a . \vec{\xi})$ parallel to $\vec{\xi}$ .Consequently we have: $(1 - \vec{\xi}\vec{\xi}) . \mathbf{S}_a = (1 - \vec{\xi}\vec{\xi}) . \mathbf{S}_a . (1 - \vec{\xi}\vec{\xi})$. This new tensor is denoted by $\mathbf{S}_{a//}$ . It is a 2D-dimensional tensor and it is called interfacial stress tensor. So the balance law for momentum is written as:

$$\rho_a \frac{\delta_s}{\delta t} \vec{V}_s = \sum_{k=1}^{2} \left\{ \dot{m}_k (\vec{V}_k - \vec{V}_s) - \mathbf{S}_k . \vec{n}_k \right\} + \vec{\nabla}_s . \mathbf{S}_{a//} + \rho_a \vec{f}_s \tag{14}$$

## 3.3 Energy interfacial balance law

For the interfacial balance of energy, we use the same method. We give for $\psi$, $\vec{J}$ and $\phi$, the quantities $e + (\vec{V}^2 / 2)$, $\vec{q} - \mathbf{S} . \vec{V}$ and $\vec{f} . \vec{V} + r$ respectively, where $e$, $\vec{q}$

and $r$ denote the internal energy per unit of volume, the heat current and the rate of energy source per unit of volume respectively. We may then put:

$$e_a = \rho_a e_s = \int_{\xi_1}^{\xi_2} \rho \left( e + \frac{1}{2}(\vec{V} - \vec{V_s})^2 \right) d\xi, \quad \rho_a \left( e_s + \frac{1}{2}\vec{V_s^2} \right) = \int_{\xi_1}^{\xi_2} \rho \left( e + \frac{1}{2}\vec{V}^2 \right) d\xi$$

$$\vec{q}_a = \int_{\xi_1}^{\xi_2} \left\{ \rho \left( e + \frac{1}{2}(\vec{V} - \vec{V_s})^2 \right)(\vec{V} - \vec{V_s}) - \mathbf{S}.(\vec{V} - \vec{V_s}) + \vec{q} \right\} d\xi$$

$$r_a = \int_{\xi_1}^{\xi_2} \rho \left( r + \vec{j}.(\vec{V} - \vec{V_s}) \right) d\xi$$

Here, $e_a$ is the surface internal energy per unit area and $e_s$ the surface internal energy per unit mass. Of course, we have used the definition (2) of $\vec{V_s}$ . The interfacial balance of energy is given by the following equation, where we only have the 2D-dimentional tensor $\mathbf{S}_{a//}$ and the tangential component $\vec{q}_{a//}$ of the interfacial heat current $\vec{q}_a$:

$$\rho_a \frac{\delta_s}{\delta t}\left( e_s + \frac{1}{2}\vec{V_s^2} \right) = \sum_{k=1}^{2} \left\{ \dot{m}_k \left[ \left( e_k + \frac{1}{2}\vec{V_k^2} \right) - \left( e_s + \frac{1}{2}\vec{V_s^2} \right) \right] - \vec{V_k}.\mathbf{S}_k.\vec{n}_k + \vec{q}_k.\vec{n}_k \right\}$$

$$+ \vec{\nabla}_s.\left( \mathbf{S}_a.\vec{V_s} \right)_{//} - \vec{\nabla}_s.\vec{q}_{a//} + \rho_a \vec{j_s}.\vec{V_s} + r_a$$

(15)

Due to the transversality property for $\mathbf{S}_a$ we have $\left( \mathbf{S}_a.\vec{V_s} \right)_{//} = \mathbf{S}_{a//}.\vec{V}_{s//}$. As usual, using the balance law for energy (15) and the balance law for momentum (13) it is possible to eliminate the kinetic energy $\rho_a \vec{V_s^2}/2$ and to write an equation for the rate of the interfacial energy $e_s$:

$$\rho_a \frac{\delta_s}{\delta t} e_s = \sum_{k=1}^{2} \left\{ \dot{m}_k \left( e_k - e_s + \frac{1}{2}(\vec{V_k} - \vec{V_s})^2 \right) - (\vec{V_k} - \vec{V_s}).\mathbf{S}_k.\vec{n}_k + \vec{q}_k.\vec{n}_k \right\}$$

$$+ \mathbf{S}_{a//}:\vec{\nabla}_s\vec{V}_{s//} - \vec{\nabla}.\vec{q}_{a//} + r_a$$

(16)

## 4. Interfacial entropy inequality

At last we propose an inequality for the interfacial entropy density. We start from the tridimensional entropy inequality written for the thin volume $\mathcal{V}$ defined in Section 2. By arguments similar to those used to obtain the equation (5) we can write:

$$\frac{\delta}{\delta t}\int_{A}\left(\int_{\xi_1}^{\xi_2}\rho s\,d\xi\right)dA - \sum_{k=1}^{2}\int_{A}\left\{\dot{m}_k s_k + \left(\vec{h}/T\right)_k.\vec{n}_k\right\}dA$$

$$+\int_{C}\left(\int_{\xi_1}^{\xi_2}\left\{\rho s(\vec{V}-\vec{V}_s)+\vec{h}/T\right\}d\xi\right).\vec{v}\,d\ell - \int_{A}\left(\int_{\xi_1}^{\xi_2}\rho(r/T)d\xi\right)dA \geq 0 \tag{17}$$

In (17), $s$ is the entropy per unit volume and $T$ is the temperature defined at each point of the transition layer. It is important to note that the entropy current vector in the transition layer is $\vec{h}/T$ and is different from $\vec{q}/T$ which is the one usually used in Newtonian fluid theory. This is because the medium within the layer is not considered to be a continuum in the classical sence. In [15], the author puts a capilarity endowed fluid and for such a fluid the two current vectors associated with the energy and the entropy are not proportionnal. We define the following surface quantities:

$$s_a = \rho_a s_s = \int_{\xi_1}^{\xi_2}\rho s\,d\xi\,, \quad \vec{h}_a^* = \int_{\xi_1}^{\xi_2}\left(\rho s(\vec{V}-\vec{V}_s)+(\vec{h}/T)\right)d\xi\,, \quad r_a^* = \int_{\xi_1}^{\xi_2}\rho(r/T)d\xi$$

Using the mass balance law, we get:

$$\rho_a\frac{\delta_s}{\delta t}s_s - \sum_{k=1}^{2}\left\{\dot{m}_k(s_k - s_s)+\left(\frac{\vec{h}}{T}\right)_k.\vec{n}_k\right\}+\vec{\nabla}_s.\vec{h}_{a//}^* - r_a^* \geq 0 \tag{18}$$

We note that the entropy current vector $\vec{h}/T$ is equal to the classical entropy current on the boundaries of the transition layer located in $\xi_1$ or $\xi_2$. In other words, $\left(\vec{h}/T\right)_k = \vec{q}_k/T_k$.

Defining the temperature $T_s$ at the interface is not a trivial task. One way is to assume that the rate $r_a$ of interfacial energy source per unit of area and the rate $r_a^*$ of interfacial entropy source per unit of area are linked by the relation $r_a^* = r_a/T_s$ by analogy to the 3D-dimensional continuous medium. If we imagine that the rate of space energy source $r$ per unit of mass are constant across the interfacial layer, we deduce the following definition for $T_s$:

$$\frac{\rho_a}{T_s} = \int_{\xi_1}^{\xi_2}\frac{\rho}{T}d\xi$$

From now on we assume that $r_a^* = r_a/T_s$ and we define $\vec{q}_a^*$ by putting $\vec{h}_a^* = \vec{q}_a^*/T_s$. Then we can write the entropy inequality (18) in the following form:

$$\rho_a \frac{\delta_s}{\delta t} s_s - \sum_{k=1}^{2} \left\{ \dot{m}_k (s_k - s_s) + \frac{\vec{q}_k}{T_k} . \vec{n}_k \right\} + \vec{\nabla}_s . \frac{\vec{q}_{a//}^*}{T_s} - \frac{r_a}{T_s} \geq 0 \qquad (19)$$

The left-hand side of (19) is the rate $\Delta$ of the irreversible production of entropy linked to the interface moving with the velocity $\vec{W}$.

Now, we assume that we can adopt the classical Gibbs relation on the interface: $de_s = T_s + \gamma d(1/\rho_a)$ where $\gamma$ denotes the surface tension. From the energy equation (16), the mass balance law (10) and the Gibbs relation written in the dynamical case, we can deduce another expression for the interfacial entropy production $\Delta$ ; it is:

$$T_s \Delta = \sum_{k=1}^{2} \left\{ \dot{m}_k \left( g_k - g_s + (T_k - T_s)s_k + \frac{1}{2}(\vec{V}_k - \vec{V})^2 \right) - \vec{q}_k . \vec{n}_k \frac{T_s - T_k}{T_k} + \mathbf{T}_k : (\vec{V}_s - \vec{V}_k)\vec{n}_k \right.$$

$$\left. - p_k (\vec{V}_s - \vec{W}).\vec{n}_k \right\} + \mathbf{T}_{a//} : \vec{\nabla}_s \vec{V}_s + \gamma(\vec{\nabla}.\vec{\xi})(\vec{V}_s - \vec{W}).\vec{\xi} - \vec{q}_{a//} . \frac{\vec{\nabla}_s T_s}{T_s} + \vec{\nabla}_s . (\vec{q}_a - \vec{q}_a^*) \geq 0 \qquad (20)$$

In (20) we have introduced the Gibbs enthalpy defined for the fluid $k$ by $g_k = e_k + p_k / \rho_k - T_k s_k$ and for the interfacial medium by $g_s = e_s - \gamma / \rho_a - T_s s_s$ , and we denoted by $\vec{\nabla}_s$ the surface gradient operator (cf appendix). We have also introduced the viscous stress tensor $\mathbf{T}_k$ and the interfacial viscous stress tensor $\mathbf{T}_{a//}$ defined by the relations:

$$\mathbf{S}_k = -p_k \mathbf{1} + \mathbf{T}_k , \qquad \mathbf{S}_{a//} = \gamma(\mathbf{1} - \vec{\xi}\vec{\xi}) + \mathbf{T}_{a//}$$

In (20) the first terms in which the mass transfer $\dot{m}_k$ , the tangential heat current $\vec{q}_k$ , and the viscous stress $\mathbf{T}_k$ appear express the lack of equilibrium between the fluid $k$ and the interface $S$. The term $\mathbf{T}_{a//} : \vec{\nabla}_s \vec{V}_s - \vec{q}_{a//} . (\vec{\nabla}_s T_s)/T_s$ expresses the non-equilibrium inside the interface. The divergence term $\vec{\nabla}_s . (\vec{q}_a - \vec{q}_a^*)$ implies that $\vec{q}_a = \vec{q}_a^*$ so that the entropy production be always positive. The two terms with the pressure $p_k$ in the fluid $k$ and the surface tension $\gamma$ are not classical; they take into account the discrepancy between the normal material velocity $\vec{V}_s . \vec{\xi}$ and the normal velocity $\vec{W}.\vec{\xi}$ . We use the relation $n_1 = -n_2 = \vec{\xi}$ to link these terms. So:

$$T_s \Delta = \sum_{k=1}^{2} \left\{ \dot{m}_k \left( g_k - g_s + (T_k - T_s)s_k + \frac{1}{2}(\vec{V}_k - \vec{V})^2 \right) - \vec{q}_k . \vec{n}_k \frac{T_s - T_k}{T_k} + \mathbf{T}_k : (\vec{V}_s - \vec{V}_k)\vec{n}_k \right\}$$

$$+ \mathbf{T}_{a//} : \vec{\nabla}_s \vec{V}_s - \vec{q}_{a//} . \frac{\vec{\nabla}_s T_s}{T_s} + \left\{ p_1 - p_2 + \gamma(\vec{\nabla}.\vec{\xi}) \right\} (\vec{V}_s - \vec{W}).\vec{\xi} \right\} \geq 0 \qquad (21)$$

We emphazise that the last term in (21) is the product of the difference between the two normal components of the velocities $\vec{V}_s$ and $\vec{W}$ , by $p_1 - p_2 + \gamma\left(\vec{\nabla}.\vec{\xi}\right)$. For an interface in thermodynamical equilibrium this last quantity is zero (Laplace's equation).If the two fluids 1 and 2 are immiscible, we have $\left(\vec{V}_s - \vec{W}\right).\vec{\xi} = 0$ and $\dot{m}_k = 0$ . Then, expression (21) for the surface entropy production reduces to the expression given in the litterature, for example in [2, 5]. If mass transfer between the fluids is allowed, and if we assume that $\left(\vec{V}_s - \vec{W}\right).\vec{\xi} = 0$, then we find the entropy production as derived by Barrère and Prud'homme [6].

## 5. A remark for an interface without mass

We now we consider the case where the interfacial mass density $\rho_a$ is zero. In order to obtain the jump conditions in this case we take the previous balance laws and we perform the limit process $\rho_a$ tending to zero. The thickness of the transition layer tends to zero, but the density $\rho$ inside is always finite. The velocities $\vec{V}$ and $\vec{V}_s$ are also finite. However, the internal energy $e$ and the entropy $s$ per unit of mass are very large inside the layer so that the interfacial internal energy $e_a = \rho_a e_s$ per unit of area and the interfacial entropy $s_a$ are finite. In the same way, the interfacial stress tensor $\mathbf{S}_{a//}$ and the interfacial heat current $\vec{q}_{a//}$ are finite. The mass balance law (10), the momentum balance law (14) with $\rho_a \vec{V}_s = 0$ and the balance law of energy lead to the following new jump conditions for an interface without mass:

$$\dot{m}_1 + \dot{m}_2 = 0 \tag{22}$$

$$\sum_{k=1}^{2}\left\{\dot{m}_k\vec{V}_k - \mathbf{S}_k.\vec{n}_k\right\} + \vec{\nabla}_s.\mathbf{S}_{a//} = 0 \tag{23}$$

$$\tag{24}$$

$$\frac{\delta_s}{\delta t}e_a + e_a\vec{\nabla}_s.\vec{W} = \sum_{k=1}^{2}\left\{\dot{m}_k\left(e_k + \frac{1}{2}\vec{V}_k^2\right) - \vec{V}_k.\mathbf{S}_k.\vec{n}_k + \vec{q}_k.\vec{n}_k\right\} + \vec{\nabla}_s.\mathbf{S}_{a//}.\vec{V}_s - \vec{\nabla}_s.\vec{q}_{a//}$$

To obtain (24) we do not use the mass balance law, but we have taken the initial equation (9). In equations (22), (23) and (24), the normal velocity $\vec{V}_{s\perp}$ does not appear. We only have the tangential velocity $\vec{V}_{s//} = \vec{W}_{//}$ and the normal geometrical velocity $\vec{W}_{\perp}$ . Thus, for an interface without mass, the distinction between $\vec{V}_{s\perp}$ and $\vec{W}_{\perp}$ is irrelevant. Finally, we give the entropy inequality and the interfacial irreversible entropy production for that limiting case:

$$\Delta \equiv \frac{\delta_s}{\delta t} s_a + s_a \vec{\nabla}_s . \vec{W} - \sum_{k=1}^{2} \left\{ \dot{m}_k s_k + \frac{\vec{q}_k}{T_k} . \vec{n}_k \right\} + \vec{\nabla}_s . \frac{\vec{q}_{a//}}{T_s} \geq 0$$

$$T_s \Delta = \left\{ \sum_{k=1}^{2} \left\{ \dot{m}_k \left( g_k + (T_k - T_s) s_k + \frac{1}{2} (\vec{V}_k - \vec{W})^2 \right) - \vec{q}_k . \vec{n}_k \frac{T_s - T_k}{T_k} + \mathbf{T}_k : (\vec{W} - \vec{V}_k) \vec{n}_k \right\} \right.$$

$$(25)$$

$$+ \mathbf{T}_{a//} : \vec{\nabla}_s \vec{W} - \vec{q}_{a//} . \frac{\vec{\nabla}_s T_s}{T_s} \geq 0$$

This last expression (25) is obtained by using the Gibbs relation $de_a = T_s ds_a$ the relation $\gamma = e_a - T_s s_a$ and equations (22), (23) and (24). The computation is classical.

# 6. Conclusion

We have introduced the "true" quantities by integration over the thickness of the layer. By logical considerations we have introduced two velocities on the interface: the material velocity of the particles belonging to the interface and the geometrical velocity of the interface. The difference between their normal components leads to an extra term in the irreversible production of interfacial entropy. We hope to have given some light on the meaning of these two velocities. It is important to notice that in the limit case of an interface without mass density, we are led to the conclusion that the distinction between the two velocities is irrelevant. This remark is also given in the papers [10, 11, 12, 13]. It is also important to notice that inside the interfacial transition layer, the fluid is not a classical Newtonian fluid. Many papers are concerned by this aspect. A similar asymptotic modelling is presented in [11, 14] by using a fluid endowed with capilarity; here an interface without mass density is obtained and described.

# Appendix

We use in a systematic way the dyadic notations that we recall here. In the oriented 3D-dimensional Cartesian space we only use orthonormal frames of the same orientation. The components of tensor $\vec{A}$ and $B$ of order 1 and 2 are denoted by $A_i$ and $B_{ij}$ where the indexes take the values 1, 2 or 3. The gradient operator is a formal vector with components $\partial_i = \partial / \partial x_i$ where $\vec{X} = (x_1, x_2, x_3)$ is the position vector in the space. The dyadic of two vectors is a tensor of order 2. The "simple product" of two tensors is denoted with one dot, and the "double product" with two dots. For example:

$$\vec{A}.\vec{B} = A_i B_i \; , \quad (\boldsymbol{A}.\boldsymbol{B})_i = A_{ij} B_j \; , \quad \vec{\nabla}.\vec{A} = \partial_i A_i \; , \quad \boldsymbol{A}:\boldsymbol{B} = A_{ji} B_{ij}$$

$$\left(\vec{A}\vec{B}\right)_{ij} = A_i B_j \; , \quad \left(\vec{\nabla}\vec{A}\right)_{ij} = \partial_i A_j$$

Finally, let be $S$ a surface and $\vec{\xi}$ a unit normal vector to the surface $S$ in the considered point. We introduce the following notations:

$\boldsymbol{P} = \boldsymbol{I} - \vec{\xi}\vec{\xi}$ , projection operator on $S$ $(\boldsymbol{P} = \boldsymbol{P}^2)$ with $\boldsymbol{I}$ identity operator,

$\vec{B} = \vec{B}_{//} + \vec{B}_\perp$ with $\vec{B}_{//} = \boldsymbol{P}.\vec{B} = \vec{B}.\boldsymbol{P}$ and $\vec{B}_\perp = \left(\vec{B}.\vec{\xi}\right)\vec{\xi}$ ,

$\boldsymbol{A}_{//} = \boldsymbol{P}.\boldsymbol{A}.\boldsymbol{P}$ where $\boldsymbol{A}$ is a tensor of order 2,

$\vec{\nabla}_s.\vec{B} = \boldsymbol{P}:\vec{\nabla}\vec{B}$ , surface divergence on $S$ ,

$\vec{\nabla}_s \varphi = \boldsymbol{P}:\vec{\nabla}\varphi$ , surface gradient of $\varphi$ .

# References

[1]  Germain, P. (1973): "Mécanique des milieux continus", *Masson et Cie.*

[2]  Delhaye, J.M. (1974): "Jump conditions and entropy sources in two-phase systems", *Int. J. Multiphase Flow 1, pp. 395-409.*

[3]  Scriven, L.E. (1960): "Dynamics of a fluid interface. Equation of motion for Newtonian surface", *Chem. Engng. Sci. 12.*

[4]  Prosperetti, A. (1979): "Boundary conditions at a liquid-vapor interface", *Meccanica 19, pp.123-136*

[5]  Bedeaux, D., Albano, A.M., Mazur, P. (1976): "Boundary conditions and non equilibrium thermodynamics", *Physica 82A, pp. 438-462.*

[6]  Barrère, M., Prud'homme, R. (1976): "Equations fondamentales de l'aérothermochimie", *Masson et Cie.*

[7]  Slattery, J.C. (1990) "Interfacial transport phenomena", *Springer-Verlag.*

[8]  Ishii, M. (1975): "Thermo-fluid dynamic theory of two phase flow" *Eyrolles.*

[9]  Landau, L.D., Lifshitz, E.M. (1959): "Statistical physics", *Pergamon Press.*

[10]  Gatignol, R. (1987): "Liquid-vapor interface conditions", *Rev. Roum. Sc. Techn.-Méc; Appl. 32, pp. 255-271.*

[11]  Gatignol, R., Seppecher, P. (1986): "Modelisation of fluid-fluid interfaces, with material properties", *J. Mécan. Th. et Appl., Numéro spécial, pp. 225-247.*

[12]  Dell'Isola, F., Kosinski, W. (1993) "Deduction of thermodynamic balance laws for bidimensional non material directed continua modelling interface layers", *Arch; Mech. 45, pp. 333-359.*

[13]  Prud'homme, R., (1995): "Balance equations for fluid curvilinear media", *this book.*

[14]  Seppecher, P., (1987): "Modélisation des zones capillaires", *Mémoire de thèse, Univ. Paris VI.*

# Surface excess momentum balances by integration across the surface of the volume balances*

A.Sanfeld[1,2] and A.Steinchen[2,3]
[1] LISA Université de Paris 7
[2] IUSTI Univ. Provence Marseille
[3] Thermodynamique Université Aix-Marseille3
Fac. des Sciences St Jérôme Bd Escadrille Normandie Niemen
13397 Marseille Cedex 20

## 1 Introduction

After the first unpublished attempts made by the authors in 1972, F.C. Goodrich [1] during his stay in Sanfeld's group published a paper on surface viscosities in which he established the basis for the understanding of surface transport properties in terms of "excess" quantities. He started from the concept of Gibbs attributing to the dividing surface a content of matter as well as a content of energy and of entropy and generalized this description to the transport coefficients. In the gibbsian method, any extensive property $Q$ of a two phase system may be split into three terms

$$Q = Q^1 + Q^2 + q \qquad (1)$$

in which $q$ is the contribution of the interface, $Q^1$ and $Q^2$ being the contribution of the bulk phases. It is evident that the sum $Q^1 + Q^2 + q$ is invariant, but that the individual values of the terms depends upon the location of the dividing surface. An important simplification already introduced by Gibbs [2] was the definition of "relative excess surface quantities" which are independant of the location of the dividing surface. These relative excess surface quantities are chosen as the excess quantities for which the absolute surface excess of one of the components is taken equal to zero. In the gibbsian approach, the adsorbed mass of the solvent in a multicomponent system was chosen as being zero to define the relative adsorptions $\Gamma_{\gamma 1}$ of the other components as well as for defining the relative surface internal energy $u^a_1$ or entropy $s^a_1$. The relation defining these relative quantities is

---

*The authors wish to dedicate the present paper to the memory of F.C.Goodrich*

$$\Gamma_{\gamma 1} = \Gamma_\gamma^a - \Gamma_1^a \frac{C_\gamma' - C_\gamma''}{C_1' - C_1''}$$

$$u_1^a = u^a - \Gamma_1^a \frac{u' - u''}{C_1' - C_1''}$$

$$s_1^a = s^a - \Gamma_1^a \frac{s' - s''}{C_1' - C_1''}$$

$$(2)$$

where $\Gamma_\gamma^a$ is the absolute surface excess of $\gamma$, $\Gamma_1^a$ is the absolute excess of the solvant 1, $u^a$ is the absolute excess surface internal energy , $u_1^a$ is the relative excess surface internal energy, $s^a$ is the absolute excess surface entropy, $s_1^a$ is the relative excess surface entropy, $C_\gamma'$ and $C_\gamma''$ are the mean concentrations of $\gamma$ in the two bulk phases, $C_1'$ and $C_1''$ the mean concentrations of the solvant 1 in each phase.

We may extend this definition of the relative surface excess to the total mass per unit area in the surface :

$$\Gamma_{T1} = \Gamma_T^a - \Gamma_1^a \frac{\rho' - \rho''}{\rho_1' - \rho_1''}$$

$$(3)$$

where $\Gamma_{T1}$ is the relative excess of total mass in the surface, while $\Gamma_T^a$ is the absolute excess of total mass in the surface, $\Gamma_1^a$ is the absolute excess of mass of the solvant in the surface, the mean mass densities in the bulk phases are $\rho'$ and $\rho''$ and the mean mass concentrations of the solvant in both phases are $\rho_1'$ and $\rho_1''$

For a pure system (one single component) the relative excess of mass is identical to the absolute excess of mass and is equal to zero. The situation is quite different if surface active materials are present in the system. If the surface excess of the solvant is chosen equal to zero, the sum of the surface excesses of the surfactants gives the value of the surface excess of total mass.

In a very simple way, Goodrich already showed that with the Gibbs convention it is possible to give a quantitative definition of the surface viscosity coefficients as capillary excess viscosities. As this problem has now taken a renewal of interest through the methods of asymptotic analysis and through the introduction of the second gradients in the developments of the local properties of the interfacial layer, (see for instance [3] )we will recall here the very elegant and simple method developed by our late friend and colleague F.C.Goodrich.

## 2 The momentum balances in a capillary layer

The capillary layer is characterized by a region of finite size in which the density varies continuously but very sharply from one bulk phase to the other bulk phase. Under static conditions, in the interior of the bulk phases adjacent to the layer, the static pressure tensor $\widehat{P}$ is isotropic while in the interfacial region it becomes an axially symmetric tensor with two independant components $p_T$ and $p_N$. If the fluids are in motion, viscous terms must be added to the static pressure tensor. In the two bulk phases the pressure tensor is

$$\widehat{P} = \begin{pmatrix} p + \pi_{xx} & \pi_{xy} & \pi_{xz} \\ \pi_{yx} & p + \pi_{yy} & \pi_{yz} \\ \pi_{zx} & \pi_{zy} & p + \pi_{zz} \end{pmatrix} \tag{4}$$

while in the interfacial region it is

$$\widehat{P} = \begin{pmatrix} p_T + \pi_{xx} & \pi_{xy} & \pi_{xz} \\ \pi_{yx} & p_T + \pi_{yy} & \pi_{yz} \\ \pi_{zx} & \pi_{zy} & p_N + \pi_{zz} \end{pmatrix} \tag{5}$$

If the fluids as well as the layer are supposed to be Newtonian, the following phenomenological equations hold in the bulk phases

$$P_{xx} = p - \left(\lambda + 2\eta\right)\frac{\partial v_x}{\partial x} - \lambda\,\frac{\partial v_y}{\partial y} - \lambda\,\frac{\partial v_z}{\partial z}$$

$$P_{yy} = p - \left(\lambda + 2\eta\right)\frac{\partial v_y}{\partial y} - \lambda\,\frac{\partial v_x}{\partial x} - \lambda\,\frac{\partial v_z}{\partial z} \tag{6}$$

$$P_{zz} = p - \left(\lambda + 2\eta\right)\frac{\partial v_z}{\partial z} - \lambda\,\frac{\partial v_x}{\partial x} - \lambda\,\frac{\partial v_y}{\partial y}$$

$$P_{xy} = -2\eta\frac{\partial v_x}{\partial y} \quad ; \quad P_{xz} = -2\eta\frac{\partial v_x}{\partial z} \quad ; \quad P_{yz} = -2\eta\frac{\partial v_y}{\partial z}$$

where $\lambda$ and $\eta$ are the Lamé coefficients. In the layer however, the axially symmetric tensor of fourth order has at most five independant components. We then have assumed the following constitutive equations

$$P_{xx} = p_T - \left(\lambda_T + 2\eta_T\right)\frac{\partial v_x}{\partial x} - \lambda_T\,\frac{\partial v_y}{\partial y} - \lambda_N\,\frac{\partial v_z}{\partial z}$$

$$P_{yy} = p_T - \left(\lambda_T + 2\eta_T\right)\frac{\partial v_y}{\partial y} - \lambda_T\,\frac{\partial v_x}{\partial x} - \lambda_N\,\frac{\partial v_z}{\partial z} \tag{7}$$

$$P_{zz} = p_N - \left(\lambda_N + 2\eta_N\right)\frac{\partial v_z}{\partial z} - \lambda_T\,\frac{\partial v_x}{\partial x} - \lambda_T\,\frac{\partial v_y}{\partial y}$$

$$P_{xy} = -\eta_T\left(\frac{\partial v_x}{\partial y} + \frac{\partial v_y}{\partial x}\right) \;\; ; \;\; P_{xz} = -\eta_N\left(\frac{\partial v_x}{\partial z} + \frac{\partial v_z}{\partial x}\right) \;\; ; \;\; P_{yz} = -\eta_N\left(\frac{\partial v_y}{\partial z} + \frac{\partial v_z}{\partial y}\right)$$

in which only four different Lamé coefficients appear : $\lambda_T$ , $\lambda_N$ , $\eta_T$ , $\eta_N$

An alternative formulation was given by Goodrich. Indeed, he proposed five different Lamé coefficients : $\lambda_T$ , $\lambda_N$ , $\eta_T$ , $\eta_N$ , $\eta_{Tz}$ and he wrote the constitutive equations

$$P_{xx} = p_T - \left(\lambda_T + 2\eta_T\right)\frac{\partial v_x}{\partial x} - \lambda_T\,\frac{\partial v_y}{\partial y} - \lambda_N\,\frac{\partial v_z}{\partial z}$$

$$P_{yy} = p_T - \left(\lambda_T + 2\eta_T\right)\frac{\partial v_y}{\partial y} - \lambda_T\,\frac{\partial v_x}{\partial x} - \lambda_N\,\frac{\partial v_z}{\partial z} \tag{8}$$

$$P_{zz} = p_N - \left(\lambda_N + 2\eta_N\right)\frac{\partial v_z}{\partial z} - \lambda_N\,\frac{\partial v_x}{\partial x} - \lambda_N\,\frac{\partial v_y}{\partial y}$$

$$P_{xy} = -\eta_T\left(\frac{\partial v_x}{\partial y} + \frac{\partial v_y}{\partial x}\right) \;\; ; \;\; P_{xz} = -\eta_{Tz}\left(\frac{\partial v_x}{\partial z} + \frac{\partial v_z}{\partial x}\right) \;\; ; \;\; P_{yz} = -\eta_{Tz}\left(\frac{\partial v_y}{\partial z} + \frac{\partial v_z}{\partial y}\right)$$

These equations are then combined with the momentum balance in the absence of external force

$$\rho\frac{d\vec{v}}{dt} = -\operatorname{Div}\widehat{P} \tag{9}$$

# 3 Integration through the layer of the momentum balances

In the capillary layer the pressure tensor is given by eq.(7) or (8) according to the description chosen. In the adjacent bulk phases it is given by eq.(6). To extend the method of Gibbs to dynamic situation one has to introduce the excess surface quantities. It means that we have to integrate across the interface the difference of the divergence of the pressure tensor in the layer and the divergence of the tensor of the two bulk phases extrapolated to the arbitrary division surface chosen. For a plane interface, integration of eq.(9) across the interface gives

$$\frac{d}{dt} \int_I^{II} \rho v_x \, dz = \Delta_s p_{xz} + \int_I^{II} \left[ \frac{\partial p_{xx}}{\partial x} + \frac{\partial p_{xy}}{\partial y} \right] dz \qquad (10)$$

$$\frac{d}{dt} \int_I^{II} \rho v_y \, dz = \Delta_s p_{yz} + \int_I^{II} \left[ \frac{\partial p_{xy}}{\partial x} + \frac{\partial p_{yy}}{\partial y} \right] dz \qquad (11)$$

$$\frac{d}{dt} \int_I^{II} \rho v_z \, dz = \Delta_s p_{zz} + \int_I^{II} \left[ \frac{\partial p_{xz}}{\partial x} + \frac{\partial p_{yz}}{\partial y} \right] dz \qquad (12)$$

Introducing now the constitutive equations (7) or (8) into the integrated balances (10) - (12) and introducing the two-dimensional divergence of $v : \theta = \dfrac{\partial v_x}{\partial x} + \dfrac{\partial v_y}{\partial y}$ we obtain either

## 3.1 Tangential balances

$$\frac{d}{dt} \int_I^{II} \rho v_x \, dz = \Delta_s p_{xz} - \frac{\partial}{\partial x} \int_I^{II} p_T dz + \frac{\partial}{\partial x} \int_I^{II} \left[ \lambda_T - \lambda_N + \eta_T \right] \theta \, dz + \nabla_s^2 \int_I^{II} \eta_T v_x dz$$

$$(13)$$

$$\frac{d}{dt} \int_I^{II} \rho v_y \, dz = \Delta_s p_{yz} - \frac{\partial}{\partial y} \int_I^{II} p_T dz + \frac{\partial}{\partial y} \int_I^{II} \left[ \lambda_T - \lambda_N + \eta_T \right] \theta \, dz + \nabla_s^2 \int_I^{II} \eta_T v_y dz$$

$$(14)$$

33

## 3.2 Normal balance

$$\frac{d}{dt}\int_I^{II}\rho v_z\,dz = \Delta_s p_{xz} + \int_I^{II}\eta_N\frac{\partial\theta}{\partial z}dz + \nabla_s^2\int_I^{II}\eta_N v_z dz \qquad (15)$$

or according to the Goodrich's choice

## 3.3 Tangential balances

$$\frac{d}{dt}\int_I^{II}\rho v_x\,dz = \Delta_s p_{xz} - \frac{\partial}{\partial x}\int_I^{II}p_T dz + \frac{\partial}{\partial x}\int_I^{II}[\lambda_T-\lambda_N+\eta_T]\,\theta\,dz + \nabla_s^2\int_I^{II}\eta_T v_x dz$$

$$(13bis)$$

$$\frac{d}{dt}\int_I^{II}\rho v_y\,dz = \Delta_s p_{yz} - \frac{\partial}{\partial y}\int_I^{II}p_T dz + \frac{\partial}{\partial y}\int_I^{II}[\lambda_T-\lambda_N+\eta_T]\,\theta\,dz + \nabla_s^2\int_I^{II}\eta_T v_y dz$$

$$(14bis)$$

## 3.4 Normal balance

$$\frac{d}{dt}\int_I^{II}\rho v_z\,dz = \Delta_s p_{xz} + \int_I^{II}\eta_{Tz}\frac{\partial\theta}{\partial z}dz + \nabla_s^2\int_I^{II}\eta_{Tz} v_z dz \qquad (15bis)$$

where $\Delta_s$ is the difference across the interface, $\theta = \frac{\partial v_x}{\partial x} + \frac{\partial v_y}{\partial y}$ and $\nabla_s^2$ is the two-dimensional Laplace operator.

# 4 The discontinuous model of surface

We now have to substract from these balances the balances obtained for the discontinuous model system. In this abrupt model of interface we write, introducing the Heaviside step function $h(z)$,

$$\lambda_T = \lambda_N = \lambda$$

$$\rho = \frac{1}{2}\left(\rho^I + \rho^{II}\right) + \frac{1}{2}\,h(z)\left(\rho^{II} - \rho^I\right)$$

$$p_T = p_N = \frac{1}{2}\left(p^I + p^{II}\right) + \frac{1}{2}\,h(z)\left(p^{II} - \rho^I\right) \tag{16}$$

$$\eta_T = \eta_N = \eta_{Tz} = \frac{1}{2}\left(\eta^I + \eta^{II}\right) + \frac{1}{2}\,h(z)\left(\eta^{II} - \eta^I\right)$$

In this discontinuous model, the velocities are continuous at the surface but the derivatives $\dfrac{\partial v_x}{\partial z}$ and $\dfrac{\partial v_y}{\partial z}$ are discontinuous as well as $\dfrac{\partial^2 v_z}{\partial z^2}$ .

We may now define like in the Gibbs treatment the following surface excess quantities :

a) $$\Gamma = \int_I^{II} \left(\rho - \frac{1}{2}\left(\rho^I + \rho^{II}\right) - \frac{1}{2}\,h(z)\left(\rho^{II} - \rho^I\right)\right) dz \tag{17}$$

which is the surface excess of mass. This quantity is zero if the system includes one single component and if the dividing surface is chosen in such a way that the integral in (17) vanishes. For a multicomponent system, the situation can be totaly different, especially if surface active substances are present. In that case, one often chose a dividing surface that corresponds to a zero surface excess of the solvent the total mass excess is then the sum of the excesses of the solutes,

b) $$\epsilon = \int_I^{II} \left(\eta_T - \frac{1}{2}\left(\eta^I + \eta^{II}\right) - \frac{1}{2}\,h(z)\left(\eta^{II} - \eta^I\right)\right) dz \tag{18}$$

where $\epsilon$ is the surface excess shear viscosity and

c) $$\kappa = \int_I^{II} \left(\lambda_T - \lambda_N + \eta_T - \frac{1}{2}\left(\eta^I + \eta^{II}\right) - \frac{1}{2}\,h(z)\left(\eta^{II} - \eta^I\right)\right) dz \tag{19}$$

where $\kappa$ is the surface excess dilational viscosity.
We now may also recall the Bakker's equation [4] defining the surface tension $\sigma$

$$\sigma = \int_I^{II} (p_N - p_T) dz \qquad (20)$$

The result of the integration across the interface of the difference of the balance equations in the layer and the balances in the discontinuous model is then

$$\frac{d}{dt} \Gamma v_x^s = \Delta_s p_{xz} - \frac{\partial \sigma}{\partial x} + \kappa \frac{\partial \theta^s}{\partial x} + \varepsilon \nabla_s^2 v_x^s \qquad (21)$$

$$\frac{d}{dt} \Gamma v_y^s = \Delta_s p_{yz} - \frac{\partial \sigma}{\partial y} + \kappa \frac{\partial \theta^s}{\partial y} + \varepsilon \nabla_s^2 v_y^s \qquad (22)$$

These equations correspond to the equations commonly used by the surface hydrodynamicists like Brenner and Wasan [5] (the inertial term is however often neglected), but the treatment has the interest to make the bridge between the surface viscosities introduced heuristically and excess properties defined in a more correct way.

For the normal balance, the problem is quite a bit different because, for a planar interface, one only finds the classical result that the difference of pressure across the surface is zero. However, for curved systems one has to handle more carrefully the normal momentum balance. Indeed it now will give rise to a generalization of the Laplace law in dynamical systems. For instance, for a sphere of radius R , one obtains

$$\frac{d}{dt} \Gamma v_r^s = \Delta_s P_{rr} - \frac{2\sigma}{R} - \varepsilon' \left[ \frac{1}{r^2} \frac{\partial}{\partial r} \left( r^2 \frac{\partial v_r}{\partial r} \right) + \frac{1}{r^2 \sin\theta} \frac{\partial}{\partial \theta} \left( \sin\theta \frac{\partial v_r}{\partial \theta} \right) + \frac{1}{r^2 \sin^2\theta} \frac{\partial^2 v_r}{\partial \phi^2} \right]_{r=R} \qquad (23)$$

where

$$\varepsilon' = \int_I^{II} (\eta_{\theta r} - \eta_r) dr \qquad (24)$$

is a surface excess shear viscosity coefficient.

If we adopt the five different viscosity coefficients like in Goodrich decomposition. But, if we adopt the phenomenological equations (7) with $\eta_r$ for the shear coefficient in the normal planes, the last term of eq.(23) vanishes. However, R.Aris [6] has introduced a surface viscosity term in his normal surface balance equation : he wrote for a surface without inertial term

$$\Delta_s P_{rr} = \frac{2\sigma}{R} + \frac{2\kappa}{R} \left[ \frac{1}{R\sin\theta} \left[ \frac{\partial}{\partial \theta} \left( v_\theta \sin\theta \right) + \frac{\partial v_\phi}{\partial \phi} \right] \right] \qquad (25)$$

where he calls κ the dilational surface viscosity coefficient. It seems thus that there exists a contradiction between our results and the equations heuristically introduced by Aris.

## References

[1] F.C. Goodrich, in "The modern Theory of Capillarity, Eds. F.C. Goodrich and A.I. Russanov, pp.19-34, Ak. Verlag Berlin (1981),.
[2] J.W. Gibbs " The Scientific Papers", I, Dover, New York, (1961)
[3] P. Casal and H. Gouin, Annales de Physique, 13, pp. 3-12 (1988)
[4] G. Bakker, in " Handbuch der experimental Physik", 6, Wien-Harms (1928)
[5] D.A. Edwards, H.Brenner and D.T. Wasan, "Interfacial Transport Processes and Rheology", Butterworth-Heineman series in Chemical Engineering, Boston, London, Oxford, (1991)
[6] R. Aris "Vectors, Tensors and the Basic Equations of Fluid Mechanics", Prentice Hall Publ., (Chap.9), Englewood Cliffs,N.J., (1962)

# Extended Irreversible Thermodynamics : Towards a Non-Local Formulation

G. Lebon

Liège University, Institute of Physics B5, Sart Tilman, B-4000 Liège 1, Belgium and Department of Mechanics, Louvain University, B-1348 Louvain-la-Neuve, Belgium

**Abstract:** In the first part of this note, one briefly recalls the motivations and the basic hypotheses underlying Extended Irreversible Thermodynamics. To illustrate the application of the formalism, one considers the problem of a viscous fluid subject to thermal heating : the relevant constitutive equations are established and their relation with the classical formulation is discussed. Extended Thermodynamics appears to be useful in treating materials involving large relaxation times (like dielectrics at low temperature, polymers,...) and high-frequency and short-wavelength processes (ultrasounds, light and neutron scattering,...). However to deal with short-wavelength phenomena, it is imperious to introduce non-locality in space. This is achieved by including supplementary variables taking the form of fluxes of the fluxes. The corresponding changes in the expression of the constitutive equations and the entropy flux are discussed.

## 1 Extended Irreversible Thermodynamics

Extended Irreversible Thermodynamics (EIT) is born out of the necessity to enlarge the range of applicability of classical irreversible thermodynamics to large frequencies and short wavelengths phenomena [1]. EIT is particularly well suited for describing ultrasonic wave-propagation, heat propagation in dielectrics at low temperature (phonon hydrodynamics), shock waves, light and neutron scattering in liquids and gases as well as materials involving long relaxation times like polymeric solutions, dielectric materials, superfluids and superconductors.

The basic idea in EIT is to extend the space of state variables by including the flux of matter, flux of momentum and flux of energy among the set of basic variables.

Denoting by $V$ the space of variables, it will be formed by the union $C \cup J$ of the set of classical variables $C$ (like mass, velocity and energy) and the flux variables $\mathbf{J}$ (like the flux of matter, momentum and energy). The classical variables $C$ obey the well-known conservation laws of mass, momentum and energy while the extra variables **J** are supposed to satisfy Markovian evolution equations of the form

$$\dot{\mathbf{J}} = -\nabla.\mathbf{F^J} + \sigma^J \tag{1}$$

a dot stands for the time derivative (either the material or any "objective" time derivative),$\mathbf{J}^F$ is the flux of flux **J** and $\sigma^J$ a source term. The closure of the set of evolution equations is ensured by expressing $\mathbf{F}^J$ and $\sigma^J$ in terms of the whole set $V$ of variables by means of constitutive equations. Since the final results have to comply with the second law of thermodynamics, one must introduce a non-equilibrium entropy. The latter is supposed to be a convex potential function of the whole set $V$ of variables and its rate of production is supposed to be non-negative.

To illustrate these general ideas, consider the motion of an incompressible mono-component fluid subject to heating. The basic variables are velocity, energy (or temperature $T$) plus the corresponding fluxes, namely the viscous stress tensor $\sigma$ and the heat flux vector **q**. The relevant evolution equations are the classical momentum and energy balance laws plus the evolution equations for the heat flux and the viscous stress tensor (for simplicity the latter is assumed to be symmetric and traceless) :

$$\dot{\mathbf{q}} = -\nabla.\mathbf{F^q} + \sigma^q \tag{2}$$

$$\dot{\sigma} = -\nabla.\mathbf{F}^v + \sigma^v \tag{3}$$

where the unknown quantities $\mathbf{F}^q$, $\sigma^q$, $\mathbf{F}^v$ and $\sigma^v$ will be given by constitutive equations.

From the positiveness of the entropy production $\sigma^s$ defined through

$$\sigma^s = \rho\dot{s} + \nabla.\mathbf{J}^s \geq 0 \tag{4}$$

wherein the entropy density $s$ and the entropy flux $\mathbf{J}^s$ are expressed by means of constitutive relations $s = s(V)$ and $\mathbf{J}^s = \mathbf{J}^s(V)$, it is shown [1] that equations (2) and (3) take the following form

$$\tau_1\dot{\mathbf{q}} = -(\mathbf{q} + \lambda\nabla T) + \beta\lambda T^2\nabla.\sigma \tag{5}$$

$$\tau_2\dot{\sigma} = -(\sigma - \eta(\nabla\mathbf{v})^{\text{sym}}) + \beta\eta T\nabla\mathbf{q} \tag{6}$$

By formulating (5) and (6), one has omitted quadratic and higher order terms in the fluxes. The undefined quantities in (4)-(6) are : $\rho$ the density, $\lambda$ the heat conductivity, $\eta$ the dynamic shear viscosity, **v** the velocity, $\tau_1$ and $\tau_2$ the relaxation times, $\beta$ a coupling parameter, the same $\beta$ appears in both relations (5) and (6) because of Onsager's reciprocity property. It is also shown that $\mathbf{J}^s$ will be given by

$$\mathbf{J}^s = T^{-1}\mathbf{q} + \beta\sigma.\mathbf{q} \tag{7}$$

while the entropy $s$ obeys a generalized Gibbs equation

$$ds = T^{-1}du - pdv - \tau_1\lambda T^2\,\mathbf{q}.d\mathbf{q} - \tau_2\eta T\,\sigma : d\sigma \tag{8}$$

wherein $p$ and $v$ denote the pressure and the specific volume respectively ; the two first terms in the r.h.s. of (8) are classical, the two last terms are new. From $\sigma^s > 0$, it is inferred [1] that

$$\lambda > 0 \qquad \eta > 0 \tag{9}$$

while the convexity property of $s$ implies that

$$\tau_1 > 0 \qquad \tau_2 > 0 \tag{10}$$

Expressions (5) and (6) contain as particular cases the classical Fourier law for heat conduction and Newton's law of hydrodynamics by letting $\tau_1$, $\tau_2$ and $\beta$ vanish. By setting in (5) and (6), the coupling parameter $\beta$ equal to zero, one recovers the Cattaneo-Vernotte and Maxwell relations, namely

$$\tau_1 \dot{\mathbf{q}} = -\mathbf{q} - \lambda \nabla T \tag{11}$$

$$\tau_1 \dot{\sigma} = -\sigma + \eta (\nabla \mathbf{v})^{\text{sym}} \tag{12}$$

However, even the rather general relations (5) and (6) reveal too crude to describe high frequency and high wavenumber phenomena as ultrasound propagation and light and neutron scattering by fluids. Moreover, it is expected that systems with very short characteristic wavelengths, like micro-devices and interfaces, require a more refined description.

## 2 Towards a Second Gradient Description

Within the framework of EIT, this can be achieved in a rather systematic way by introducing additional variables ; these extra variables may take the form of fluxes of fluxes. For instance in the problem of heat conduction, Cattaneo-Vernotte equation will be generalized in the form

$$\tau^{(1)} \dot{\mathbf{q}}^{(1)} = -\nabla . \mathbf{q}^{(2)} - \mathbf{q} - \lambda \nabla T \tag{13}$$

where $\mathbf{q}^{(2)}$ is a tensor of rank two representing the flux of the heat flux $\mathbf{q}$. Next, equation (13) will be complemented by evolution equations for the fluxes $\mathbf{q}^{(2)}$, $\mathbf{q}^{(3)}$, ... , $\mathbf{q}^{(n)}$:

$$\tau^{(2)} \dot{\mathbf{q}}^{(2)} = -\nabla . \mathbf{q}^{(3)} - \sigma^{(2)} \tag{14}$$

$$\tau^{(n)} \dot{\mathbf{q}}^{(n)} = -\nabla . \mathbf{q}^{(n+1)} - \sigma^{(n)} \tag{15}$$

wherein $\sigma^{(2)}$ , ... $\sigma^{(n)}$ represent the corresponding source terms. For simplicity, consider only $\mathbf{q}^{(1)}$ and $\mathbf{q}^{(2)}$ as variables and express $\sigma^{(2)}$ and $\mathbf{q}^{(3)}$ by means of constitutive equations. At the lowest order of approximation, one has

$$\sigma^{(2)} = a\mathbf{q}^{(2)} , \quad \mathbf{q}^{(3)} = b\mathbf{q}^{(1)}\mathbf{I} \tag{16}$$

wherein $a$ and $b$ are arbitrary scalar functions of the temperature while $\mathbf{I}$ stands for the identity tensor. Assuming $\tau^{(2)} << \tau^{(1)}$, and eliminating $\mathbf{q}^{(2)}$ between (13) and (14), it is finally found that equation (13) reads as

$$\tau \dot{\mathbf{q}} = -\mathbf{q} - \lambda \nabla T - \gamma \nabla^2 \mathbf{q} \qquad (17)$$

wherein we have dropped superscript (1). Expression (17) contains a non-local term in $\nabla^2 \mathbf{q}$ and is referred to in the literature as Ginzburg-Landau or Guyer-Krumhansl type equation [2] ; such a relation has been widely used to study heat conduction in dielectric crystals at very low temperature between 5° K and 20° K.

It should be mentioned that non-locality can be introduced into EIT without using higher order fluxes, whose physical meaning is not always easy to assign, at least at the macroscopic level of description. One may for instance express the fluxes $\mathbf{F}^q$ and $\mathbf{F}^v$ defined through equations (2) and (3) in terms of the gradient of the whole set of variables. To be explicit, if $\mathbf{F}^q$ contain a term in $\nabla \mathbf{q}$ and $\mathbf{F}^v$ a term in $\nabla \sigma$, then the evolution equations (5) and (6) for $\mathbf{q}$ and $\sigma$ will be generalized as

$$\tau_1 \dot{\mathbf{q}} + \mathbf{q} + \lambda \nabla T = -\lambda T^2 \beta \nabla . \sigma - L_2 \nabla . \nabla \mathbf{q} \qquad (18)$$

$$\tau_2 \dot{\sigma} + \sigma - \eta (\nabla \mathbf{v})^{\text{sym}} = -\eta T \beta \nabla \mathbf{q} - l_2 \nabla^2 \sigma \qquad (19)$$

which contain "second gradients" of $\mathbf{q}$ and $\sigma$.

It is worth emphasizing that the above results (18) and (19) refer to weak non-locality as non-local effects are truncated at second order terms in the gradients. It is nevertheless hoped that such an approximation will be sufficient for describing a great majority of non-local effects.

# References

1.    D. Jou, J. Casas-Vazquez and G. Lebon, *Extended Irreversible Thermodynamics*, Springer-Verlag, Berlin, 1st edition 1993, 2nd edition 1996
2.    T. Dedeurwaerdere, J. Casas-Vazquez, D. Jou and G. Lebon, Phys.Rev. E, 1996 (in press)

Acknowledgements. The author wishes to thank Profs. A. Steinchen (Marseille) and A. Sanfeld (Paris) for valuable comments.

# Transient Surface Properties of Liquid Bridge and Pendant Drop Menisci in Gravity and Low Gravity

J F PADDAY

Nether Crutches, Jordans, Beaconsfield, Bucks, HP9 2TA UK.

**Abstract**. Liquid bridges and pendant drops, whether in a force field or in a neutral environment are all menisci. Those of greatest interest are axisymmetric in shape, which when perturbed or stretched, tend to break into three parts one of which is a free floating satellite drop. In this study an attempt has been made to understand the transient surface properties that control the processes of bifurcation and breakage of the liquid system.

The paper demonstrates that a trifurcation process occurs similarly with both high and low viscosity liquids, and both lead to the formation of a connecting rod of liquid between the two halves of the bridge. At low viscosity the rod appears to break simultaneously (in 0 g) at both ends by Rayleigh instability of the type attributed to the break up of jets. At high viscosity, the rod continues to thin and stretch until it disappears evanescently. In terrestrial gravity breakage occurs similarly and a satellite drop is formed.

A new type of instability was found using two liquids of widely different viscosity. A bridge formed from two miscible half zones, one of high and one of low viscosity, formed a satellite drop impaled on a thread of the high viscosity liquid.

Keywords. Liquid bridges, pendant drop, menisci in low gravity, microgravity, capillary instability, drop breakage

## 1    Introduction

The process of crystal growth from the melt involves forming a liquid bridge between the solid crystal and the free surface of the molten salt [1]. Satisfactory growth requires that the bridge does not break and also that, in the earlier stages, the diameter of the crystal be increased from the few mms of the seed to 200 or more mm. This expanding growth is achieved by withdrawing the solid crystal upwards at a rate determined by the surface properties of the stable bridge. Thus a knowledge of the stability limit provides the crystal grower with boundary conditions within which stable growth may be obtained.

In printing, an ink is applied to a solid surface (Image areas), from an inking roller. As the inked off-set printing surface seperates from the inking roller many liquid bridges are broken and in that process minute droplets of ink are formed in the surrounding atmosphere (ink fly). The study of the process of liquid bridge rupture thus reveals  how much liquid ink is transferred and the source and size of the droplets released into the air.

Ink application takes place in liquid bridges that are 1mm or less in length. On this scale surfaces forces and viscosity predominate with gravity playing a minor or second order role. Also the scale is so small that very little detail can be observed. Also, dynamic processes of breakage of the very small system occur within  microseconds - thus necessitating the use of ultra high speed recording. It has been shown that by studying much larger liquid bridges in reduced gravity [2] the main features of the microscopic system becomes observable and the characteristic time for bifurcation and breakage is increased by three orders of magnitude to the millisecond range which is now within the range of normal video and cine cameras.

a)      Break up of a liquid zone with an unduloid neck

b)      Break up into equal parts with satellite drop

c)      Break up of amphora with neck and bulge into twop unequal parts

Figure 1      Symmetrical and unsymmetrical trifurcation of a
               liquid bridge in low gravity

The study of the trifurcation and breakage of menisci both in terrestrial and zero gravity has only recently become of interest. It has been known from the studies of Lord Rayleigh [3] that low viscosity water emerging from a tap very soon breaks up into droplets whilst very high viscosity liquids such as syrup form an ever thinning thread which does not break up into droplets at all.

The explanation is to be found in understanding the type of Laplace shape the liquid bridge assumes prior to it reaching the point of critical stability [4]. The stability of liquid bridges in zero gravity has attracted wide attention in recent years but most investigators did not investigate the shapes of such critically stable bridges. Here it is proposed to detail this shape in a generalised form.

This study sets out the main features of the stabiltiy limit and the subsequent bifurcation and breakage processes of a bridge of the much larger system performed in reduced gravity ( 0.02 g ). The study and understanding of the shape of the system prior to rupture provides an indication of the amount of liquid remaining each side of the broken bridge. The bifurcation and breakage of a pendant drop, occuring in 1 g, corresponds to a type of liquid bridge breakage and has been studied previously. These data have been compared with the stability studies of the low gravity system.

Consider the formation and breakage of a liquid bridge formed in a low gravity field such as is seen in the two examples of figure 1. Both liquid bridges have been formed with a low viscosity liquid held betwwen two end plates of equal diameter. The upper two photos show that with a relatively small amount of liquid breakage occurs symmetrically but in the third photo, with a somewhat larger volume of liquid, bifurcation and breakage is clearly asymmetric with a larger volume on the left hand side. If one is to understand the process of bifurcation an explanation of this asymmetric must first be sought.

## 2    Liquid Bridge Shape Prior to Bifurcation in Zero g

A liquid bridge is defined as an axisymmetric meniscus supported between two rotationally symmetric solid surfaces separated by a known distance apart measured along the axis of symmetry. There are many types of stability to be considered but here the stability of a fixed volume maintained at a known distance of separation is the only type to be analysed in detail.

Liquid bridge shape is obtained by integrating Laplace's capillary equation, which relates the sum of the principal curvatures to the ratio of the capillary pressure to the surface tension. Most investigators reduce the dimensions to some unit length that here is the radius of curvature of the plates. In the event of the two plates being of unequal size then the radius of the smaller plate is taken as unit length. The shapes of any given bridge depends solely on the ratio, C, defined in equation 1,

$$[\ 1/\ R_v + 1/\ R_h\ ] = P/\ \sigma\ = C \qquad ..................(1)$$

where $R_v$ and $R_h$ are the principal radii of curvature of the liquid - fluid interface, P, the capillary pressure and $\sigma$, the surface tension.

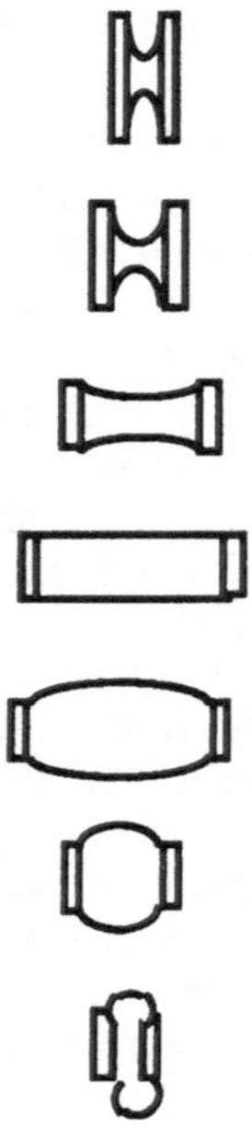

| Shape | Pressure | Constant C |
|---|---|---|
| Invert Nodoid | - ve | < 0 |
| Catenoid | 0 | = 0 |
| Unduloid | + ve | 1 <C> 0 |
| Cylinder | + ve | = 1 |
| Unduloid | + ve | 2 <C> 1 |
| Sphere | + ve | = 2 |
| Nodoid | + ve | C > 2 |

Figure 2    Laplace Capillary Shapes applied to liquid bridges. Each bridge type is derived from the locus of foci of rolling conic sections.

Integration of equation 1 produces a set of shapes that are solely dependent on the value of C. They are shown in figure 2. In principle each shape is generated by the locus of the focal point of a conic section rolling along the axis of symmetry of the liquid bridge. A full treatment of these geometric identities is given by Bouasse [5].

In this study Laplace's capillary equation has been integrated using 4 th order Runge Kutta but with a floating argument chosen according to the end data being sought. The choice of arguments for the increments were:

$\Sigma$ F, distance along profile: normal integration process

$\Sigma$ ø, angle subtended at the axis by the radius: preparing tables at set angles

$\Sigma$ X, the distance of the surface normal to the axis of symmetry

$\Sigma$ V, the total volume of the liquid bridge

Also it was necessary to adjust the increment in F to much smaller values when passing through the inflexion points. A second summation process was performed simultaneously in order to derive the total energy of the liquid bridge, using the equation:

$$E = \sum \{\sigma.2.\pi.X.dF\} + \sum \{P.\delta V\} \quad ........(2)$$

where E is the total energy, made up of two components, the work done increasing the surface and the work performed against capillary pressure. When the capillary pressure is zero as with catenoid shapes, the second term is zero and for invert nodoid systems P is negative and can exceed the value of the first term. It is now possible to set criteria for equilibrium and stability for the particular liquid bridge bounded by completely wetted end plates and for a fixed volume and fixed separation distance between the end-plates. The criteria are:

Stable   equilibrium

$$\delta E / \delta \varphi = 0 --- \& --- \delta^2 E / \delta \varphi^2 = +ve \quad ............(3)$$

Unstable equilibrium

$$\delta E / \delta \varphi = 0 --- \& --- \delta^2 E / \delta \varphi^2 = -ve \quad ............(4)$$

Critical equilibrium

$$\delta E / \delta \varphi = 0 --- \& --- \delta^2 E / \delta \varphi^2 = 0 \quad .......(5)$$

Non-equilibrium

$$\delta E / \delta \varphi = \pm ve \quad .......(6)$$

where $\delta\phi$ is a perturbation of shape. Though the perturbation may be any stimulation or artificially induced change of profile, those that are most damaging may be considered as those that are of lowest energy change. Such perturbations must be axisymmetric and usually involve the capillary pressure changing linearly along the axis of symmetry or otherwise as a sine wave perturbation as used by Lord Rayleigh [3].

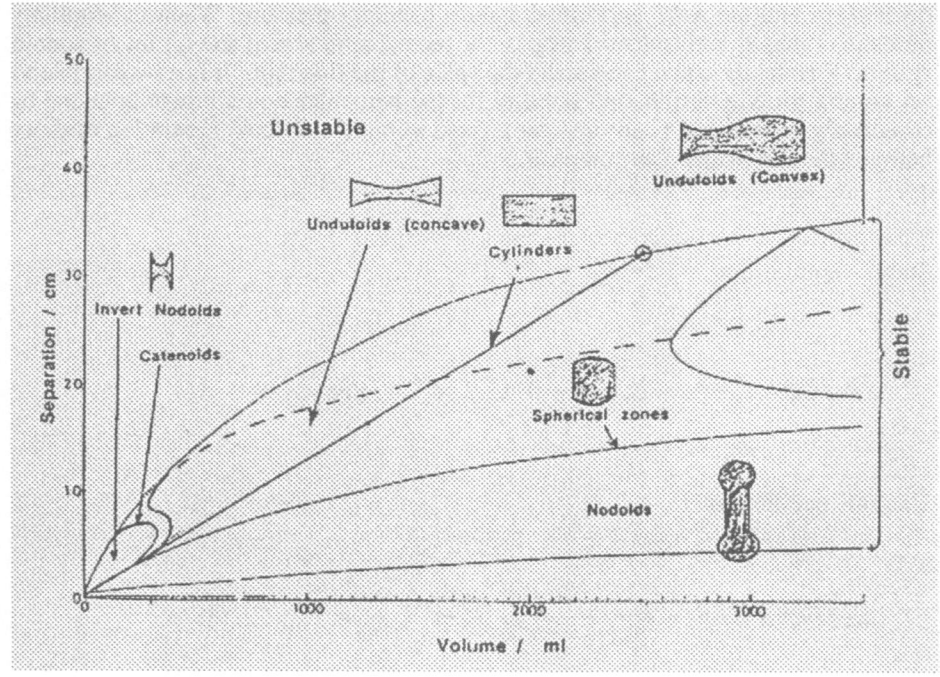

Figure 3    Liquid Bridge Shapes Formed Between Two Equal End Plates

Consider a family of liquid bridges formed between two equal end plates of radii 50mm [6].The separation between the plates  is plotted in figure 3 (figure 9 of ref [6]), as a function of the bridge volume of liquid for a wide variety of bridge shapes. To give clarity, all data points have been omitted and only boundary lines are shown. The upper curved line sets the boundary between all Laplace equilibria; stable {eq.3}, unstable {eq.4} and critical {eq.5}.  Shapes which lie below the line  and those that are either not Laplace shapes or otherwise are non-equilibrium shapes {eq.6}, lie above the line. In addition there exists a second boundary of stability, the lowest of the curves, on which lay all critically stable shapes and above which are contained once again, all stable and unstable Laplace equilibrium shapes. Thus the two boundary curves form enveloping borders encompassing all equilibrium shapes.

The upper boundary denotes the first type of stability limit at which the liquid bridge bifurcates and eventually breaks into two halves whilst the lower limit denotes a second, totally different type of instability, whereby the compressed toroid of liquid becomes asymmetric and a droplet escapes from the system leaving the diminished volume once more at stable equilibrium. This study considers instabilities of the first type at the upper boundary only.

In figure 3 a straight line extends from the origin and denotes all the combinations of separation and volume that form a cylindrical bridge between the two end plates. Up to the critical stability limit all cylinders are stable. Rayleigh [3] and others have shown that, at the critical limit of a rod, the separation, Z, and consequently the volume, V between successive necks are given by

$$Z = 2.\pi.X \qquad\qquad \dots\dots\dots\dots(7)$$

$$V = 2\pi^2.X^3 \qquad\qquad \dots\dots\dots\dots(8)$$

The data of figure 3 were plotted on the basis of a real value of X equal to 50mm thus the critical cylinder length is 314.159mm and the critical volume is 2.4674011 litres. Let this point be designated the Plateau point.

All critical shapes with volumes greater than that of the Plateau point will be of the unduloid convex type and will posses both a bulge and a neck. This shape is sometimes referred to as an amphora. Such bridges bifurcate asymmetrically as would be expected. At relatively small volumes catenoids are formed and all such shapes must lie on an isobaric line as marked in the figure. Within the area of the graph enclosed by the catenoid curve all liquid bridges posses negative capillary pressure and are uniquely invert nodoid shaped. Here invert is used to indicate that the liquid (phase of greater density) lies outside the generating curve. Liquid bridges with liquid volumes greater than those of catenoids and yet lying to the LHS of the cylinder line form concave unduloids as shown. Whereas all catenoids and invert nodoids bifurcate and break symmetrically, it is found experimentally that concave unduloids do not behave uniformly. Those critical volumes lying near to the Plateau point appear to bifurcate unsymmetrically and those nearer to the catenoid curve to break symmetrically. Somewhere between these two positions lies the unique point at which critical instability changes from symmetry to asymmetry.

The corresponding stability limits for liquid bridges have been widely investigated by others and a stability diagram is presented in reference 6 as figure 8. Obviously all liquid bridges that posses an inbuilt asymmetry must inevitably bifurcate unevenly. Even so, they are classed into two sets, those that follow a bifurcation process similar to bridges with a bulge and those that correspond to bifurcation of a catenoid wheteher or not between equal or unequal end-plates.

Consider now what happens in physical terms once critical stabiltiy is reached and bifurcation commences.

# 3     Bridge Stability, Bifurcation and Breakage of a Liquid Bridge in Zero g

Although the stability criteria for liquid bridges has been widely studied, The bifurcation of a liquid bridge leading to it breakage has not received such wide attention. Those studies that have been attempted have tended to be theoretical based on computer modelling [7].

In the course of extensive experimental studies the author has shown that the bifurcation process - that is the irreversible flow of liquid away from the neck of the bridge - occurs in two stages and if the liquid viscosity is sufficiently low, is closely followed by total breakage of the bridge. Figure 6 of reference [4] shows bifurcation and breakage very clearly.

The stages are as follows:

Bifurcation   - 1 -

The liquid bridge cascades through a series of non-equilibrium axisymmetric Laplace shapes with the surface area continuously diminishing and a capillary pressure gradient being generated along the axis with the greatest pressure being located at the central position of the neck.

Trifurcation  - 2 -

The next stage is for the bridge to become stressed so that each half zone attached to the end plates adopts a pyramidal shape connected at the neck by a cylinder of liquid. The cylinder or rod of liquid increases in length and at the same time diminishes in diameter until a critical point is reached at which each end of the rod breaks away from the supporting pyramids of liquid.

Breakage

The process of breakage is fast and catastrophic and is completed within 30 milliseconds. The rod of liquid released at both ends simultaneously contracts into a single satellite drop which free floats between the two half zones.

Recoil

The release of stress in the two half zones following breakage is seen as each half pulsating in and out along the axis of symmetry until finally the new equilibrium position is reached. The final shape assumed is, of course, that of a spherical cap.

Experimentally it was not possible to record the liquid bridge in its transient state immediately prior to breakage therefore it became rather a matter of luck for this shape to be recorded. At present, attention will be confined to the ratio of the length of the rod to its mean radius at this point as it is very likely that the actual breakage will follow a pattern similar to that of the break up of jets [3,5].

In figure 4 below three frames of the breakage sequence show that breakage takes place very rapidly, the consequetive frames being seperated by 30 milliseconds. The centre frame in fact shows a silhouette of the shape immediately prior to indentation of each end of the rod of liquid leading to breakage.

The diameter of the rod in the centre frame is about 1.0 mm diam and 2,75 mm long, and the satellite drop in the third frame is about 1.5 mm diam. The magnification of the photographs in figure 4 is 4.0 Thus the comparison of volumes produces:

Volume of neck rod = 0.00216 ml

Volume of satellite drop = 0.00178 ml

The comparison of the volumes is only correct to an order of magnitude because the satellite drop diameter is not sufficiently clear to provide an accurate measurement. The main conclusion is that breakage at the end of each side of the rod is very likely to be simultaneous.

In Rayleigh's treatment of the break up of a jet into droplets [3] he derived the total energy of a section of the jet perturbed by an axisymmetric wave perturbing the rod or jet. He assumed that the energy of perturbation is due solely to surface area changes and such changes were constrained at constant volume per unit wavelength of undisturbed rod. His equation for energy change is given by:

$$E - E_0 = (\pi.\alpha^2 / 2X).(k^2.X^2 - 1) \qquad \ldots\ldots\ldots(9)$$

using the symbols of this study with $\alpha$ being a very small fraction displacement of the radius $X$, and $K = 2.\pi.1$ where 1 is the wavelength. Thus for stability the energy increase must be positive, for critical rupture zero and for instability, negative.

A similar perturbation at constant volume of the rod forming the neck region, can be considered on the condition that perturbation and breakage takes place very much more rapidly than the slow evolution of the rod thinning process. Furthermore, one assumes that the wave form of the perturbation is that as measured from from neck to neck with the bulge of the wave at the centre of the rod. The value of $2.\pi..X$ for the rod at breakage equals 3.125 mm as compared with the estimated length of 2.75 mm.

It is thus concluded that there is a prima facie case for considering liquid bridge breakage at low viscosity as a type instability driven solely by surfaces forces. The reason that the process occurs so quickly is related to the very small amounts of liquid that needs to be moved very short distances in order to effect the break . Thus

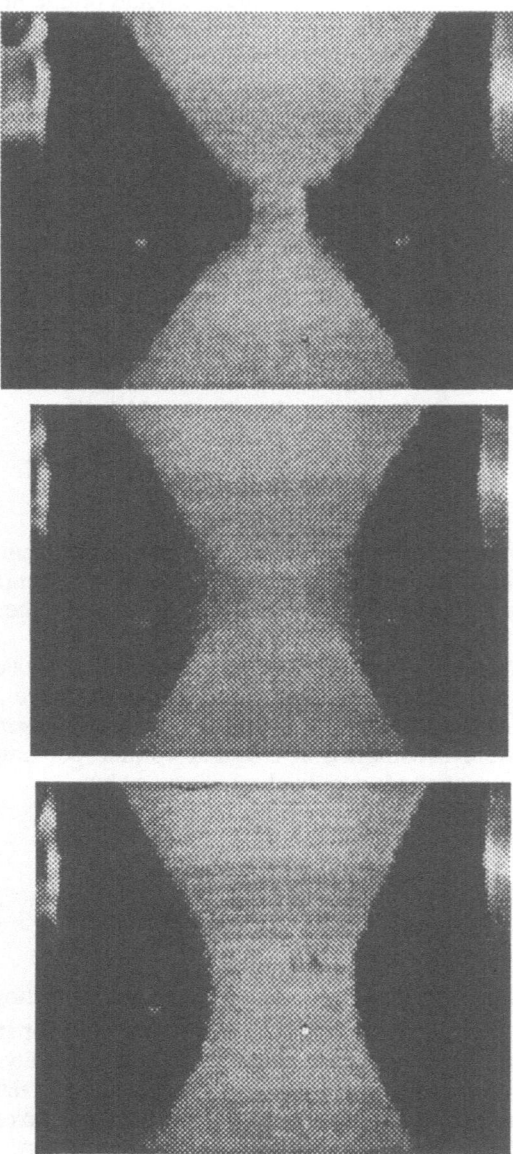

Figure 4  Liquid Bridge bifurcation and breakage
with satellite drop formation

one would expect the rate process to be controlled by an inertia term that is best analysed by classical hydrodynamics.

The data of figure 4 were obtained from an experiment conducted at about .02 g during parabolic flight. The end plates, not completely visible in the pictures, were 30mm diam. And the liquid itself was a mixture of 90% water with 10% glycerol. Recording was in PAL using a Sony Video 8mm recorder. The single frames were then copied photographically and the prints scanned electronically to produce the prints shown in the figure.

# 4      Pendant Drop Stability, Trifurcation and Breakage in 1 g

The bifurcation and breakage of a pendant drop in a terrestrial field has also been studied using an ultra high speed cine camera taking 2500 frames / second.

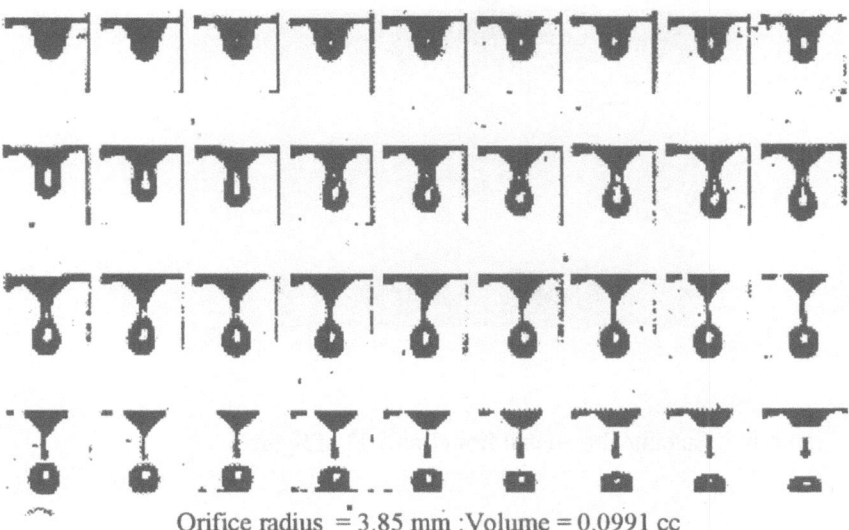

Orifice radius = 3.85 mm : Volume = 0.0991 cc

Figure 5      The Bifurcation and Breakage of a Pendant Drop
in Terestrial Gravity

The pendant drop of figure 5 adopts its critically stable shape in the first frame of the top row. The orifice radius was 3.85mm and the liquid was water. The time interval between each frame was thus 400 μs. The bifurcation and breakage occurred with remarkable similarity to the bridge behaviour in zero gravity with the exception that the satellite drop, once released, produced an unduloid structure prior to colapsing into a single round drop.

The periodic structure of the satellite dropof figure 6 is explained in terms of the gravitational hydrostatic pressure gradient inherited by the liquid rod prior to its breakage. The rod of liquid in this experiment was found to be slightly cigar shaped prior to rupture and it is very clear from the bottom row of frames of figure 5 that the

Figure 6    Satellite drop detail from frame 33 of figure 5

main drop breaks away from the connecting rod of liquid well before the upper end of the rod breaks away from the liquid attached to the orifice.

The principal radii of curvature of the surface showing this characteristic periodicity are obviously smaller at the top than at the bottom suggesting that the capillary pressure is greater at the top than at the bottom. The probable reason for this is that breakage occurred at the lower end and the collapsing process, which relieves the capillary pressure is more advanced than at the top.

# 5     A New type of Two Component Liquid Bridge Instability in Reduced Gravity

It has already been reported that bifurcation of liquid bridges formed with very high viscosity liquids do not break in the same manner as those with low viscosity liquids. It has been shown [4] that, at high viscosity (60Pas) the rod of liquid forming the narrow part of the thinning neck does not break when the length reaches πxdiameter, but rather the diameter continues to stretch and thin whilst the

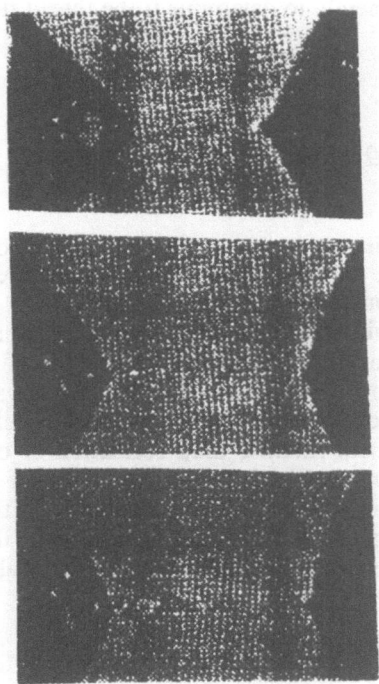

Figure7           Composite Liquid Bridge Showing
New Type of Instability

two half zone relax towards the shape of spherical caps whilst still being connected.

The thread never seemed to break at all it just seems to disappear altogether without recoil. Accordingly experiments were performed using two miscible liquids such that the left hand end plate held a spherical cap of low viscosity and the right hand side, a cap of very high viscosity of the thread forming type. These two zones under low gravity conditions were brought together to form a liquid bridge and then moved apart so that the critical length was reached.

The first stages of bifurcation were followed as though the bridge were formed of a low viscosity liquid . When the ratio of length to diameter exceeded $\pi$ , The breakage began to occur fairly slowly , as shown in frame 1 of figure 7. Thereafter some of the liquid broke away from each end towards the centre leaving a thread connecting the two sides.

The outer casing of of liquid receded from each end of the rod to form a satellite drop impaled on the high viscosity thread as seen in frames 2 and 3 of figure 7. This new type of instability demonstrates that both thread formation and breakage to form a satellite drop can occur simultaneously and are separate physical processes. Also the characteristic relaxation of the two half zones took place as expected with thread formation. All these experiments were carried in the low gravity of parabolic flight which had the limitation of only 20 seconds of low gravity. Thus it was not possible to age the the thread sufficiently to determine if the thread finally retracted into the drop or the two half zones.

# 6    Discussion and Conclusions

The main purpose of this investigation was to investigate both experimentally and theoretically the physics of the process of bifurcation and breakage of liquid bridge system in zero and terrestrial gravity.

It is clearly shown that bifurcation of the system is driven solely by surface forces accompanying a reduction of the total area of the system. At first the systems cascade through a series of axisymmetric Laplace shapes that are not at equilibrium and then competing surface and viscous forces appear to govern the pathway which eventually leads either to breakage with satellite drop formation or otherwise leads to the formation of an ever thinning thread. Both these types of bifurcation can occur simultaneously with a system formed of two liquids of very different viscosity.

The next stage of investigation is to produce an account of the liquid movement as the non-equilibrium Laplace shapes evolve to the pyramidal half zones connected by a cylindrical rod of liquid.

# Acknowledgements

The experiments performed in parabolic flight were carried out in two campaigns during 1990 and 1993. The author is indebted to the considerable in-flight assistance given by Drs Richard Davies, Nigel Wildman , Christophe Corneujols and Vladimir Pletzer all of whom have acted as experimental flight assistants. Also the theoretical work was performed within the framework of the European Network programme contract    HCM ERBCHRXCT 940481

# References

| 1 | Martinez I and Eyer A | "Liquid Bridge Analysis of Silicon Crystal Growth Experiments under Microgravity" J Cryst. Growth 75(3), 535-544, 1986 |
|---|---|---|
| 2 | Carruthers J R and Grasso M | "Studies of Floating Zones in simulated Zero Gravity ", J. App. Phys., 43(2), 436-445, 1972 |
| 3 | Lord Rayleigh | "On Liquid Jets - The Length of the Continuous Part " Proc. Roy. Soc. London ; June 15, p137 1882: "On the Instability Jets", Proc. London Math. Soc., X,4-13, 1879. |
| 4 | J F Padday | "The Formation and Breakage of Liquid Bridges Under Microgravity", Microgravity Quarterly; Vol 2, No4, 239-249. 1992 |
| 5 | H Bouasse | "Capillarité Phénoménes Superficiels ",Delgrave,Paris 1924. Chapter III |
| 6 | J F Padday | "Capillarity in Microgravity". In "Lecture Notes in Physics" No386 Eds Pétré and Sanfeld. Springer-Verlag, p90, 1991 |
| 7 | R M S M Schulkes | "Non-linear Dynamics of Liquid Columns: A Comparative Study" submitted to Physics of Fluids A - |

# Bénard-Marangoni Instability in a Rotating Liquid Layer with a Deformable Free Surface

G. Lebon [1], P. Cerisier [2], O. Dupont [3]

[1]Université de Liège, Institut de Physique B5, Sart Tilman, B-4000 Liège 1, Belgium and Département de Mécanique, Université Catholique de Louvain, B-1348 Louvain-la-Neuve, Belgium
[2]IUSTI, UMR-CNRS 139, Université d'Aix-Marseille 3, F-13397 Marseille 20, France
[3]MRC, Chimie Physique EP, C.P. 105, Université Libre de Bruxelles, B-1050 Bruxelles, Belgium

**Abstract:** This work deals with Bénard-Marangoni convection in a thin horizontal fluid layer heated from below. The lower surface of the layer is in contact with a uniformly heated rigid plate and the upper face is deformed and submitted to a temperature-dependent surface tension. Two situations are analyzed : in the first one, the container is at rest ; in the second one, it is uniformly rotating about a vertical axis. The role of the combined effects of surface deflection and rotation is examined. Two problems receive a particular attention : overstability and the observation that in pure surface-tension driven instability, the system is unconditionally unstable with an infinite critical wavelength.

## 1 Introduction

There are two causes which are responsible for the onset of convection in a thin fluid layer subject to a temperature gradient, namely buoyancy forces (Rayleigh-Bénard instability) and surface tension gradients (Marangoni instability). The problem was first examined by Nield [1] who assumed that the layer was heated from below with a non-deformable upper free surface ; in addition, the experimental set up containing the fluid was at rest : no rotation of the support was allowed. The effect of a deformable upper surface was investigated by Scriven and Sternling [2] and Smith [3] in the case of surface-tension driven instability (Marangoni problem), these works were revisited by Takashima [4,5] who considered in particular the problem of a layer heated from above and that of a layer whose underside is a free surface. Scriven and Sternling [2] showed that disturbances of sufficiently short wave numbers are unconditionally unstable, even for a vanishing temperature difference between bounding surfaces. This unpleasant result was circumvented by Smith [3] who argued that gravity waves may play an important role at short wave numbers. By including the gravity waves in his description, Smith showed that instability at small wave numbers occurs for a small but non-zero temperature gradient. It was supposed by Sternling and Scriven as well as by Smith that convection starts under the form of steady motion. This result was justified by Vidal and Acrivos [6] in the case of Marangoni instability in layers with a non-deformable upper surface. But as soon as the surface is allowed to be deformed, oscillatory convection cannot be excluded a priori as shown by Takashima [4,5]. The above discussion refers to pure capillary driven instability. When this effect is coupled with buoyancy, the hypothesis of exchange of stability was still proved to remain valid [7] at least in absence of surface deflections. The problems resulting from the presence of a deformable upper surface were examined by Perez-Garcia and Carneiro [8]. These authors determine under which conditions which mode, either oscillatory or steady, will be observed. In the present work, we go one step further by allowing the system to rotate as a whole : the upper surface is deformable and the fluid is heated from below. The paper is organized as follows. In Sect. 2, the basic equations governing the instability problem are established. The main results obtained by Smith [3], Takashima [4,5] and Perez-Garcia and Carneiro [8] are analyzed in Sect. 3. The effects resulting from the coupling of rotation and surface deflections are examined in Sect. 4 in both cases of steady and oscillatory motions; the competition between these two modes receives a particular attention. In Sect. 5, conclusions are drawn and open questions are raised. Disturbances are supposed to be of infinitesimal amplitude so that the analysis is strictly linear.

## 2 Mathematical Formulation

Consider a thin horizontal liquid layer heated from below ; the following hypotheses are introduced : the liquid is newtonian, incompressible and boussinesquian, the layer extends horizontally to infinity and rotates at a uniform angular velocity about a vertical axis. The lower surface of the layer is in contact with a rigid perfectly heat conducting plate. The upper surface is free and deformed with a deflection $\delta(x, y, t)$ with respect to the mean height $d$, it is also assumed that heat transfer between the upper surface and the motionless gas in contact with it is negligible. In the unperturbed reference state, the fluid is at rest with respect to the rotating container and heat propagates by conduction : the corresponding velocity and temperature fields $\mathbf{v}_r$ and $T_r$ are given by

$$\mathbf{v}_r = 0 \;, \;\; T_r = T_H - \beta z \left( \beta = \frac{\Delta T}{d} > 0 \right)$$

where $T_H$ is the temperature at the hot lower face while $\Delta T$ is the temperature difference between the lower and upper surfaces, $z$ is the vertical component pointing in opposite direction of the gravity acceleration $\mathbf{g}$. After that motion has set in, the temperature disturbance $\theta (= T - T_r)$ and the velocity field $\mathbf{v}(u, v, w)$ obey the following balance equations :

$$\nabla \cdot \mathbf{v} = 0 \tag{1}$$

$$(\partial_t + \mathbf{v} \cdot \nabla)\, \mathbf{v} = -\nabla\left(\frac{\mathbf{P}}{\rho_0} - \frac{1}{2}\,|\Omega \times \mathbf{r}|^2\right) + \alpha g \theta \mathbf{e_z} + \nu\, \nabla^2 \mathbf{v} + 2\mathbf{v} \times \Omega \mathbf{e_z} \tag{2}$$

$$(\partial_t + \mathbf{v} \cdot \nabla)\, \theta = \kappa \nabla^2 \theta \tag{3}$$

$p$ is the pressure, $\rho_0$ a constant reference density, $\Omega$ the constant vertical angular velocity, $\mathbf{r}(x, y)$ the horizontal position vector, $\alpha$ the volume expansion, $\nu$ the kinematic viscosity and $\kappa$ the heat diffusivity ; in absence of rotation, the two terms in Eq. (2) involving $\Omega$ will be cancelled. Cartesian coordinates will be used with the origin located at the lower plate. The corresponding boundary conditions are :

$$at\ the\ lower\ surface\ (z\ =\ 0)\ :\ w\ =\ \partial_z w\ =\ \theta\ =\ 0 \tag{4}$$

$$at\ the\ upper\ surface\ (z\ =\ d + \delta)\ :\ w\ =\ \partial_t \delta\ ,\ \partial_z \theta\ =\ 0 \tag{5}$$

$$\rho\nu\, (\partial_x w\ +\ \partial_z u)\ =\ -\gamma\, (\partial_x \theta\ -\ \beta\partial_x \delta) \tag{6}$$

$$\rho\nu\, (\partial_y w\ +\ \partial_z v)\ =\ -\gamma\, (\partial_y \theta\ -\ \beta\partial_y \delta) \tag{7}$$

plus a condition on the pression which will be of no use here, the two last equations express the equilibrium between the viscous and surface tension forces, $\gamma$ denotes the variation of the surface tension $\xi$ with temperature :

$$\gamma\ =\ -\frac{\partial \xi}{\partial T}\ \ (\gamma > 0) \tag{8}$$

The boundary conditions (5)-(7) are transferred from $z = d + \delta$ to $z = d$ after peforming a Taylor expansion and omitting second order corrections. From now on, the procedure is classical. The continuity equation and the pressure gradient are eliminated by applying twice the rot operator on Eq. (2). This procedure results in three differential equations (which are made dimensionless) for the three unknown quantities $w$ (the z-component of velocity), $\theta$ (the temperature disturbance) and $\zeta$ (the z-component of the vorticity). Expressing $w, \theta, \zeta$ and the deformation $\delta$ in terms of normal modes, namely

$$(w, \theta, \zeta, \delta)\ =\ [W(z)\ ,\ \Theta(z)\ K(z)\ ,\ E]\ \exp\ [i(k_x x\ +\ k_y y)\ +\ \sigma t] \tag{9}$$

where $\mathbf{k}(k_x, k_y)$ is the horizontal wave number and $\sigma$ the (complex) time growth and substituting in (1)-(3), one obtains the following equations for the amplitudes $W(z), \Theta(z)$ and $K(z)$ :

$$(D^2\ -\ k^2\ -\ \sigma)\,(D^2\ -\ k^2)\,W\ -\ Ta\ DK\ =\ Ra\ k^2 \Theta \tag{10}$$

$$(D^2\ -\ k^2\ -\ \sigma)\,K\ =\ -DW \tag{11}$$

$$\left(D^2\ -\ k^2\ -\ \text{Pr}\ \sigma\right)\Theta\ =\ -W \tag{12}$$

where $D = d/dZ$ while the dimensionless numbers $Ta$ and $Ra$ are defined by :

$$Ta \ = \ 4 \ \frac{\Omega^2 d^4}{\nu^2} \quad (Taylor \ number)$$

$$Ra \ = \ 4 \ \frac{\alpha g \Delta T d^3}{\nu \kappa} \quad (Rayleigh \ number)$$

For further purpose, we introduce also the Marangoni, crispation, Bond, and Prandtl numbers :

$$Ma \ = \ (\gamma \Delta T d) \ / (\rho_0 \nu \kappa) \ (Marangoni \ number)$$

$$Cr \ = \ (\rho_0 \nu \kappa) \ / (\xi d) \ (crispation \ number)$$

$$Bo \ = \ (\rho_0 g d^2) \ / (\xi) \ (Bond \ number)$$

$$\mathrm{Pr} \ = \ \nu \ / \ \kappa \ (Prandtl \ number)$$

The corresponding boundary conditions are :

$$at \ z \ = \ 0 \ : \ W \ = \ \Theta \ = \ K \ = \ DW \ = \ 0 \tag{13}$$

$$at \ z \ = \ 1 \ : \ D\Theta \ = \ DK \ = \ 0 \tag{14}$$

$$W \ = \ \sigma \ \mathrm{Pr} \ E \tag{15}$$

$$(D^2 \ + \ k^2) \ W \ = \ - \ k^2 \ (\Theta \ - \ E) \ Ma \tag{16}$$

$$Cr \ [(D^2 \ - \ 3k^2 \ - \ \sigma) \ DW \ - \ Ta \ K] \ = \ (B_0 \ + \ k^2) \ k^2 \ E \tag{17}$$

The set of equations (10)-(17) constitutes an eigenvalue problem which is solved numerically by generalizing a method proposed by Nield 51] ; it consists essentially in developing the unknown quantities in terms of Fourier expansions and forcing them to obey the boundary conditions (13)-(17) ; for details, the reader is referred to references [9] and [10]. At neutral stability ($Re \ \sigma \ = \ 0$), the stability parameter reduces to its imaginary part ; if in addition the hypothesis of exchange of stability is satisfied, one has simply $\sigma \ = \ 0$. In the next Sect., we assume that the fluid is not rotating (Ta = 0) : two subcases will be examined, respectively pure Marangoni instability ($Ra \ = \ 0 \ , \ Ma \ \neq \ 0$) and afterwards the more general Bénard-Marangoni problem ($Ra \ \neq \ 0 \ , \ Ma \ \neq \ 0$).

# 3 Non-Rotating Fluids with a Deformable Surface

## 3.1 Marangoni instability

The eigenvalue problem yields an expression of the Marangoni number in terms of the crispation number Cr, the Bond number Bo and the wave number $k$ :

$$Ma = \ Ma(Cr, Bo, k)$$

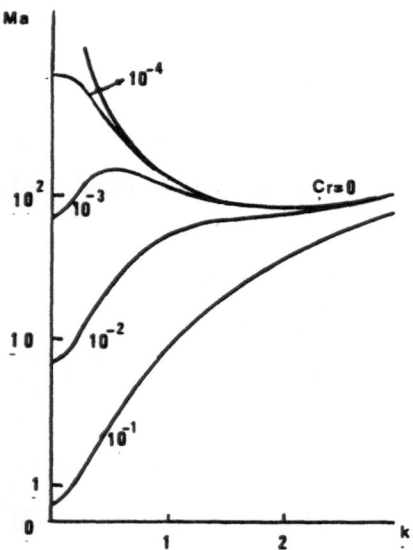

**Fig. 1.** Neutral stability curves Ma versus k, for several values of the crispation number (Ra = 0, Bo = 0.1) (reproduced from Takashima [4])

The Prandtl number and the imaginary part of $\sigma$ do not appear if exchange of stability is supposed to be satisfied, which is the case that we examine first. Neutral stability curves corresponding to Bo = 0.1 and several values of Cr are given in fig. 1, reproduced from Takashima's paper [4]. The case Cr = 0 corresponds to a non-deformable surface and was treated by Pearson [11] : the critical Marangoni and wave number take the well-known values $Ma^c \approx 80$ and $k^c \approx 2$. For $Cr \neq 0$, the neutral stability curve is characterized by two minima, one at k = 0, the other at k $\approx$ 2. In Scriven and Sternling [2] work, Bond's number was not introduced and it was found that the lowest minimum of the marginal curve was located at $Ma^c = 0$, $k^c = 0$, from which follows that the system should be unconditionally unstable. This unphysical result was circumvented by Smith [3] who introduced gravity waves through a non-zero Bond number. Smith showed that the critical Marangoni number was no longer zero at k = 0 and that the critical $Ma^c$ increases when the crispation number is lowered (see Fig. 1). It was proved by Takashima [4] that for $Cr < Bo / 120$, the critical values $Ma^c$, $k^c$ are these found by Pearson, which means that surface deformations are unimportant in Marangoni problem for Cr-values smaller than Bo/120.

Beyond this value, the critical $Ma^c$ is sensitive to the values taken by Bo and Cr while the corresponding critical wave number is equal to zero. Let us now assume that exchange of stability is not verified ; interesting results are then displayed. It was shown by Takashima [5] that overstability is only possible for $Ma < 0$ ; negative Marangoni values mean either that one heats from above a fluid with $\partial \xi / \partial T < 0$ (the usual case) or that one heats from below a fluid with $\partial \xi / \partial T > 0$s. Critical Ma-values as a function of Cr are

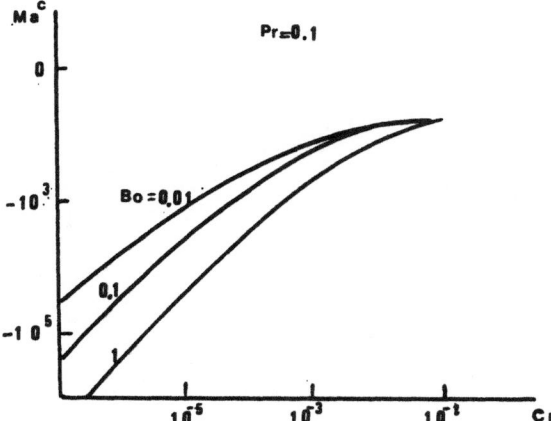

**Fig. 2.** Critical Marangoni number as a function of the crispation number for several Bond number values (Ra = 0). (Reproduced from Takashima [5])

represented in fig. 2. The regions above the curves represent stable situations. It is essentially shown that deformations have a destabilizing role while large values of Bo promote stability. An important difference with the steady convection is that in the present case, the critical wave number is not zero and grows with the crispation number.

Hitherto we have neglected the influence of gravity : this effect will be incorporated into the analysis in the next sub-section.

## 3.2 Rayleigh-Bénard-Marangoni instability

We now examine to which extent the previous results are modified by including buoyancy forces.

Interesting results have been proposed by Perez-Garcia and Carneiro [8] and Gouesbet et al. [12]. The physical situation is well described in a $Ma^c - Ra^c$ plane (see Fig. 3). In absence of surface deformation (Cr = 0), the locus of the critical points $Ma^c - Ra^c$ is a straight line which was first established by Nield [1], the region below this straight line represents stable states while the points above these curves correspond to unstable states. It should also be emphasized that for a given fluid and a given layer depth, the Ma and Ra numbers are not independent but related by

$$Ma = \Gamma Ra \qquad (18)$$

wherein

$$\Gamma = \frac{\frac{\partial \xi}{\partial T}}{\rho_0 g \alpha d^2}$$

this relation represents a straight line crossing the origin O. Since $\Gamma$ varies like $d^{-2}$, the slope of OP is close to $\pi/2$ for very thin layers and tends to zero when the layer becomes thicker and thicker. The critical temperature difference is

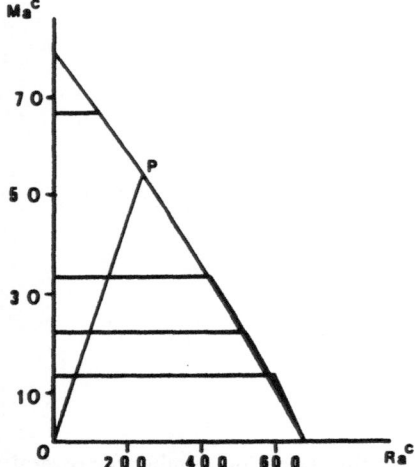

**Fig. 3.** Critical $Ma^c - Ra^c$ loci in the plane $Ma^c - Ra^c$ ($B_0 = 0.1$), the horizontal lines correspond to $Cr = 6.10^{-3}$, $3.10^{-3}$, $2.10^{-3}$, $10^{-3}$ respectively starting from the bottom

proportional to the distance between O and the crossing point P at the interSect. of the line $Ma = \Gamma Ra$ and Nield's line.

Typical results corresponding to steady convection are given in fig. 3. As in Marangoni problem, there exist two critical modes, one with $k^c = 0$, the other with $k^c \approx 2$. The horizontal plateaus of fig. 3 describe regions where convection starts with $k_c = 0$ ; in these regions, $Ma^c$ does not vary with $Ra^c$ but its value depends strongly on the crispation number. $Ma^c$ increases when Cr is decreased and below a critical value of Cr $\left(Cr < 8.10^{-4}\right)$, one recovers Nield's straight line, which means that surface deformations cease to be effective. It is also observed in Fig. 3 that for a given Cr, the portion of the curve corresponding to $k^c \approx 2$ lies beyond Nield's curve : this indicates that for sufficiently large values of Ra, surface deformation is stabilizing ; recall that in pure Marangoni convection, surface deflection is always destabilizing.

The next question to be asked is under which condition overstability may occur. Perez-Garcia and Carneiro [8] have shown that oscillatory convection is not possible for positive Ma-values, overstability is only observable for negative Ma's. This is seen on Fig. 4, where the regions above the dotted lines represent stable states. It is worth stressing that the results are only significant for Ra-values satisfying $Ra < B_0/Cr$ ; this inequality results from Boussinesq approximation requiring that $\alpha \Delta T < < 1$. As a consequence, it is only in a small region of fig. 4 corresponding to $Ra \lesssim 100$ that the results are relevant. It is thus observed that overstability is only predicted in regions with relatively high negative values of Ma ($|Ma| \gtrsim 1000$). In the limiting case of Cr tending to zero, the critical $Ra^c$ is infinite from which is concluded that no overstability is expected in absence of surface deformation.

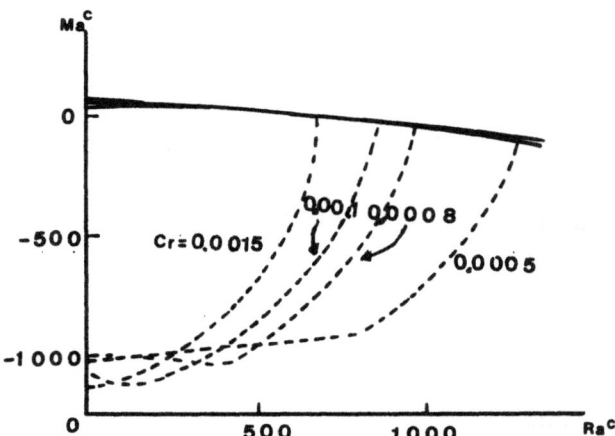

**Fig. 4.** Critical oscillatory states for different values of the crispation number in the $Ma^c - Ra^c$ plane (reproduced from [8])

## 4 Rotating Fluids with an Upper Deformable Surface

We have now to deal with the complete set of equations (10)-(17) with $Ta \neq 0$ and $Cr \neq 0$. Several authors, among which Chandrasekhar [13], Veronis [14], Rossby [15], have studied buoyancy driven instability in rotating systems but only a few authors (Vidal and Acrivos [16], Mc Conaghy and Finlayson [17]) have treated Marangoni convection ; the coupled buoyancy and surface-tension driven instability was examined by Sarma [18] but we have checked that his results are in error. One of our objective was therefore to correct and to generalize Sarma's work by including deflections of the upper free surface. In a first stage we consider only steady convection : exchange of stability is taken for granted. In very thin layers or in microgravity conditions (Ra = 0), it is found, as represented in Fig. 5, that the Marangoni number is practically unaffected by the surface deflection. In contrast, $Ma^c$ increases considerably when the angular velocity of the fluid is increased ; similarly, it is shown that the critical wave number increases with the angular velocity. Clearly, as expected, rotation plays a strong stability role. The same conclusion can be drawn when in addition buoyancy is present, as shown in Fig. 6. The region of stability (i.e. the region below the representative curves) is becoming larger and larger when Ta is increased. But now the effect of a surface deformation becomes more perceptible, as exhibited by Fig. 7 which represents Nield's curves in the $Ma^c - Ra^c$ plane for a fixed value of the Taylor number and different values of the crispation number. Surface deformation has a double effect : it produces a curvature of the Nield straight line (see Fig. 7) and may appreciably modify the critical temperature difference at which motion sets in ; this effect becomes relatively important for large crispation numbers $(Cr > 10^{-3})$. We now focus our attention on the problem of overstability. In this case, the stability parameter $\sigma$ is no longer zero at marginal stability but

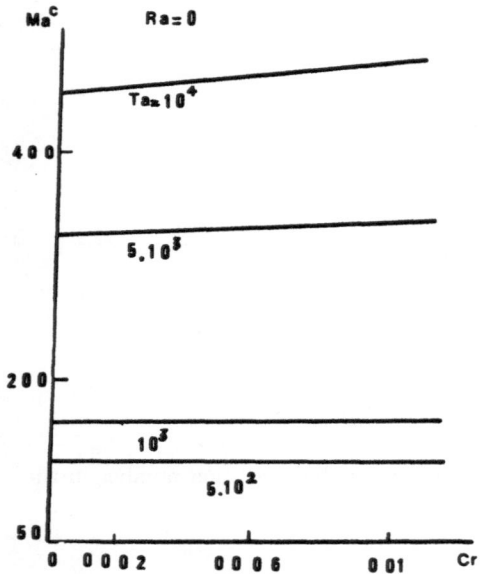

**Fig. 5.** Critical Marangoni number versus the crispation number for several values of Taylor number

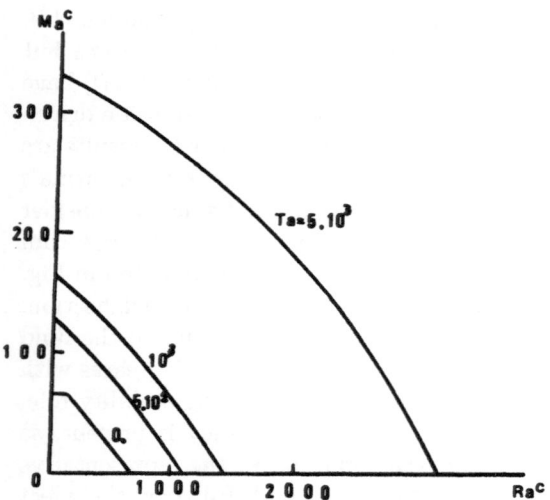

**Fig. 6.** Neutral stability curves in the $Ma^c - Ra^c$ plane for different values of Ta : Ta $= 0$ corresponds to a non-rotating fluid $\left(Cr = 10^{-3}\right)$

it takes imaginary values $(Im\ \sigma \neq 0)$. The solutions of the dispersion relation expressing Ma in terms of $Im\ \sigma$ and the relevant parameters, namely Ra, Ta, Pr, Cr, $B_0$ and k are now complex :

$$Ma = (ReMa) + i(ImMa)$$

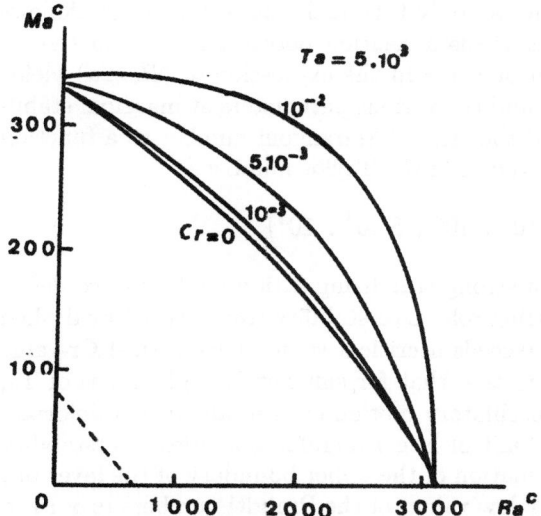

**Fig. 7.** Neutral stability curves in the $Ma^c - Ra^c$ plane for several values of Cr and $Ta = 5.10^2$. The dashed straight line corresponds to zero values of Ta and Cr

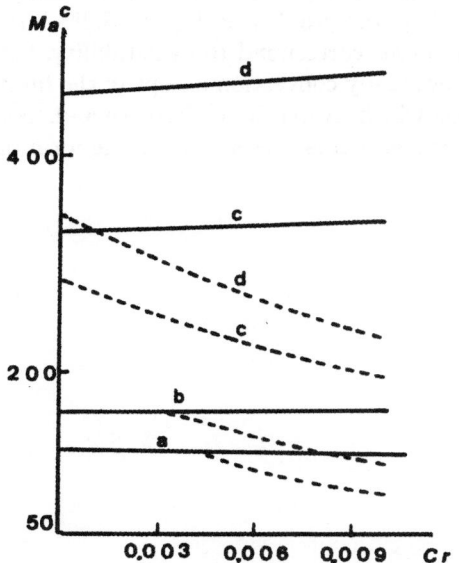

**Fig. 8.** Critical Marangoni number $Ma^c$ vs the crispation number Cr for Ra = 0, Pr = 0.1 and four different values of the Taylor number : $Ta = 5.10^2$ $(a)$ , $Ta = 10^3$ $(b)$ , $Ta = 5.10^3$ $(c)$, $Ta = 10^4$ $(d)$. Overstable solutions are represented by dashed curves

Since Ma is a physical quantity, it can only take real values from which follows that $ImMa = 0$. This conditions yields a relation between $Im\ \sigma$ and the remaining parameters. Substitution of $Im\ \sigma$ in the expression of $(ReMa)$ yields the required relation between Ma and the various parameters at marginal stability. In Fig. 8, we have represented the critical Marangoni number as a function of the crispation number for four values of the Taylor number

$$\left(Ta\ =\ 5.10^2\ ,\ 10^3\ ,\ 5.10^3\ ,\ 10^4\right)$$

It is seen that Coriolis force has a strong stabilizing action while surface deformation plays generally a destabilizing role. Overstability (represented by dashed lines) is observed as soon as Cr exceeds a critical value. This critical Cr-value decreases when Ta grows and it is seen that for sufficiently high values of Ta, instability occurs in the form of oscillatory motion even in absence of deformation, i.e. for Cr = 0. Within the limit of larger angular velocities, overstability is unavoidable whatever the deformation of the upper boundary of the layer but this is only observed at sufficiently low values of the Prandtl numbers (say Pr < 5). At higher Prandtl numbers, we have shown that the hypothesis of exchange of stability is justified whatever the angular velocity and the surface deformation. The same conclusion is valid when buoyancy is included (Ra $\neq$ 0). When both surface tension and buoyancy effects are acting, the results of the analysis are conveniently represented in a $Ma^c - Ra^c$ graph (see Fig. 9).Solid lines describe steady convection while dotted curves correspond to overstability. For small Ta's (Ta < 500), instability occurs as steady convection except in the limit of very small Ra's (0 $\leq$ Ra $\leq$ 120) for which overstability is predicted. For larger rotation velocities $\left(Ta\ \geq\ 10^3\right)$, convection is present under the form of

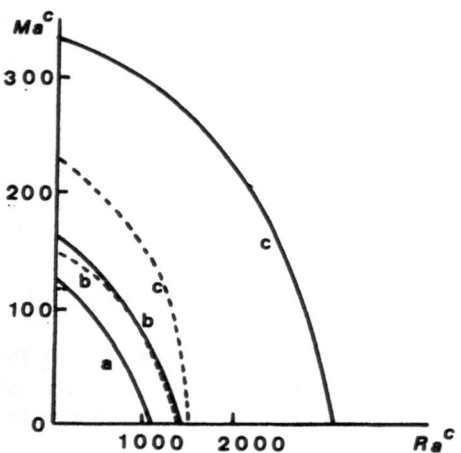

**Fig. 9.** Critical Marangoni number $Ma^c$ as a function of the critical Rayleigh number $Ra^c$ for $Cr\ =\ 5.10^3$ , $Pr\ =\ 0.1$ , $Bo\ =\ 0.1$ and values of the Taylor number $Ta\ =\ 500\ (a)$ , $Ta\ =\ 10^3\ (b)$ , $Ta\ =\ 5.10^3\ (c)$. Solid lines describe steady instability, dashed lines correspond to oscillatory instability

oscillatory motions whatever the value of Ra ; indeed, the dotted curves (oscillations) are lower than the solid lines (stationary convection). After that the angular velocity has been fixed, it is observed that the presence of overstability is promoted by increasing the surface deformation. This is not surprising and was still observed in non-rotating fluid layers. It can be concluded from the above analysis that overstability will only be observed for small Prandtl numbers (Pr < 5) and that both surface deflection and rotation are factors that promote the appearance of overstability.

## 5 Conclusion and Open Problems

The present study is concerned with a linear analysis of thermoconvection in respectively non-rotating and rotating fluids with a deformable upper free surface ; the fluid layer is horizontal and of infinite lateral extent. For pure surface-tension driven instability in a non-rotating layer, it is observed that for relatively small values of the crispation number ($Cr < B_0/120$), instability appears in the form of steady convection with $Ma^c = 80$ and $k^c = 2$ while deformation of the upper surface is of no influence on the critical values. In contrast, for larger Cr-values ($Cr > B_0/120$), surface deformation is destabilizing ($0 < Ma^c < 80$) and a long wave instability mode ($k^c = 0$) is predicted. In presence of buoyancy effects, the above conclusions remain in form : for low Cr's, one recovers the classical results established by Nield while for high values of Cr, steady convection will occur with an infinite wave length ; surface deformation is still destabilizing at least for sufficiently small values of the Rayleigh number. At high Rayleigh numbers corresponding to a thickness of the layer larger than a critical value, deflection plays a stabilizing role. It is also known that oscillatory convection can only be produced when the Marangoni number is negative, i.e. by heating from above or alternatively by heating from below a fluid whose surface tension increases with temperature. If in addition, one superimposes a rotation around a vertical axis, it is shown that rotation plays a stabilizing role. This stabilizing effect is still reinforced by surface deflections at least for rather high values of the angular velocity $\Omega$ ($Ta > 10^3$) and large Pr-values (Pr > 5). In contrast, at low angular velocity, the surface deformation is destabilizing. Oscillatory motion is predicted at low Prandtl (Pr < 5) even for fluid layers heated from below, i.e. with positive Ma numbers.

All the results reported in the present paper have been established for a positive Bond number $B_0$, which means that the free surface of the layer is upside. It would be interesting to investigate the case of a negative $B_0$, when the underside of the layer is the free surface. To our knowledge, the case $B_0 < 0$ has only been treated for the sole Marangoni problem by Takashima [4,5]. Another interesting open problem is to study the time-growth and the correlation length of disturbances with a large wavelength ($k^c \approx 0$). Moreover, as mentioned by Takashima [4], the situation with $k^c = 0$ is not significative, as for $k^c = 0$, the eigenvalue problem predicts that the velocity $w$ is identically zero, at least for Marangoni instability. In practice however, the presence of lateral walls would impose a non-zero value for $k^c$ and therefore, it would be useful to introduce side-walls into the description. Finally, we recall that the present analysis as well as the references quoted in this work refer to a linear approach : of course, a non-linear description is desirable as it provides information about the behaviour of the fluid far from the threshold but it helps also to elucidate the important question of pattern formation. Moreover, by departing from the threshold, structural disorder progressively appears [19]. The occurrence of disorder in convective patterns is closely related to the problem of wavelength selection, which is of crucial interest in the onset of turbulence and transition to chaos.

Acknowledgements : The authors wish to thank the EC program Human Capital and Mobility under contract ERB CHRCX-CT-94-0481 for financial support. Partial financial support from SSTC, Belgian Science Policy Programming, under contract PAI 21 is also acknowledged.

# References

1.      D.A. Nield: J.Fluid Mech. **19** 1341 (1964)
2.      L. Scriven, C. Sternling: J.Fluid Mech. **19** 321 (1964)
3.      K.A. Smith, J.Fluid Mech. **24** 401 (1966)
4.      M. Takashima: J.Phys.Soc.Japan **50** 2745 (1981)
5.      M. Takashima: J.Phys.Soc.Japan **50** 2751 (1981)
6.      A. Vidal, A. Acrivos: Phys.Fluids **9** 615 (1966)
7.      M. Takashima: J.Phys.Soc.Japan **20** 810 (1970)
8.      C. Perez-Garcia, G. Carneiro: Phys.Fluids A **3** 292 (1991)
9.      A. Kaddame, G. Lebon: Appl.Sc.Research **52** 295 (1994)
10.     A. Kaddame, G. Lebon: Microgravity Q. **4** 69 (1994)
11.     J.R.A. Pearson: J.Fluid Mech. **4** 489 (1958)
12.     G. Gouesbet, J. Maquet, C. Roze, R. Darrigo: Phys.Fluids A **2** 903 (1990)
13.     S. Chandrasekhar, *Hydrodynamic and Hydromagnetic Stability*, Oxford, London, 1961
14.     G. Veronis: J.Fluid Mech. **24** 545 (1966)
15.     H.T. Rossby: J.Fluid Mech. **36** 309 (1969)
16.     A. Vidal, A. Acrivos: J.Fluid Mech. **26** 807 (1966)
17.     G. Mc Conaghy, B. Finlayson: J.Fluid Mech. **39** 49 (1969)
18.     G. Sarma, in *Physico-Chemical Hydrodynamics*, M. Velarde, ed., Plenum, pp. 271-289, 1988
19.     P. Cerisier, H. Nguyen Thi, B. Billia: Physica D **61** 113 (1992)

# BALANCE EQUATIONS AND THE PROBLEM OF CONSTITUTIVE RELATIONS IN VARIED DIMENSIONS CURVILINEAR MEDIA

Roger Prud'homme

Laboratoire de Modélisation en Mécanique, Université Pierre et Marie Curie & C.N.R.S., Case 162, 4, place Jussieu, F75252 Paris Cedex 05, France.

This paper summarizes and completes a previous one [1]. An important problem to be solved is to obtain constitutive relations for material whose thickness is small compared to their radius of curvature. These material media are by definition moving, deformable and present -more or less far from equilibrium-irreversible phenomena. This question comes up with fluid lines or interfaces.

Complex phenomena such as stability changes can affect this sort of material medium.

Thus, considering evaporation under a microwave field [2], see Fig. 1, one may ask whether a simpler approach might not be possible to develop. This new description should provide the behaviour of the actual thick interface which is submitted to very strong constraints. Is it possible to reach, in some situations, correct simplified balance equation systems and constitutive relations ?

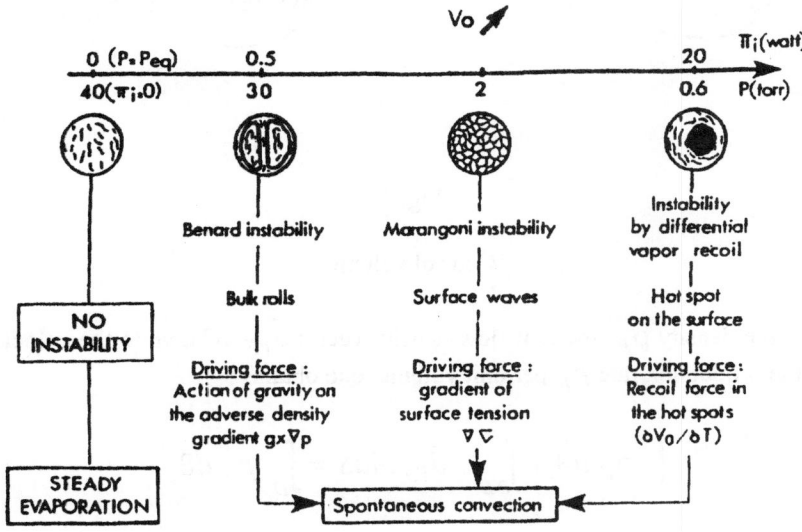

Fig. 1.

Hydrodynamic instability of liquids under low pressure or/ and under microwave irradiations.

70

The same kind of question may be asked concerning interfaces between two-phases and monophasic media (for example, how to define the surface tension of a ferro-fluid ?), concerning 1D curvilinear media (common line between capillary surfaces, fluid flows inside flexible pipe), concerning near-critical-point interfaces.
We first look at the question of 3D and 2D bulk and interface equations.
The 3D equation of a property F may be symbolically written :

$$\frac{d_v F}{dt} + \Phi_{VF} = P_F \tag{1}$$

for an extensive quantity $F$ which is contained, see Fig. 2, into a closed convex domain $(D)$ with boundary $(\partial D)$. The continuum $(D)$ is assumed to move with an arbitrary local velocity $\vec{V}$. The flow of $F$ through $(\partial D)$ and the production of $F$ are represented by $\Phi_{VF}$ and $P_F$ respectively.

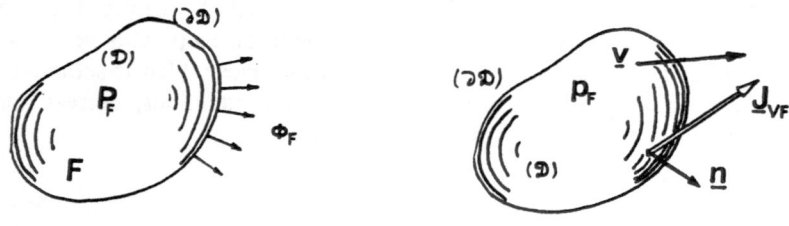

Fig. 2

Control volume

Use of the density $\rho_F$, the unit flow density vector $\vec{J}_{VF}$ relative to the velocity $\vec{V}$ and of the production rate $p_F$ per unit volume, one obtains :

$$\frac{d_v}{dt}\int_D \rho_F \, d\vartheta + \int_{\partial D} \vec{J}_{VF}.\vec{n}dS = \int_D p_F \, d\vartheta \tag{2}$$

$$\frac{d_v \rho_F}{dt} + \rho_F \vec{\nabla}.\vec{V} + \vec{\nabla}.\vec{J}_{VF} = p_F \tag{3}$$

At the level of a moving and flexible interface, we have, see Fig. 3 :

$$\frac{dF_s}{dt} + \Phi_F^S + \Phi_F^\pm = P_F^S \tag{4}$$

$$\frac{d_v}{dt} \int_S \rho_F^S \, dS + \int_{\partial S} \vec{J}_{VF}^S \cdot \vec{\eta} \, d\ell + \int_{S^+} \vec{J}_{VF}^+ \cdot \vec{\xi} \, dS - \int_{S^-} \vec{J}_{VF}^- \cdot \vec{\xi} \, dS = \int_S p_F^S \, dS \tag{5}$$

$$\frac{d_v \rho_F^S}{dt} + \rho_F^S \, \vec{\nabla}_{//} . \vec{V} + \left[ J_{VF\perp} \right] + \vec{\nabla}_{//} . \vec{J}_{VF}^S = p_F^S \ . \tag{6}$$

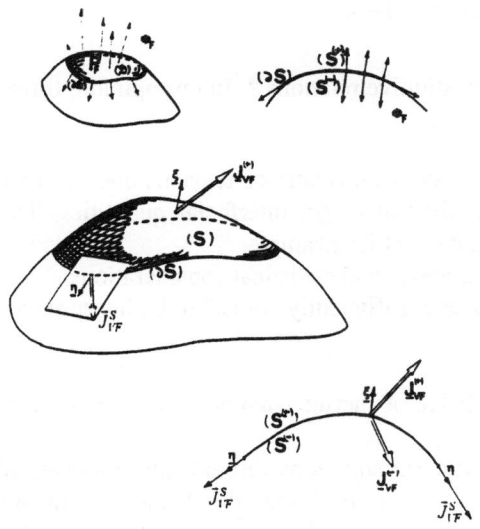

Fig. 3

Control surface

These 2D expressions have the same meaning as the 3D previously stated but with surface quantities and two kinds of fluxes : the tangential fluxes between two interfacial zones on the one hand and the fluxes between the interface and the 3D adjacent bulks on the other. The local surface quantities curently used are excess ones [1]. Now that the system of local equations is obtained, one is faced with the problem of the closure of this system. In general, we know the equations of state and the constitutive relations for 3D media satisfying the equation 3.

The equations of state interlink the quantities $\rho_F$, the additional laws enable the flux $\vec{J}_{VF}$ to be expressed in terms of the density gradients $\vec{\nabla}\rho_F$ and the production rates $p_F$ in terms of the $\rho_F$ [3].

At the interface, the problem may be posed at and away from equilibrium. For some non equilibrium interface phenomena, there may not exist any reference equilibrium configuration. This is the case for flames where we are, straight off, strongly away from equilibrium [4]. Unusual relations may then exist between surface densities $\rho_F^S$, volume density $\rho_F$, and their gradients. These relations can be obtained by integrating the terms present in equation 3, along the normal coordinate through the interface.

This example shows that the equations, the method to obtain them and the choice of the control volume velocity field $\vec{V}$, which has to be unchanged along the normal coordinate, are very important questions which must be very carefully considered.

## Interface constitutive relations

How can we obtain the constitutive relations ? In our opinion, three methods may be used :

1 - Via experiment : one obtains surface tensions and even surface viscosities, but sometimes it is very difficult to get interfacial quantities. That is the case of surface tensions under field effect for example.

2 - From molecular theories and statistical considerations.

3 - When the interface is sufficiently "thick" to be interpreted as a microscopic 3D continuous medium :

. from the knowledge of thermodynamic and mechanical properties of this medium.

. by integrating the 3D equations through the thickness of the interface ; however, in this case, we have to know the behaviour of this 3D medium, therefore its equations of state and others constitutive relations which is not always obvious.

Two situations can happen with this third method :

3.1 - 3D media with classical properties

That is the case of high activation energy premixed flames, diffusion flames and some boundary layers. One then finds non classical constitutive relations for the 2D interface, for which it would be very interesting to be able to elaborate a coherent theory. This would then allow us to obtain directly this type of relations, without having to integrate along the normal coordinate.

3.2 - Non classical properties 3D medium

The most common example is that of capillary surfaces, but many other situations can be considered.

Four approaches seem to be particularly interesting for interfaces and have been used for this purpose, so far :

A. Theories involving non spherical pressure tensor (see G.B. Bakker, 1928 [5], J.G. Kirkwood and F.P. Buff, 1949 [6], B. Widom, 1965 [7], A. Steinchen et al, 1993 [8]).

B. Theories based on "second gradient" concept (see Cahn and Hilliard model [9] and R. Gatignol [10], P. Seppecher [11], H. Gouin [12] works).

C. "Extended irreversible thermodynamics" due to D. Jou et al [13].

D. "Excess work" theories of J. Ross et al [14] and of M. Velarde et al [15].

## Balance equation for a fluid line

What we call fluid line is simply a 1D curvilinear medium which is fluid or which contains a fluid.

The common line between capillary surfaces is an example of such a fluid line. But in some cases, we may consider similarly a fluid flow into a flexible pipe or a continuous fluid jet, or even a fluid bridge between moving drops.

Such fluid lines have been studied by other authors with slightly different points of view [16, 17].

We shall assume that the considered medium thickness is small compared to the local line curvature and that there is a quasi uniform velocity on any normal cross section. A local equation is then obtained for an arbitrary extensive property $F$, in the following form :

$$\frac{d_v \rho_F^\ell}{dt} + \rho_F^\ell \, \vec{\nabla}_{//} . \vec{V} + \varphi_{VF}^\ell + \vec{\nabla}_{//} . \vec{J}_{VF}^\ell = p_F^\ell \tag{7}$$

where the various linear quantities have been defined in [1].

Note the similarity between equation 7 and the interface equation 6, but notice too also that the subscript ($//$) no longer specifies the tangent plane to the interface but the tangent unit vector $\vec{t}$ of the line. Note also that the flux $\varphi_{VF}^\ell$ results from integration of the normal flux density along the boundary ($C$) of the line cross section ($S$), see Fig. 4. Thus, the moving and straining fluid line may be considered as a singular line analogous to sources, sinks or vortex lines of theoretical hydrodynamics.

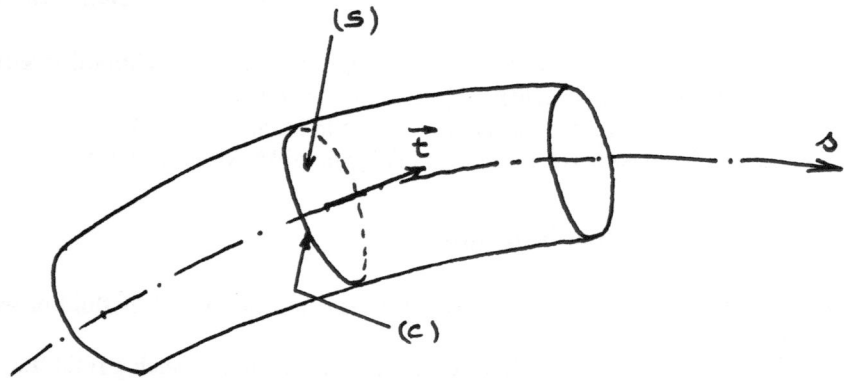

Fig. 4

Fluid line feature

The case of a pipe flow has already been presented [1]. The equations for mass and momentum have been given. We will give here the results of momentum in the Frenet reference frame. Starting from a parametric definition of the fluid line :

$$\begin{cases} \vec{x} = \vec{f}(s,t) \\ s = s(\alpha,t) \end{cases} \tag{8}$$

and from the parametric definition of the pipe :

$$\begin{cases} \vec{x} = \vec{f}(\sigma,t) \\ \sigma = \sigma(\alpha,t) \end{cases} \tag{9}$$

one obtains for the velocities :

$$\begin{cases} \vec{v} = \dot{s}\,\vec{t} + \dfrac{\partial \vec{f}}{\partial t} & \dot{s} = \dfrac{\partial\,s(\alpha,t)}{\partial\,t} & \vec{t} = \dfrac{\partial \vec{f}(s,t)}{\partial\,s} \\[2mm] \vec{v}^{s} = \dot{\sigma}\,\vec{t} + \dfrac{\partial \vec{f}}{\partial t} & \dot{\sigma} = \dfrac{\partial\,\sigma(\alpha,t)}{\partial\,t} & \vec{t} = \dfrac{\partial \vec{f}(\sigma,t)}{\partial\,\sigma} \\[2mm] \vec{v} - \vec{v}^{s} = (\dot{s} - \dot{\sigma})\vec{t} & & \end{cases} \qquad (10)$$

This allows us to find an expression for the friction force $p_f$ usually in terms of $(\dot{s} - \dot{\sigma})$.

Writing :

$$\begin{cases} \vec{V} = \vec{v} \\[1mm] \vec{p}_f^{\,\ell} = \rho^{\ell}\,\vec{v} \end{cases} \qquad (11)$$

the components of the momentum equation in Frenet formulation are :

$$\begin{cases} \rho^{\ell}\left( \ddot{s} + \vec{t}.\dfrac{\partial^2 \vec{f}}{\partial t^2} \right) + p_{//} + \dfrac{\partial p^{\ell}}{\partial s} = \rho^{\ell}\left( \vec{g} - \vec{\gamma}_e - \vec{\gamma}_c \right).\vec{t} \\[3mm] \rho^{\ell}\left( 2\dot{s}\,\vec{n}.\dfrac{\partial^2 \vec{f}}{\partial s\partial t} + \dfrac{\dot{s}^2}{R} + \vec{n}.\dfrac{\partial^2 \vec{f}}{\partial t^2} \right) + \vec{p}_\perp.\vec{n} = \rho^{\ell}\left( \vec{g} - \vec{\gamma}_e - \vec{\gamma}_c \right).\vec{n} \\[3mm] \rho^{\ell}\left( 2\dot{s}\vec{b}.\dfrac{\partial^2 \vec{f}}{\partial s\partial t} + \vec{b}.\dfrac{\partial^2 \vec{f}}{\partial t^2} \right) + \vec{p}_\perp.\vec{b} = \rho^{\ell}\left( \vec{g} - \vec{\gamma}_e - \vec{\gamma}_c \right).\vec{b} \end{cases} \qquad (12)$$

Recall that :

$$\ddot{s} = \dfrac{\partial^2 s(\alpha,t)}{\partial t^2}, \quad \dfrac{1}{R} = \left| \dfrac{\partial^2 \vec{f}}{\partial s^2} \right|, \quad \vec{n} = \dfrac{\partial^2 \vec{f}}{\partial s^2} \Big/ \left| \dfrac{\partial^2 \vec{f}}{\partial s^2} \right|, \qquad (13)$$

$$\vec{b} = \vec{t} \wedge \vec{n}$$

We also have the wellknown relations :

$$\frac{\partial \vec{t}}{\partial s} = \frac{\vec{n}}{R} \quad , \quad \frac{\partial \vec{n}}{\partial s} = -\frac{\vec{t}}{R} - \frac{\vec{b}}{T} \quad , \quad \frac{\partial \vec{b}}{\partial s} = \frac{\vec{n}}{T} \tag{14}$$

where $1/R$ is the curvature and $1/T$ the torsion.

In the case of common lines between capillary surfaces, the inertial terms disappear [16].

So, we have written the momentum equation in a form which is convenient for a class of problems.

In addition, we can now be interested by the energy balance equation. This is important for non isothermal problems along the line, as well as when there exists heats exchanges between the line and its environment. Then we introduce the linear heat flux $\vec{q}^{\ell}$ along the line and the quantity $Q$ which indicates the local heat exchange between the line and the bulk.

Then, writing the total energy equation in the form of equation 7 and deducing the kinetic energy equation from the momentum equation, we obtain by sub'traction the internal energy balance equation in the following form :

$$\frac{d_v \rho_E^{\ell}}{dt} + \rho_E^{\ell} \, \vec{\nabla}_{//} \cdot \vec{v} + Q + \vec{\nabla}_{//} \cdot \vec{q}^{\ell} = \left( \vec{v} - \vec{v}^s \right) p_f \vec{t} - p^{\ell} \, \vec{\nabla}_{//} \cdot \vec{v}. \tag{15}$$

To solve a given real fluid line problem, we have to know the expression of $Q$ and $\vec{q}^{\ell}$ as functions of thermal gradients and jumps and other possible quantities. The resulting constitutive relations depend upon the choice by hypothesis for the thermal behaviour of the considered media.

## Polarised media interface problems in the Minkowski space-time

Use of the Minkowski space-time is not compulsory in order to obtain non relativistic polarized media equations when an electromagnetic field is present. Nevertheless, the use of 4D formulation is convenient because the different quantities remain in tensorial form, resulting in the flux being easily identifiable to name one.

The tensors and operators appearing in the writing of the Maxwell equations of a polarized medium are the following :

$$\underline{H} = \begin{bmatrix} 0 & H_z & -H_y & -iD_x \\ -H_z & 0 & H_x & -iD_y \\ H_y & -H_x & 0 & -iD_z \\ iD_x & iD_y & iD_z & 0 \end{bmatrix} , \quad \underline{F}^* = \begin{bmatrix} 0 & -E_z & iE_y & B_x \\ iE_z & 0 & -iE_x & B_y \\ -iE_y & -E_x & 0 & B_z \\ -B_x & -B_y & -B_z & 0 \end{bmatrix} ,$$

$$\underline{J} = \begin{bmatrix} I_x/_c \\ I_y/_c \\ I_z/_c \\ i\tilde{\rho} \end{bmatrix} \quad , \quad \underline{\nabla} = \begin{bmatrix} \partial/_{\partial x} \\ \partial/_{\partial y} \\ \partial/_{\partial z} \\ -\dfrac{i}{c}\,\partial/_{\partial t} \end{bmatrix} \tag{16}$$

where $\underline{H}$ and $\underline{D}$ are the magnetic and electric displacement vectors, $\underline{B}$ and $\underline{E}$ the magnetic and electric fields, $\tilde{\rho}$ the classical electric charge per unit volume and $\underline{I}$ the current vector.

The Maxwell equations can be written as follow :

$$\begin{cases} \underline{\nabla}.\underline{\underline{H}} = \underline{J} \\ \underline{\nabla}.\underline{\underline{F}}^* = \underline{O} \end{cases} \tag{17}$$

For a medium at rest, we have for the relations between fields and displacements :

$$\underline{D} = \underline{\underline{\varepsilon}}.\,\underline{E}$$
$$\underline{H} = \underline{\underline{\mu}}^{-1}.\,\underline{B} \tag{18}$$

These relations verify form-invariance under Galilean transformation. ($\underline{\underline{\varepsilon}}$ and $\underline{\underline{\mu}}$ are the dielectric and permeability tensors respectively). This last point is very important because we have to follow the fluid motion to express the fluxes of the diverse quantities in a moving frame in the non relativistic case $(v\,/\,c << 1)$.

For any quantity $F$, one can write :

$$\underline{J_F} = \begin{bmatrix} J_{Fx} \\ J_{Fy} \\ J_{Fz} \\ i\rho_F \end{bmatrix} \quad , \quad p'_F = \frac{1}{c}\,p_F \tag{19}$$

78

where $J_{Fx}$, $J_{Fy}$ and $J_{Fz}$ are the components of the flux $F$ through a fixed surface in the 3D space and where $p_F$ is the usual production rate of $F$.

The equation of a quantity $F$ will be of the form :

$$\underline{\nabla} \cdot \underline{J_F} = p'_F \tag{20}$$

and will be applied to mass, electric charge and energy-momentum tensor [18, 19].

Before establishing the balance equations for a flexible moving interface, we first have to express the interface in the Minkowski space-time and to deduce the universe velocity :

$$\underline{V} = \begin{bmatrix} V_x/\alpha \\ V_y/\alpha \\ V_z/\alpha \\ ic/\alpha \end{bmatrix} \quad , \quad \alpha = \sqrt{1 - \frac{V^2}{c^2}} \tag{21}$$

where $V_x, V_y, V_z$ are the classical 3D velocity components and $V$ is the velocity modulus. This step does not present any particular difficulty. After that we must write the 4D equations in a suitable form to integrate along normal direction 4-vector $\vec{\chi}$ to the interface. This is not easy, except in the case of media at rest.

Then, we will have to establish the following interface equations for a polarized moving medium via an integration in 4-vector $\vec{\chi}$ direction :
. Maxwell equations
. Mass balances
. Electric charge balance
. Energy-momentum balance

(The problem of separation between field terms and mass terms is, as usual, difficult to solve).

This part of the work is difficult to achieve, for media at rest as well as in equilibrium situation [2, 8].

An other way consisting of deducing directly the form of the interface equations, without integrating along the normal direction to the interface [19] does not seem to be applicable in the general case. This is because the methodology described here is not suitable to handle easily the level of complexity that would result. Hence a new approach would be required for a general case.

7779

# References

[ 1] Prud'homme, R. (October 14-15th 1994) : "Balance equations for fluid curvilinear media". 1st Meeting on Basic Constitutive Relations for Surfaces. EEC Network "Dynamics of multiphase flows across interfaces". *Contract HCM ERB CHRCT 940481. CNRS Meudon Bellevue. In these proceedings.*

[ 2] Steinchen-Sanfeld, A., Lallemant, M., Courville, P. and Bertrand, G. (1988) : "Volume or surface instabilities during liquid evaporation under reduced pressure or/and microwave irradiation", *Proc. NATO ASI and EPS Summer-School and conference on PCH "Interfacial phenomena", Plenum corp., Ed. M.G. Velarde, p.387.*

[ 3] De Groot, S. R. and Mazur, P. (1961) : "Non-equilibrium thermodynamics", *North-Holland Publishing Company.*

[ 4] Prud'homme, R. (1988) : "Fluides hétérogènes et réactifs : écoulements et transferts", *Lecture Notes in Physics 304, Springer-Verlag.*

[ 5] Bakker, G.B. (1928) : "Kapillarität und Oberlächen-Spannung", *in Handbuch Exp. Physik VI (Akad. Verlagsgeselschaft, Leipzig).*

[ 6] Kirkwood, J.G. and Buff, F.P. (1949) : *J. Chem. Phys., 17, p. 338.*

[ 7] Widom, B. (1965) : *J. Chem. Phys., 43, p. 3892.*

[ 8] Liggieri, L., Sanfeld, A. and Steinchen, A. (1994) : "Effects of magnetic and electric fields on surface tension of liquids". *Physica A, 206, pp. 299-331.*

[ 9] Cahn, J.W. (1977) : "Critical point wetting". *J. Chem. Phys., 66, n° 8, P. 3667.*

[10] Gatignol, R. and Seppecher, P. (1986) : Modelisation of fluid-fluid interfaces with material properties". *J. de Mécanique Théor. et Appl., n° spécial, pp. 225-247.*

[11] Seppecher, P. (1988) : "Etude d'une modélisation des zones capillaires fluides : interfaces et lignes de contact", *Thèse de l'Univ. P. et M. Curie, Paris.*

[12] Casal, P. and Gouin, H. (1985) : "Relation entre l'équation de l'énergie et l'équation du mouvement en théorie de Korteveg de la capillarité". *C.R. Acad. Sci., Paris, 300, Série II, pp. 231-234.*

[13] Jou, D., Casas-Vazquez, J. and Lebon, G. (1993) : "Extended irreversible thermodynamics", *Springer-Verlag.*

[14] Ross, J., Chu, X., Hjelmfelt, A. and Velarde, M.G. (1992) : "Thermodynamic and stochastic theory of transport processes far from equilibrium". *The J. of Phys. Chemistry, 96, n° 26, pp. 11054-11065.*

[15] Velarde, M.G., Chu, X., Ross, J. (1994) : "Toward a thermodynamic theory of hydrodynamics : the Lorenz equations". *Phys. of Fluids, 6, n° 2, pp. 550-563.*

[16] Slattery, J.C. (1990) : "Interfacial transport phenomena", *Springer-Verlag.*

[17] Coutris, N. (1993) : "Balance equations for fluid lines, sheets, filaments and membranes". Int. J. Multiphase Flow. Vol. 19, n° 4, pp. 611-637.

[18] Eringen, A.C. and Maugin, G.A. (1990) : "Electrodynamics of continua - tomes 1 et 2", *Springer-Verlag.*

[19] Prud'homme, R. and Dudeck, M. (mai 1976) : "Equations de bilan d'un milieu conducteur avec champ électromagnétique". *Rapport 76-2 du Laboratoire d'Aérothermique du CNRS.*

# Some Preliminary Results About Equilibrium Surface Model

F. Cuadros [1] ; A. Mulero [1] and W. Okrasinski [2]

[1] Departamento de Física, Universidad de Extremadura, 06071 Badajoz, Spain
[2] Institute of Mathematics, University of Wroclaw, Wroclaw, Poland

**Abstract.** From some time ago we have been interested on a procedure to study the adsorption phenomena from a molecular point of view using first principles. The ideas of Defay *et al* can be used to consider the normal density profile $\rho(z)$ into a porous as a sucession of layers with different densities at the same temperature. Some authors have studied this problem and they have obtained density profiles, $\rho(z)$, by means of computer simulations and/or Density Functional Theory (DFT). By using $\rho(z)$ computer simulations results and following the Defay procedure, we are sure to solve definitely the adsorption problem into a porous. For this we need an accurate and, if possible, straightforward equation for Helmholtz free energy, equation of state, etc., of 2D-LJ fluids. The work is about some preliminary results on this subject.

**Keywords.** Adsorption, Steele's theory, computer simulation, thermodynamic properties

## 1. Steele's theory of the adsorption [1]

The partition function of a system adsorbed onto a solid wall can be writen as

$$z^{ads.} = \int_{v^{ads.}} \cdots \int \exp[-\beta\phi(\vec{r})]d\vec{r}^N \qquad (1)$$

being $\beta = \frac{1}{K_B T}$, and

$$\phi(\vec{r}) = \sum_{i=1}^{N} u_{sf}(\vec{r}_i) + \sum_{i<j}^{N} \sum_{j}^{N} u_{ff}(r_{ij}) \qquad (2)$$

where $u_{sf}(\vec{r}_i)$ is the potential between the solid and fluid at $\vec{r}_i$, and $u_{ff}(r_{ij})$ is the potential (assumed to be central) fluid-fluid.

The expression (1) can be writen in other form,

$$z^{ads} = z^{(o)} \left\{ 1 + z^{(1)} + z^{(2)} + \dots \right\} \tag{3}$$

being

$$z^{(o)} = \int_{\mathcal{V}^{ads.}} \cdots \int \prod_i e_0(z_i) \exp \left[ -\beta \sum_{j<k}^{N} \sum_k^N u_{ff}(r_{jk}) \right] d\vec{r}^N \tag{4}$$

the partition function for the adsorbate onto homogeneous (without periodicity) surfaces which properties are given by $e_0(z_i)$, and taking into account that the adsorbed monolayer dynamics is given by $u_{ff}(r_{jk})$.

The terms $z^{(1)}, z^{(2)}, \dots$ contain the periodicity effects of the solid surface.

## 1.1 First approximation

$z^{ads} = z^{(0)}$

This is the simplest, and the adsorbate can be considered as a two-dimensional system.

On the basis of this approximation we have been able [2] to determine the adsorption isotherm as

$$\ln p = \ln \frac{\pi \rho}{4} + \frac{\mu_0}{T} - 2\frac{\rho}{T}\alpha_{CM}(T, \rho) - \frac{\rho^2}{T}\frac{\partial \alpha_{CM}(T, \rho)}{\partial \rho} \tag{5}$$

as well as the isosteric heat,

$$\Delta q_{st} = -T^2 \left[ \frac{\partial (\mu_0/T)}{\partial T} \right] - T^2 \frac{\partial}{\partial T} \left\{ \frac{\partial [\rho^2 \alpha_{CM}(T, \rho)]}{\partial \rho} \right\} \tag{6}$$

where in both equations $\mu_0$ is the chemical potential of the WCA reference system (repulsive forces only), and $\alpha_{CM}(T, \rho)$ is a function that gives the contribution of the attractive forces treated as a perturbation.

## 1.2 Second approximation

Adsorption of fluids onto homogeneous surfaces with periodicity.

The basic idea is to consider this system as a "mixture" between the fluid and the periodic solid adsorbent.

For adsorbates Lennard-Jones (LJ)-type (simple fluids), the interaction solid-fluid, $u_{sf}$, is considered as LJ-type too, and the expression for the potential interaction between adsorbate-adsorbent is given by

$$u_{sf}(z) = 2\pi\epsilon_{sf}\rho_s\sigma_{sf}^2\Delta\left[\frac{2}{5}\left(\frac{\sigma_{sf}}{z}\right)^{10} - \left(\frac{\sigma_{sf}}{z}\right)^4 - \frac{\sigma_{sf}^4}{3\Delta(z+0.61)^3}\right] \quad (7)$$

Here, $z$ is the distance between a molecule of the fluid and the solid surface, $\Delta$ is the distance between solid layers and $\epsilon_{sf}$ and $\sigma_{sf}$ are the LJ parameters of the solid-fluid interaction obtained through the corresponding "mixing rules".

## 1.3 Multilayer adsorption

Some procedures are in use:

- Serie expansion on bulk density.
- Density Functional Theories (DFT) with different approximations [3]. This procedure es the most used.

## 1.4 Computer simulation

They are a few computer simulation results about the multilayer adsorption phenomena. For example the MC computer simulation of $N_2$ adsorbed into a wide graphite porous obtained by Lastoskie *et al* [4].

## 1.5 Heterogeneus surfaces

The interaction solid-fluid depends on the $(x, y)$ point of the solid surface.

To modelize this, Molina-Sabio *et al* [5] have considered that the potential depth of the solid-fluid interaction, $\epsilon_{sf}$, is a random variable.

## 2. A new procedure

Following the ideas of Defay, Sanfeld and Steinchen [6] we can suggest a new procedure to attack the homogeneous adsorption problem by dividing the density profiles of the adsorbate into a sucession of two-dimensional layers with separation $\delta z$.

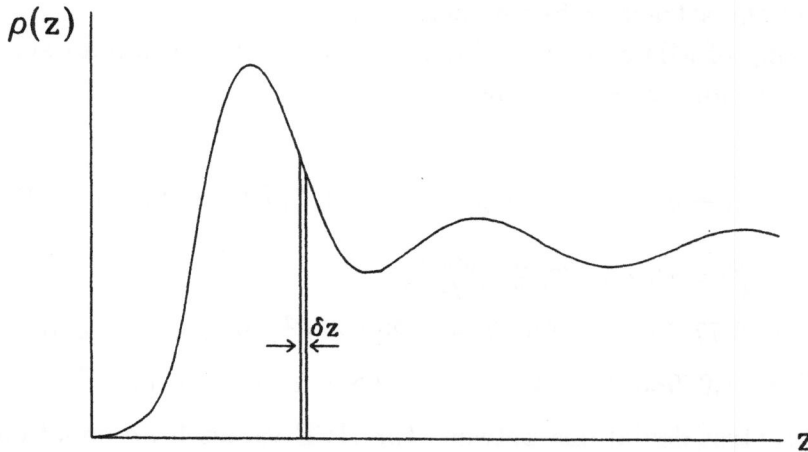

Figure 1: Picture of the procedure we will follow.

To apply this model we need:

### 2.1 The density profile results

Now we have the computer simulation results from Steele about the density profiles of $CH_4$ in microporous carbons.

### 2.2 Good models about the two-dimensional thermodynamics

Referring to this point, we will show some results.

#### 2.2.1 Helmholtz free energy (HFE)

Recently we have proposed [7] an analytical expression to calculate the HFE of two-dimensional LJ system using the WCA separation of the intermolecular potential:

$$\frac{A}{T} = \frac{A_0}{T} - \frac{\rho}{T}\alpha_{CM}(T,\rho) \tag{8}$$

Here and in the following equations all the properties are expressed in reduced LJ units. $A_0$ is the HFE of the reference system (repulsive forces only), $\rho$ and $T$ are the density and temperature, respectively, and $\alpha_{CM}(T,\rho)$ is a function that gives the contribution on HFE of the attractive forces (perturbation).

By fitting of MD computer simulation results we obtain an analytical expression for $A$ as follows:

$$\frac{A}{T} = (1-y)^{-1} - \ln(1-y) - 1 - \frac{\rho}{T}[C_1(T) + C_2(T)\rho] \tag{9}$$

where $y = \frac{\pi\rho d^2}{4}$ with $d = \frac{0.3837+1.068/T}{0.4293+1/T}$,

$C_1(T) = 3.77711 - 3.39554\,T + 2.20294\,T^2 - 0.46687\,T^3$ and

$C_2(T) = -1.00985 + 4.83792\,T - 3.19864\,T^2 + 0.68183\,T^3$.

The greatest deviations between the HFE values from equation (9) and the computer simulation results are less than 5% in the fluid region.

### 2.2.2 Vapour-liquid equilibrium (VLE) for two-dimensional LJ systems

Using equation (9) we can obtain the two-dimensional pressure, $P$, and the chemical potential, $\mu$, as follows:

$$\frac{p}{\rho T} = \frac{p_0}{\rho T} - \frac{p}{T}\alpha_{CM}(T,\rho) - \frac{\rho^2}{T}\frac{\partial\alpha_{CM}(T,\rho)}{\partial\rho} \tag{10}$$

and

$$\frac{\mu}{T} = \frac{\mu_0}{T} - \frac{2\rho}{T}\alpha_{CM}(T,\rho) - \frac{\rho^2}{T}\frac{\partial\alpha_{CM}(T,\rho)}{\partial\rho} \tag{11}$$

where $p_0$ and $\mu_0$ give the pressure and the chemical potential of the reference system, respectively, and $\alpha_{CM}(T,\rho)$ gives the contribution of the attractive forces to these quantities.

On the other hand, the thermodynamics requeriments for VLE are:

$$p_L = p_v \text{ (Mechanical equilibrium)}$$

$$\mu_L = \mu_v \text{ (Chemical equilibrium)} \tag{12}$$

By using the Eqns. (10) and (11) and the equalities (12) we have been able to determine the VLE curve for two-dimensional LJ fluids.

Moreover, because our computer simulations were carried out for five values of the perturbative WCA parameter: $\lambda = 0$, 0.25, 0.50, 0.75 and 1, we have been able to determine the VLE curve when the intensity of the attractive forces is changed (see Fig.2).

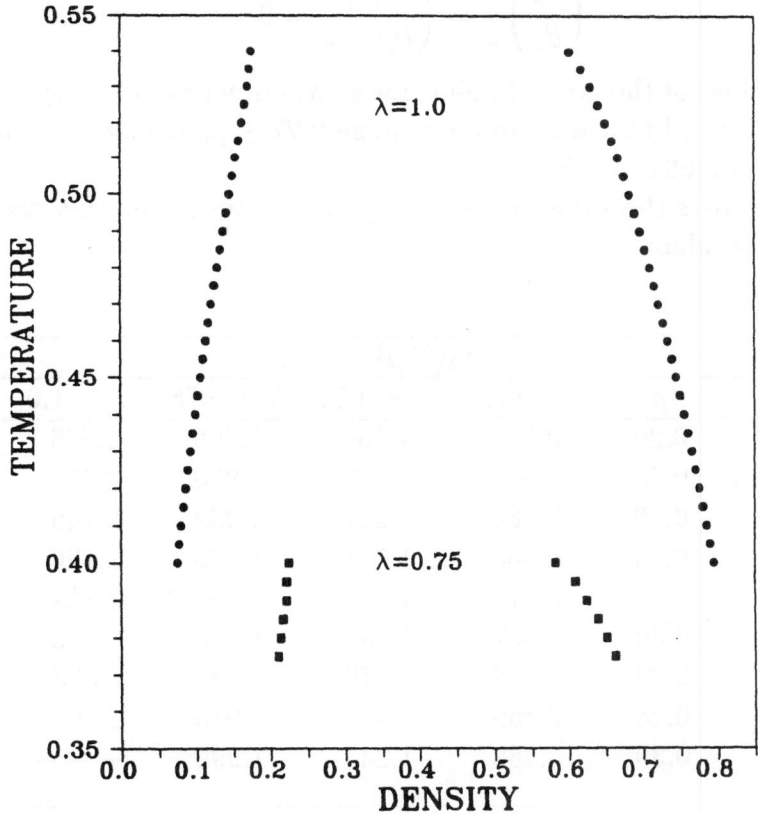

Figure 2: VLE curve for two-dimensional LJ fluids for two values of the perturbative parameter, $\lambda$, of the WCA theory.

More computer simulations are needed to determine more exactly the VLE curve for $\lambda < 1$, because as we will see the critical point is extraordinarily depending with the "charge" of the attractive forces.

### 2.2.3 Critical point of two-dimensional LJ systems

On the basis of our computer simulation results of $\alpha_{CM}(T, \rho)$ for two-dimensional LJ systems, we have developed a method to determine – with an error below 1% – the location of the critical point by using the thermodynamic requeriment,

$$\left(\frac{\partial p}{\partial \rho}\right)_{T_c} = \left(\frac{\partial^2 p}{\partial \rho^2}\right)_{T_c} = 0 \tag{13}$$

The values of the critical point, for a two-dimensional LJ system, reported in the literature are on a range $0.27 \leq \rho_c \leq 0.38$ for $\rho$ and $0.60 \leq T_c \leq 0.6844$ for $T$.

Table 1 lists the values of $\alpha_{CM}(T, \rho)$ for states inside this region of the phase plane.

**Table 1**

| T | $\rho$ | $\alpha_\lambda(T,\rho)$ | | | |
|---|---|---|---|---|---|
| | | $\lambda = 0.25$ | $\lambda = 0.50$ | $\lambda = 0.75$ | $\lambda = 1.0$ |
| 0.625 | 0.20 | 0.547 | 1.159 | 1.385 | 2.628 |
| | 0.25 | 0.567 | 1.202 | 1.930 | 2.774 |
| | 0.30 | 0.584 | 1.227 | 1.939 | 2.745 |
| 0.60 | 0.20 | 0.550 | 1.171 | 1.888 | 2.736 |
| | 0.25 | 0.572 | 1.214 | 1.954 | 2.814 |
| | 0.30 | 0.587 | 1.231 | 1.955 | 2.802 |
| 0.55 | 0.20 | 0.539 | 1.170 | 1.930 | 2.833 |
| | 0.25 | 0.559 | 1.185 | 1.916 | 2.805 |
| | 0.30 | 0.588 | 1.243 | 1.990 | – – – |

As we can see the values of $\alpha_{CM}(T, \rho)$ are approximately constant for each temperature around the critical one.

Taking into account the values of $\alpha_\lambda(T, \rho)$ obtained from computer simulations and given in Table 2, we may calculate the location of the critical points given in the same Table 2 ($\rho_c = 0.27$ for all values of $\lambda$).

**Table 2**

| $\lambda$ | $\alpha_\lambda$ | $T_C$ | $P_C$ |
|------|------|------|--------|
| 0.25 | 0.7 | 0.15 | 0.0147 |
| 0.50 | 1.4 | 0.30 | 0.0294 |
| 0.75 | 2.1 | 0.45 | 0.0442 |
| 1.00 | 2.8 | 0.60 | 0.0589 |

**LINEARITY!** An important result, but not surprising, because in a previous paper [8] we have analyzed the linearity of the thermodynamic properties with the perturbative parameters of WCA theory.

# 3. Preliminary three-dimensional results

As a marginal point of this work we would like to present the first results about the use of this procedure for a 3D LJ system.

Fig.3 shows the values of $\alpha_{CM}(T, \rho)$ for $\lambda = 1$ and for different temperatures and densities around the critical point taken of the literature (the critical point for LJ systems is between $0.26 \leq \rho_c \leq 0.40$ and $1.58 \leq T_c \leq 1.25$). As we can see in Fig.3, the values of $\alpha_{CM}(T, \rho)$ are very similar for all temperatures and they can be fitted by the linear function

$$\alpha(T, \rho) = 2.334\rho + 5.395 \tag{14}$$

Following the same procedure that in the 2D case, we have been able to determine the critical point of 3D-LJ systems.

We pretend to follow the same studies in 2D as well as in 3D to determine the different location of the critical point and the change of the ELV curve when the intensity of the attractive forces is changed.

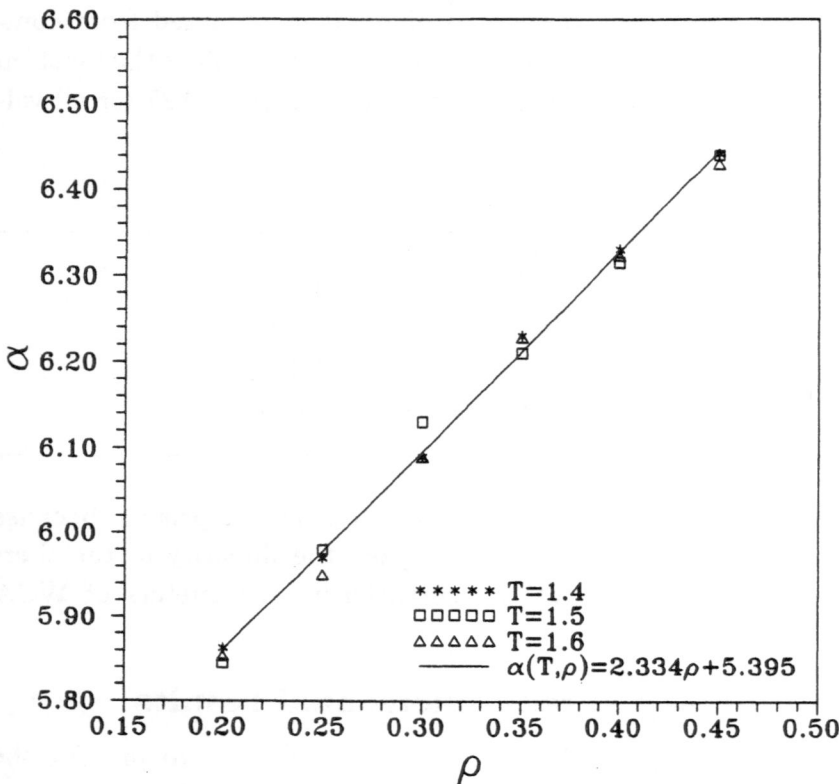

Figure 3: Critical point and values of $\alpha$ and fitting of $\alpha$ as function of $T$ and $\rho$ for 3D LJ systems. Critical point: $\rho_c = 0.31$; $T_c = 1.52$; $p_c = 0.33$. Mean absolute deviation for $\alpha$: 0.47%

## 4. Some questions

1. The results for 3D systems, could be used to attack the variation of the critical point and ELV curve when we have an external attractive field?

2. Ferrofluids?

3. Change on the surface tension in function of the attractive forces?

## Acknowledgements

We would like to express our gratitude to the European Union for financial support through the European Network "Dynamics of Multiphase Folws across Interfaces", contract ERB CHRXT 940481 and to the DGICYT of the Spanish Government for financial support through the grant SAB 94-0197.

## References

[1] W.A. Steele; *J. Chem. Phys.* **65**, 5256 (1976)

[2] A. Mulero; Doctoral Thesis. University of Extremadura (1994)

[3] S. Sokolowsky and J. Fischer; *J. Chem. Soc. Faraday Trans.* **89** (5), 789 (1993)

[4] C. Lastoskie, K.E. Gubbins and N. Quirke; *Langmuir* **9** (10), 2693 (1993)

[5] M. Molina-Sabio, F. Rodríguez-Reinoso, D. Valladares and G. Zgrablich; Carbon'94. Granada (Spain) 246 (1994)

[6] R. Defay, A. Sanfeld and A. Steinchen; *J. Chimie Phys.* **9**, 1380 (1972)

[7] F. Cuadros and A. Mulero; *Chem. Phys.* **177**, 53 (1993).

[8] F. Cuadros, A. Mulero and W. Okrasinski; *Physica A*, accepted for publication (1995).

# Experimental study of the convective phenomena during the evaporation of aqueous solutions of sucrose

Blaise Simon

Institut Universitaire des Systèmes Thermiques Industriels ,

Université de Provence, U M R 139, Centre de S$^t$ Jérôme,

13397 Marseille, France.

## 1 Introduction

This work was initiated in order to study the convection in a liquid layer cooled at the top by evaporation instead of being heated from below. The water-sucrose system has been chosen because solutions can be prepared in a wide range of viscosities. It was expected to see the regular lattice of convective cells of the Bénard- Marangoni convection. This turned out differently, and in the course of the work it was found that the evaporation of shallow layers of aqueous solutions of sucrose in the free air of the laboratory, a simple enough phenomenon, is nevertheless a rich field of investigation .

A shallow liquid layer ( usual depth between 0,5 mm and 3 mm ) evaporating in the free air of the laboratory is placed in a horizontal glass container with side walls made up of glass or of teflon. It is illuminated from above by the lamp of a binocular microscope, and the transmitted image ( shadowgraph ) is observed on a screen at 5 cm under the container, where it can be recorded directly on photographic paper. The temperature and humidity of the air are recorded, but are not imposed ( the usual values are t =20° , H = 40% ) .

The system is not strictly stationary, since during the evaporation the thickness decreases and the viscosity increases, but this evolution is slow : The figure 1 shows the mass of a liquid layer during the evaporation : in usual conditions of the laboratory the initial evaporation rate is of the order of $10^{-5}$ kg m$^{-2}$ s$^{-1}$, and this is constant in a wide range of concentrations, in spite of the increase of the concentration and viscosity .

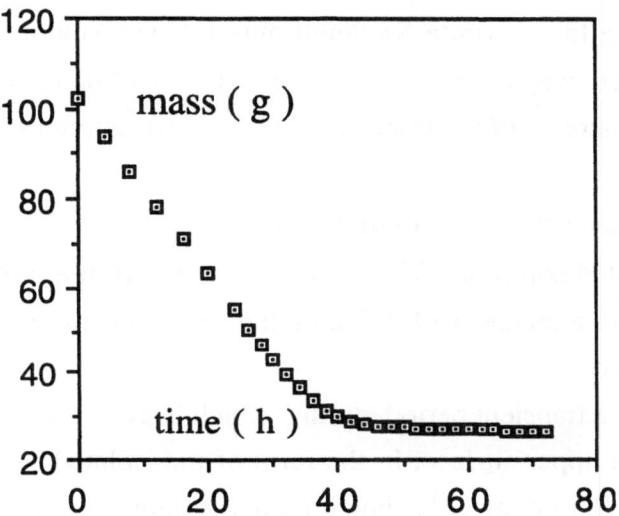

Fig. 1 Mass of an evaporating solution with an initial concentration of 28% sucrose. The concentration at $t = 28$ h is 59% .

The duration of a typical run is a few hours, and in a time interval of ten minutes the system can be considered as stationary.

The evaporation of water at the upper surface induces there an increase of density , due firstly to the evaporative cooling, and secondly to the increase of the sucrose concentration. As we shall see later, the effects of surface tension do not play any influence, so that the instability can be described in principle as thermosolutal.

From the the mass transfer of water evaluated from the figure 1 one can calculate the thermal- and the solutal Rayleigh numbers. For solutions containing 20% of sucrose and more, the solutal Rayleigh number is usually about $10^3$ times greater than the thermal one , so that the convection can be considered as solutal.

In the course of an experiment the Rayleigh number decreases : For instance in the case of the figure 1 , the initial value is Ra = 2.2 $10^6$ , and at t = 28 h , Ra = 5.9 $10^3$ .

Up to now it has not been possible to detect a convective threshold corresponding to a definite Rayleigh number. The smooth departure from the linear evaporation rate beginning at t = 28 hours corresponds rather to the increase of the viscosity of the sucrose solutions.

## 2    Free convection at  small depths [1]

A concentrated solution ( 50% sucrose ) is placed in a circular glass container, with a diameter of 240 mm, the initial thickness of the fluid being 1,35 mm .

There is first a transient period, during which heavier parts of the fluid sink from the upper surface, in the form of the isolated " streamers " described by Berg et al. [2] , showing up as bright lines on the screen. In half an hour, these streamers join in an irregular array of non-closed polygons. Progressively, there appears a lattice of closed polygons with straight edges. The overall planform is definitely organized at = 5 hours ( figure 2 )

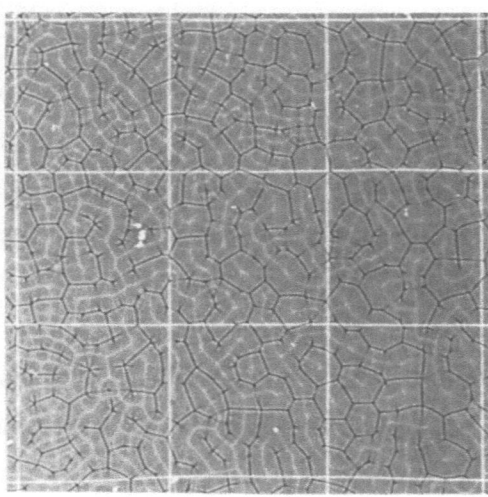

Fig. 2    Convective cells in an evaporating solution of sucrose (concentration 60%, thickness 1 mm ). The side of a  square is 2 cm.

Later, there are only  slight rearrangements, a few sides disappearing on the spot, so that the number of polygons decreases somewhat. Later, as the evaporation proceeds, the surface becomes still more deformed, the polygons  do not disappear, but are chilled as the layer takes the consistence of a honey. Still later, as the layer hardens the contrasts vanish  and the free surface becomes flat.

Each of these polygons is a convective cell, in which the fluid is sinking along the edges, and rising along the cell center. The free surface of the liquid is not flat, but each side of  a polygon is marked on the surface by a  narrow groove, which can be seen with the naked eye ( figure 3 ).

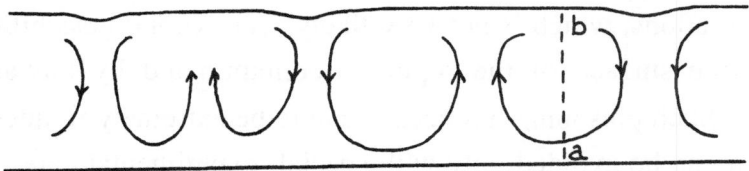

Fig. 3 Fluid flow in a convective layer.

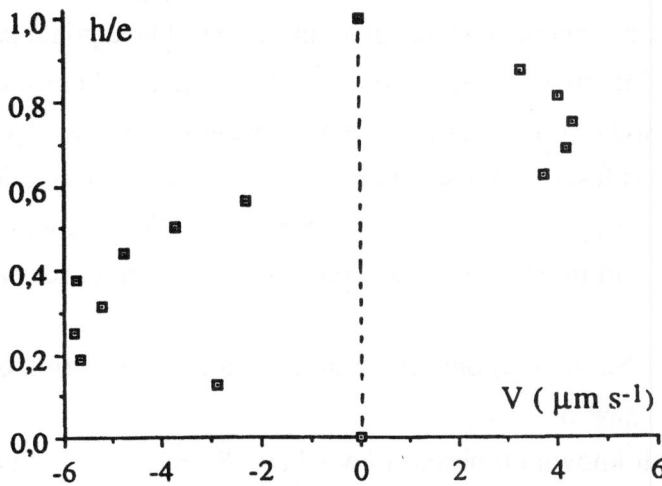

Fig. 4 Horizontal  velocity measured along the line
ab of the preceding figure.

The fluid motion is extremely slow : the displacement of latex spheres (with a density slightly less than that of the liquid) being observed with a microscope, it is possible to measure a velocity profile, as indicated in the figure 4. The maximum velocity of the order of 5 $\mu$m s$^{-1}$. The important fact is that the free surface is immobile in first approximation (i.e. as measured with the microscope) : the liquid seems to move between two rigid boundaries. Now it is known that for pure aqueous solutions of sucrose $d\sigma / dT < 0$ and $d\sigma / dC > 0$ .Therefore the free surface should move in the direction of the sinking streamers. Since it is not so, there are two possibilities : the first is that the surface tension forces for ordinary (not pure) solutions oppose the buoyancy forces at all concentrations, which is not very likely. The second possibility is that the free surface of the liquid is contaminated by unwanted impurities. Such poisoning has been found to be extremely frequent in water, and can be avoided only with special experimental care. This can lead to a "strengthening of the surface to such an extent that it will become completely stagnant" [3]. This is most probably the case here.

The bases of the convective cells are polygons with straight edges, but their arrangement is not at all the classical hexagonal packing of the Bénard-Marangoni convection [1] . Here the number of sides is comprised between three and eight. The most numerous polygons are those with five sides, then those with four, then those with six sides (the mean number of sides being < n > = 4,94). The corners of the polygons may join three, four, and even five sides without rearranging to three.

The mean area $S_n$ of polygons with n sides is a linear function of n, as shown in the figure 5.

This is the well known empirical Lewis law, $S_n = an + b$ , which has been observed in extremely different systems, such as the basaltic prisms of the Giant's Causeway, the lattice of epithelial cells in the skin of cucumbers, the packing of soap froths[4]. Another more recent example is the system of convective cells during the crystallization of Pb-Tl alloys [5].

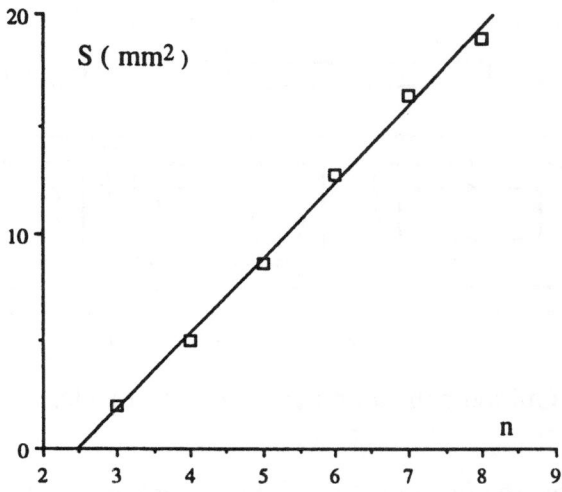

Fig. 5 Mean area of the cells with n sides as a function of n.

Rivier [6] has demonstrated that in random stackings this law should read

$$S_n = S \lambda n + S ( 1 - <n> \lambda )$$

Where S is the mean area of the polygons, and $\lambda$ a constant.

Fitting this law to the experimental line, with $S = 8.683$ mm$^2$ the constant $\lambda$ can be evaluated in two independant ways, firstly from the slope of the curve, and secondly from the ordinate at $n = 0$. The values thus obtained are respectively 0,405 mm$^{-2}$ and 0,405 mm$^{-2}$, i.e. they are equal. So the Lewis-Rivier law is accurately observed, indicating a random arrangement of cells. Furthermore, there is no tendency to order.

# 3    Guided convection

## 3.1 Guided convection as a system of rolls

Placing above the liquid a set of parallel glass plates leaving between them areas where the evaporation is more active, one imposes a convection as a system of rolls or of regular cells [7].

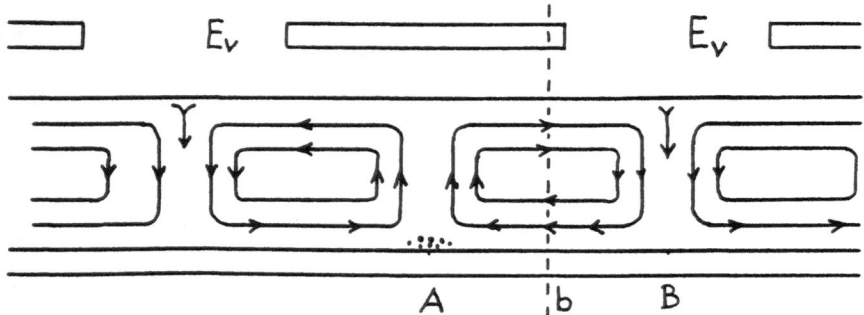

Fig. 6 Guided convection as a system of rolls.

In such conditions, the heavier fluid sinks at regular intervals, so that a system of parallel rolls takes place. This can be obtained for all concentrations of the solutions, even with pure water.

The velocity profiles in such rolls are quite similar to that shown in the figure 3, but with higher maximum velocities : 30 $\mu$m s$^{-1}$ for pure water , 20 $\mu$m s$^{-1}$ for a sucrose concentration of 40%. Here again, the free surface is immobile in first approximation. The guided evaporation of the liquid is thus an easy way to get the fluid moving regularly at a small velocity.

In such a system of rolls it is interesting to study the transport and sedimentation of solid particles more dense than the liquid. Suspensions of sulphur grains have been used, with diameters ranging from 2 $\mu$m (small ) to 10 $\mu$m (large grains). It is clear from the figure 6 that grains near the bottom are swept in the direction of the stagnation line A, where they must accumulate. This is shown in the figures 7.

A direct simulation of this transport has been made by R. Occelli and gives a good agreement with the experiment [8]. For small grains the simulation predicts a sedimentation as a double bank along the stagnation lines (figure 8b): this unexpected result has been confirmed later experimentally (figure 7b).

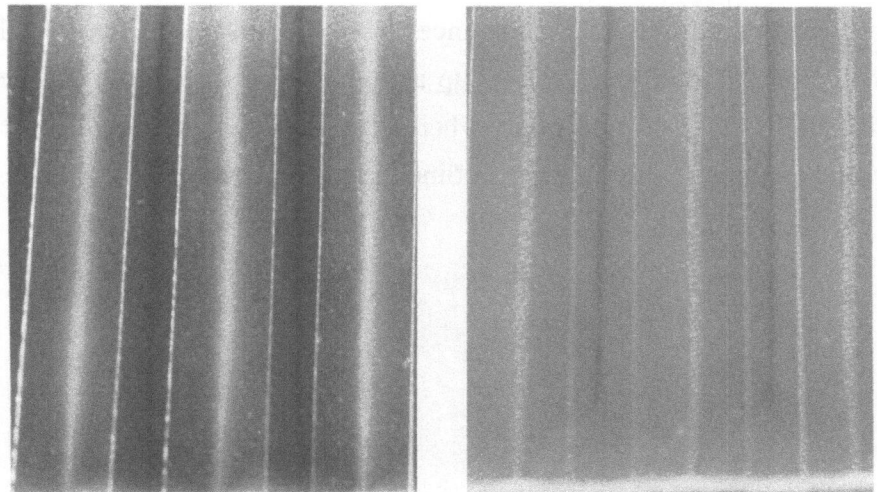

Fig. 7 . a ) Sedimentation of large grains.
b ) Small grains

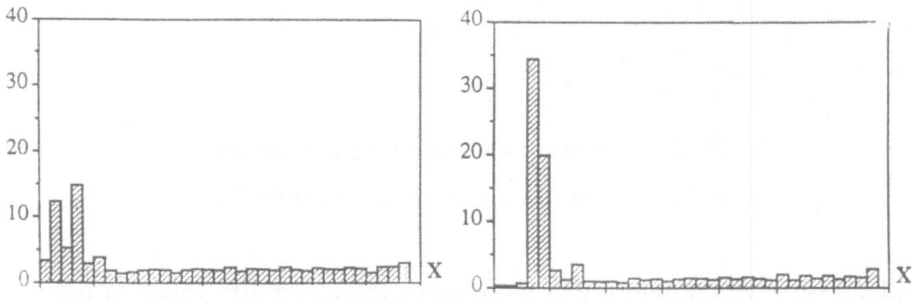

Fig. 8 . Direct simulation of the density of sedimented grains
a ) Large grains b) small grains

Suspensions of still smaller grains ( less than 2 µm) are stable in a vertical column and exhibit the Tyndall effect during months. When such suspensions are put to convect as a system of rolls, the sedimentation along lines occurs in about 24 hours. This apparatus is thus working like a centrifuge ( an ecological one, in fact ).

98

## 3.2 Guided convection as a system of cells

When placing above the liquid a perspex mask drilled with a set of holes, where the evaporation is more active, there results a convection as a system of cells. For instance, with holes in a hexagonal arrangement, the contrasts show up as indicated in the figure 9. The striae on the surface appear only where the liquid is exposed to the free air, and the narrow bright lines outline the convection cells .

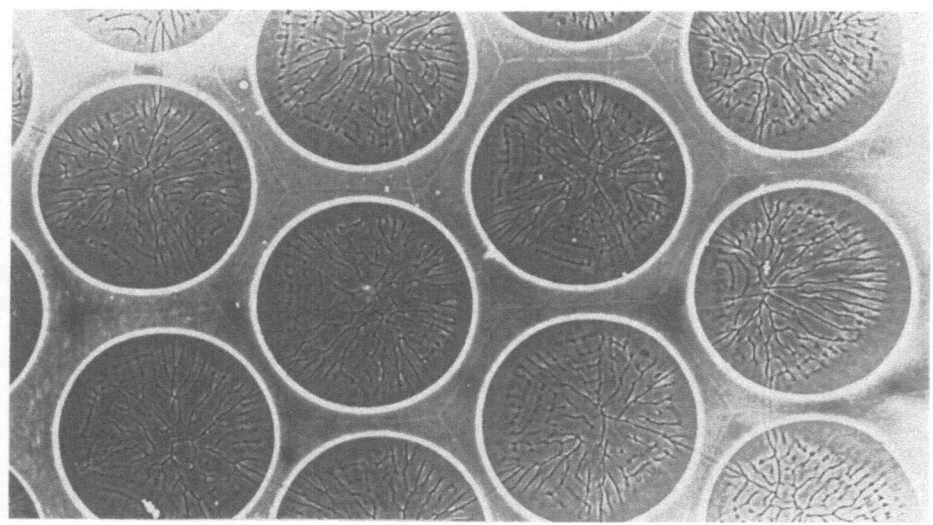

Fig. 9   Convection as a system of hexagonal cells.
The diameter of the holes in the mask is 2 cm

With an infrared camera it was shown that there is a difference of half a degree between the cold zones (center of convective cells) and the hot zones (walls of the cells).

In such a system of cells the sedimentation of grains on the bottom of the container occurs as described in rolls, i.e. along the stagnation lines, with the interest that here this sedimentation is an accurate means to determine exactly the limits of the convective cells and the influence of the confinement (figure 10). It is seen that in such a confined container only the central cell is hexagonal.

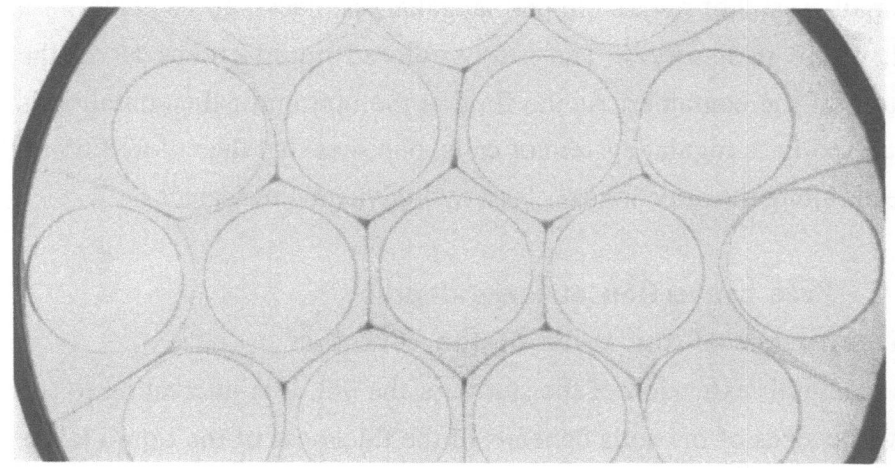

Fig. 10 . Sedimentation of grains ( dark lines) in a confined systems;
for a hexagonal lattice of holes only the central cell is hexagonal

Let us describe the transport of an inert component along the bottom
of a container where there is a system of convective cells (figure 11).

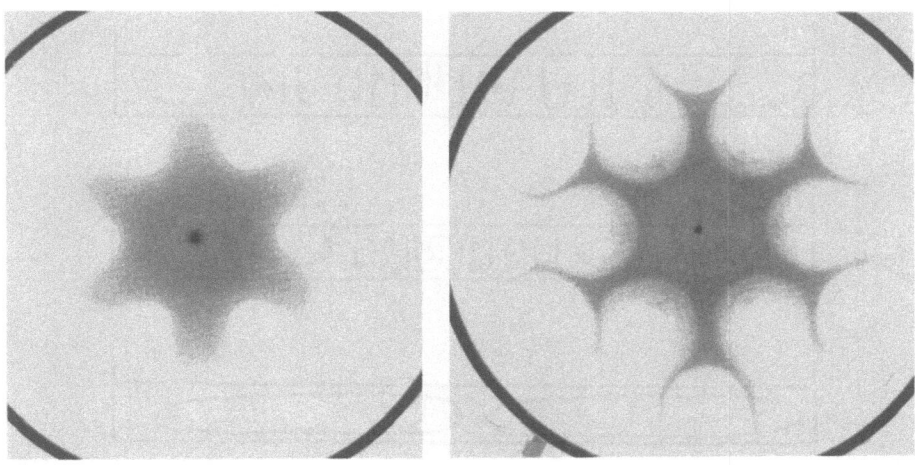

Fig. 11.  Spreading of a heavy liquid  obtained by the dissolution of
a grain of  potassium permanganate, at t = 30 min, and at  t = 1,5 h

A small crystal of potassium permanganate is placed at the center of a cell, where it dissolves. The heavy colored liquid creeps along the bottom of the container. As the fluid at the bottom of the container is organized in a regular system of cells, one sees that the colored liquid spreads along the directions of least hydrodynamic resistance.

## 4. Free convection at large depths
### 4.1 Influence of the walls of the container

As a natural extension of the study of the § 2, it is interesting to see how the sizes of the cells depends of the thickness of the liquid layer. At the concentration of 50% and at depths greater than 1.9 mm the streamers do not rearrange into the network of well defined polygonal cells. Moreover it is found that increasing the depths of the liquid amounts to increase the influence of the lateral walls of the container, and for small aspect ratios this precludes the observation of the convective cells.

A 50% sucrose solution is placed in a circular container 100 mm in diameter, the thickness of the liquid being 3 mm. The figure 12 shows the evolution of the hydrodynamic regime.

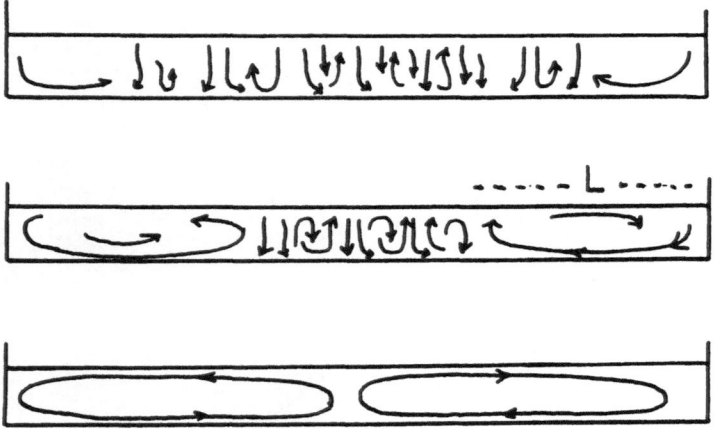

Fig. 12. Large depths: the roll initiated at the wall invades the whole volume of the container.

There appears first the usual chaotic system of streamers, but these do not rearrange in a system of convective cells; at the same time there appears a roll near the walls of the container, with the sense indicated on the figure 12.

Progressively this roll invades the whole space and the chaotic zone disappears in about six hours. Two stages of this process are shown in the figure 13 .

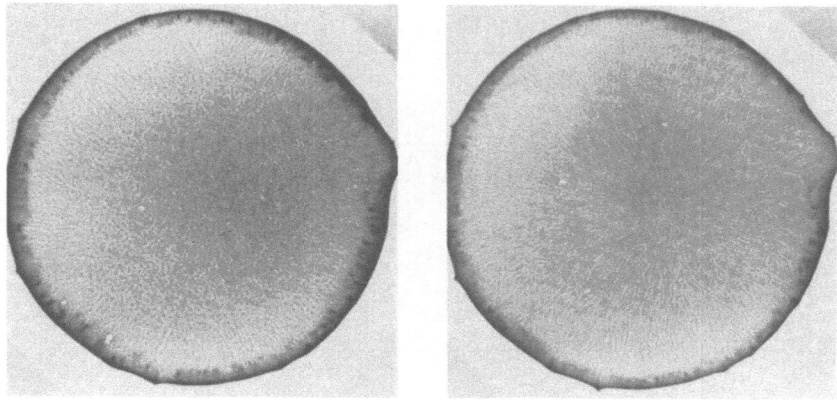

Fig. 13. Two stages of the invasion of the volume by the laminar flow.

The reason for this is as follows: when a streamer far from the walls of the container is sinking, it spreads at random along the bottom of the container. But near the wall a sinking streamer must spread on the bottom in the direction of the center of the container, so that a laminar motion is initiated here. By viscosity effects, this laminar motion forces the adjacent streamers to spread also in the direction of the center, so that the laminar zone progressively invades the whole container. The distance from the wall of this limit between the chaotic and the laminar zone is proportional to $t^{0.5}$ .

## 4.2 Convection far from the walls of the container

To avoid the influence of the walls of the container the following experiment has been devised. A perspex cover with a large hole is placed above the liquid, so that the convecting zone is expected to exist in an ocean of immobile fluid . The figures 14 and 15 show the evolution of the hydrodynamic regime.

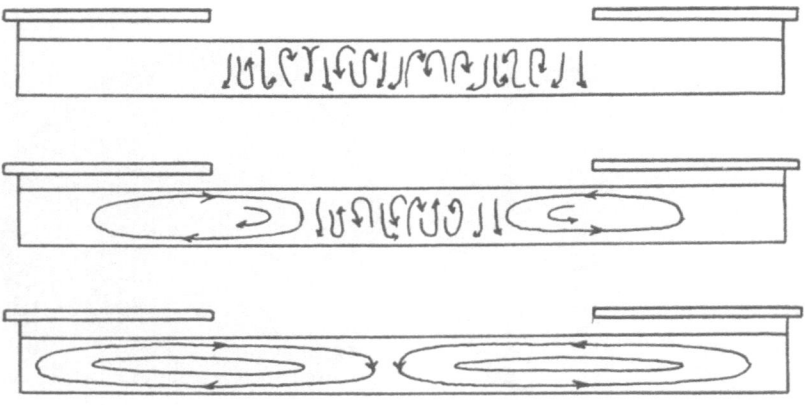

Fig. 14 . Evolution of the convection regime for a preferential evaporation under a large hole in the mask

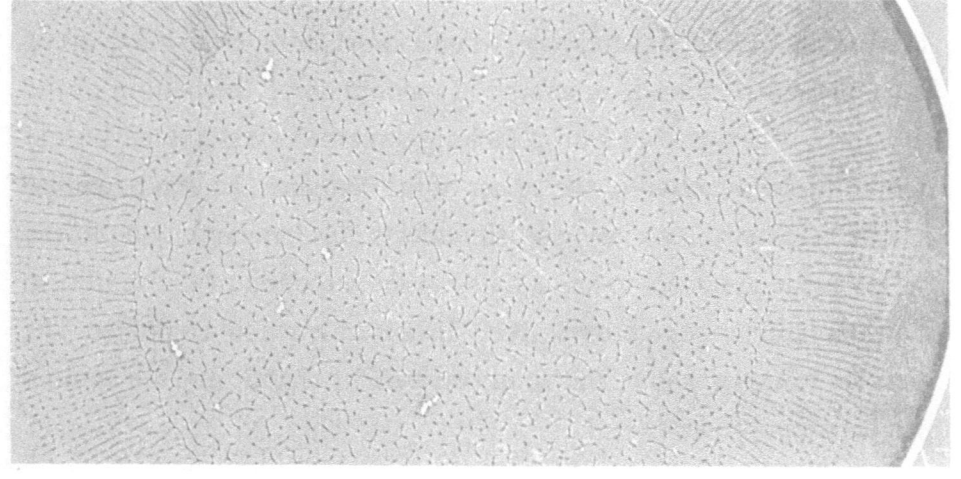

Fig. 15 Intermediate view of the liquid layer. The bright circle is the rim of the circular mask( diameter 14 cm).

Indeed it is found that the convective chaotic zone initially exists in the part of the fluid exposed to the free air. But progressively there appears at the periphery a system of radial contrasts, indicating a laminar motion here (figure 15). This system invades the whole container, as in the experiments with solid walls. But now, the laminar motion is divergent along the bottom of the container; furthermore, the motion is not limited to the zone where the optical contrasts are seen. The striae on the surface exist only where this is exposed to the free air, but there is a motion below the covered zone. In fact the fluid is in movement in the whole volume of the container .

The sense of the motion in the laminar zone can be explained as follows (figures 14 ) : consider the sinking of a streamer at the center of the container:  it spreads along the bottom  in random directions; consider now a streamer forming at the periphery of this zone: this will spread rather on the bottom in the direction of the periphery, initiating there a laminar motion . By viscosity this laminar motion will put in motion the whole immobile zone. Now the adjacent more central streamer will follow, so that, as in the case of small depths previously described the laminar zone invades the chaotic zone, and makes it disappear  .

## Conclusion

We have shown that the evaporation of sucrose solutions is at the origin of many unexpected phenomena. The fluid velocities are very low, so that long periods of observation are necessary to detect the consequences of the motion.

At small depths the free convection occurs as a system of random cells; this is due probably to the high values of the Rayleigh number, and maybe also to the fact that the experiments are performed with decreasing Ra, so that hysteresis phenomena may play a role. At greater depths (much higher Rayleigh numbers) there is a system of irregular streamers, and the influence of the confinement becomes important.

Much remains to be done to explore systematically this field of evaporative hydrodynamics. Let us cite the influence of the smaller depths and of the concentration on the morphology of the cellular regime. In particular, efforts will be made to work with extremely pure systems, and in better controlled atmospheres. In this case, lower values of the Rayleigh number should be interesting to study.

**References**

1  B. Simon , M. Belmedani  "Convection cellulaire dans des couches minces de solutions aqueuses de saccharose: loi de Lewis". C.R. Acad. Sci. Paris, t. 319, Série II , p. 865 - 871 (1994)

2   J. C. Berg, A. Acrivos, M. Boudard  "Evaporative convection". Adv. Chem Eng.  $\underline{6}$ p. 61 - 123  (1966)

3   A. G. Kirdyashkin, "Thermogravitational and thermocapillary flows in a horizontal liquid layer under the conditions of a horizontal temperature gradient", Int. J. Heat Mass Transfer, Vol.27, n° 8,  p. 1205 -1218  (1984)

4   N. Rivier "Order and disorder in packings and froths". In "Disorder and granular media", D. Bideau and A. Hansen (editors), Elsevier Science Publishers B.V., p. 55 -102  (1993)

5   H. Jamgotchian, B. Billia, H. Nguyen Thi, "Organisation des réseaux bidimensionnels de cellules de solidification dirigée". Ann. Chim. Fr., $\underline{16}$ p. 229 - 235  (1991)

6   N. Rivier,  "On the structure of random tissues or froths and their evolution", Phil. Mag. B , $\underline{47}$ , n° 5,  p. L45 - L49  (1981)

7   B. Simon et Y. Pomeau, "Free and guided convection in evaporating layers of aqueous solutions of sucrose".  Phys. fluids A, $\underline{3}$ , p. 380 - 384  (1991)

8   B. Simon, R. Occelli, "Sédimentation de grains dans un système de rouleaux convectifs", J. de Physique III, $\underline{4}$ , p. 1411 - 1420  (1994 )

# Experimental Study of the Competition Between Convective Rolls in an Enclosure

P. Cerisier and S. Rahal

IUSTI, UMR CNRS 139, University of Provence, 13397 Marseille Cedex 20, FRANCE

**Abstract.** An experimental study was carried out to describe the interaction and competition between Rayleigh-Bénard roll pattern and a roll induced by a lateral heating in a horizontal or in an inclined enclosure.

## 1 Introduction

A great number of studies have been presented on natural convection in enclosures because of its great importance in many engineering applications such as design of buildings, electronic devices, solar collectors, thermal storage devices, cooling systems for reactors in the nuclear industry, etc.....

The most studied case, the Rayleigh-Bénard convection, is that of a layer submitted to a vertical difference of temperature ($\Delta T$) which organizes itself into a roll pattern when $\Delta T$ is beyond a critical value $\Delta Tc$. This configuration is an exemplary case for the study of non-linear fluid dynamics and transition to turbulence. A certain number of works have been devoted to the evolution and the stability of such structures for supercritical conditions and for various thermal and/or geometrical conditions.

An experimental investigation of the planform of Rayleigh-Bénard convection in bounded circular, square or rectangular containers was performed by Koschmieder [1]. From these experiments, Koschmieder concluded that the form of the pattern in a bounded fluid layer is determined by the shape of the lateral walls (in rectangular containers, straight rolls whith axes parallel to the short side of the container are observed; in circular vessels, concentric rolls, etc..). These results were confirmed experimentally by Stork and Müller [2] and theoretically by Davis [3] and Charlson and Sani [4].

For a fixed value of the Prandtl number (Pr), the Rayleigh-Bénard convection undergoes a number of discrete transitions, remaining in each regime for a finite range of Rayleigh number (Ra), which is the characteristic nondimensional number in this problem. A large amount of studies were carried out to answer questions : in what region of the (Ra,Pr) space does one observe two dimensional rolls? Are they steady or time dependent? Are there three dimensional patterns? What is the nature of instabilities which lead two dimensional rolls to three dimensional patterns?

A non-linear analysis of the stability problem [5] gave a region of stable rolls which is often referred as the "Busse ballon" named after F. Busse who identified the many secondary instabilities (zig-zag, cross-rolls, knot, etc...) beyond which different types of convecting states are observed. These states are either more complicated steady roll patterns or time-dependent states which may be periodic or non periodic.

A precise experimental test of the predicted stability diagram is not easy to carry out since the ideal geometry of the theory cannot be realized, but the predictions were confirmed experimentally at least semiquantitatively [6-11], both in the approximate location of the stability boundaries and in the nature of each instability.

Other works have been devoted to convective motions driven by lateral temperature gradients [12-18]. Generally, in this case, the flow consists of a main circulation in which fluid rises along the hot wall, sinks along the cold wall.

Many papers have been written on convection in a horizontal fluid layer induced by vertical or horizontal temperature gradients, whereas few papers [19,20] have been devoted to convection induced by inclined temperature gradient. Besides the theoretical importance, this configuration is useful for modeling a number of physical systems, examples of such systems can be found in the paper of Drummond and Korpela [19].

Finally, another configuration which is of great importance in engineering applications is that of enclosures inclined to the direction of gravity where we need to consider the effects of both the tangential and normal components of the buoyancy force relative to the differentially heated walls. Review of theoretical, experimental and numerical studies on this subject until 1980 is given by Schinkel [21]. Other investigations [22-31] were performed later to study the influence of the nature of fluid, the angle of inclination, aspect-ratios of enclosures on the patterns and the transitions between them, the measurement of heat transfer, etc...

It seems that no investigation has been devoted until today to the problems of interface and competition between patterns driven by different heatings in a horizontal or in an inclined enclosure, more precisely when the Rayleigh-Bénard pattern is disturbed by a lateral heating parallel to the short wall of the box. Such experiments have been performed to understand how a roll driven by a lateral heating interacts with a Rayleigh-Bénard pattern. At the interface, the two flows can be parallel or antiparallel. Indeed, in that case, shear effects can be expected. This paper shows the first results obtained by our team.

The outline of this paper is the following : experimental procedure is described in section 2, typical experimental results are given in section 3; finally, some conclusions are gathered in section 4.

## 2 Experimental Procedures

The enclosure is a rectangular cavity ($1 \times 3 \times 12$ cm$^3$) filled with silicone oil Rhodorsil 47V100 (Pr=880 at T = 25°C)(Fig. 1), made of polycarbonate "Lexan" which is transparent to the ligth and has about the same thermal conductivity (0.22 w/m.K) as the oil (0.16 w/m.K). Therefore the horizontal surfaces ($3 \times 12$ cm$^2$) C and C' limiting the liquid layer are moderate heat conductors. The fluid is limited by two small lateral vessel walls ($1 \times 3$ cm$^2$), A and A' and two lateral walls ($1 \times 12$ cm$^2$), B and B'. A, A', C and C' walls have the same thickness : 0.3 cm. That of B or B' is 1 cm. A and A' are made of copper and separated from the atmosphere by a polycarbonate wall. They are isotherm and can be kept at fixed or variable temperature or at the ambiant temperature. In our experiments, the ambient room temperature could be smaller or larger than the average temperature of both C and C'. This fact is important because it imposes the flow directions along A and A' : upward or downward. C' and C are respectively in contact with a cooling and a heating water flow at controlled temperature. The inclination angle of the box is $\phi$ ($\phi < 0$ when A is at the lower part of the vessel and $\phi > 0$ when it is at the upper part). The experiments were limited to $|\phi|$ values smaller than 6°.

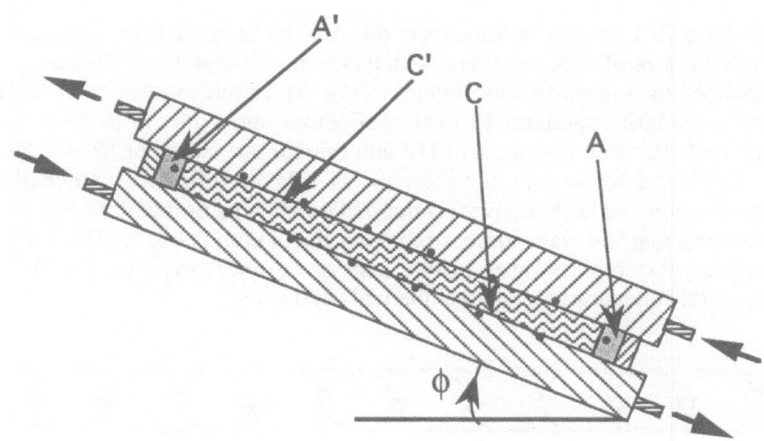

**Fig.1.** Schematic of the apparatus

The temperatures in A and A' and on C and C' are measured with fourteen thermocouples (diameter : 0.5 mm).

The flows are visualized in vertical planes and in planes parallel to C by means of a laser beam. The streamlines are made visible using small aluminium flakes in the liquid.

Five experiment series were performed :

Series 1 - It is the classical Rayleigh-Bénard convection but between moderately thermal conducting horizontal boundaries : a vertical temperature gradient greater than the critical one is applied to the liquid layer. A roll pattern parallel to the small side walls (A, A') is observed in the fluid layer.

Series 2 - It is the classical Rayleigh-Bénard experiment in an inclined box. Experiments are performed for Ra = 3.1 Rac (Rac : critical value of Ra).

Series 3 - The two horizontal limiting surfaces (C and C') are at the same temperature. The temperature of A (TA) is increased step by step. A roll is induced along A. This experiment was performed respectively for horizontal and inclined fluid layer.

Series 4 - For the horizontal configuration, we superposed the two heatings : when a stable Rayleigh-Bénard roll pattern is established, A is heated and disturbs the roll pattern. The size of the induced roll increases whereas the number of the other rolls decreases.

Series 5 - C and C' are differentially heated in inclined configuration. Once the Bénard pattern established, the lateral heating is switched on in the A wall. Then, the steady regime being reached, a new temperature for A is imposed.

For each series, the steady state being reached, the sizes of the rolls and the temperatures of the limiting surfaces are measured.

## 3 Results and Discussion

Figure 2 shows three patterns corresponding to three values of the temperature difference (and Ra). It can be observed that the number of rolls decreases (and consequently the size of rolls increases) with increasing temperature difference.

We calculated the marginal stability curve (Fig. 3) for our experimental conditions (horizontal boundaries moderately heat conductors and with finite thickness, the critical value of Ra : Rac is equal to 1318 and critical wavenumber Kc = 2.6 instead of Rac = 1705 and Kc = 3.11 for the case of a layer infinitely extended in the horizontal direction and between perfect thermal conductors horizontal boundaries). It can also be seen that the wavenumber of supercritical convective motions decreases with increasing Rayleigh number in our case as in other experiments [1, 32-36] where horizontal boundaries are perfect thermal conductors

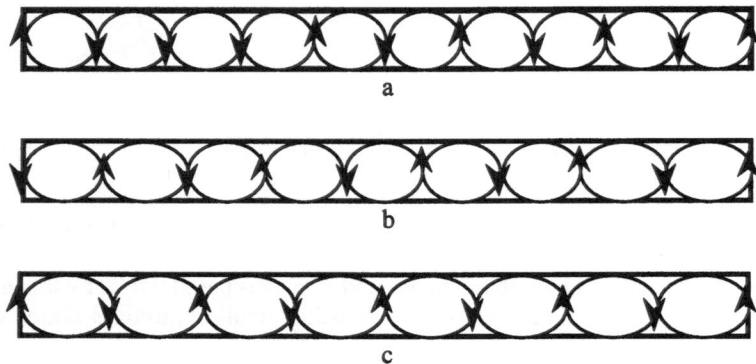

**Fig.2.** Series 1 : Rolls observed in the vertical median plane (Drawn from photos)
a : Ra = 2.4 Rac, b : Ra = 3.3 Rac, c : Ra = 5 Rac

As discussed by Gershuni and Zhukhovtskii [37], longitudinal rolls represent the preferred form of convection at high Pr (let us recall that for our experiment Pr = 880) for almost all inclinations as long as there exists a finite component of the temperature gradient opposite to the direction of gravity. Figure 4 shows two longitudinal rolls and two transverse rolls near A and A'. We could think that the presence of these transversal rolls is due to the non adiabaticity of A and A', but in a study of the stability of natural convection in an inclined fluid layer performed by Masuoka and Shimizu [38], it was found, experimentally and analytically, that for small angles of inclination, the effects of lateral walls can give rise to a secondary flow in the form of stationary transverse rolls with the horizontal axes parallel to the shorter side. That transverse rolls are not regular cylinders, they " drive in wedges" between the two longitudinal rolls because the flows are all dowward in the central

109

part but they have opposite directions close to B and B' : the shear effect is important in that regions.

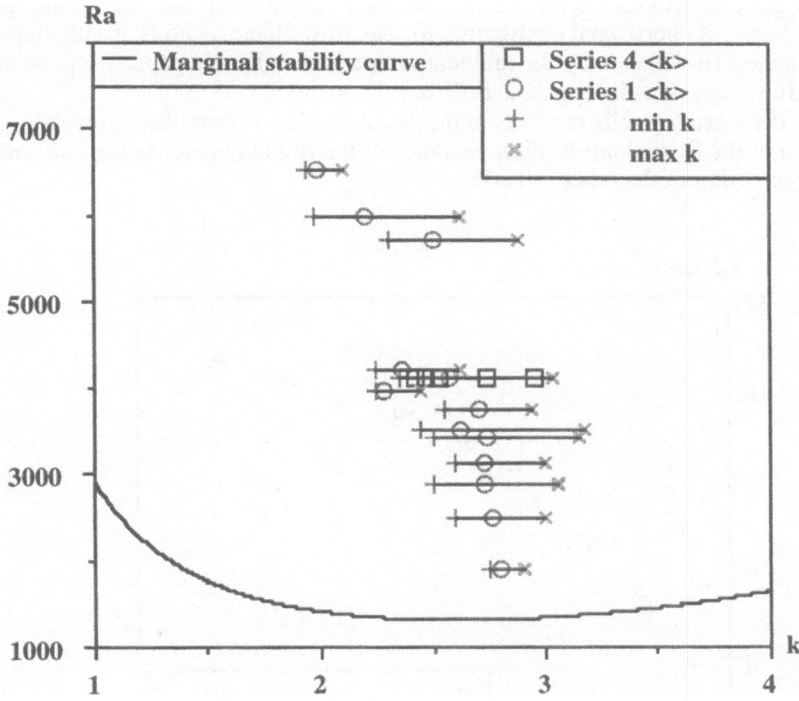

**Fig.3.** Series 1, Series 4 : Stability diagram
Ra : Rayleigh number, k : wave number

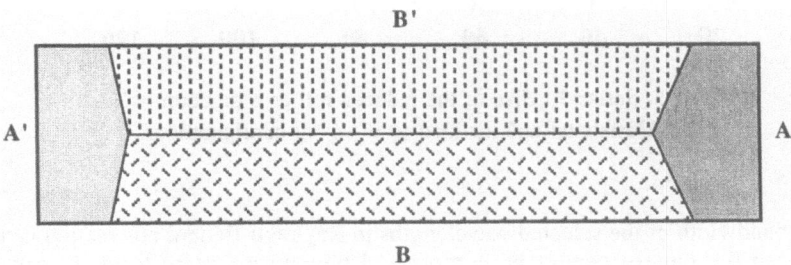

**Fig.4.** Series 2 : Rolls observed in the horizontal median plane (Drawn from photo)
($\phi = +3°$), Ra = 3.1 Rac

In Fig. 5 we compare the length of the induced roll as a function of TA for series 3. It can be seen that they have the same trend and that the size of the induced roll looks smaller for the inclined configuration ($\phi = -4.5°$) than in the horizontal one.

The pattern in Fig. 6 corresponding to series 4 is the result of competition between the roll pattern without lateral heating (series1) and the roll induced by the lateral heating (series 3 : horizontal configuration). The flow along A and A' is either upward or downward (it depends on the ambient temperature). In this paper we present the case of flows downward along A and A'.

First, there are ten rolls in the pattern; the heating of A provokes a reversing of a direction of the flow along A, disappearance of the roll neighbour to the induced roll and compression of the other rolls (Fig. 7).

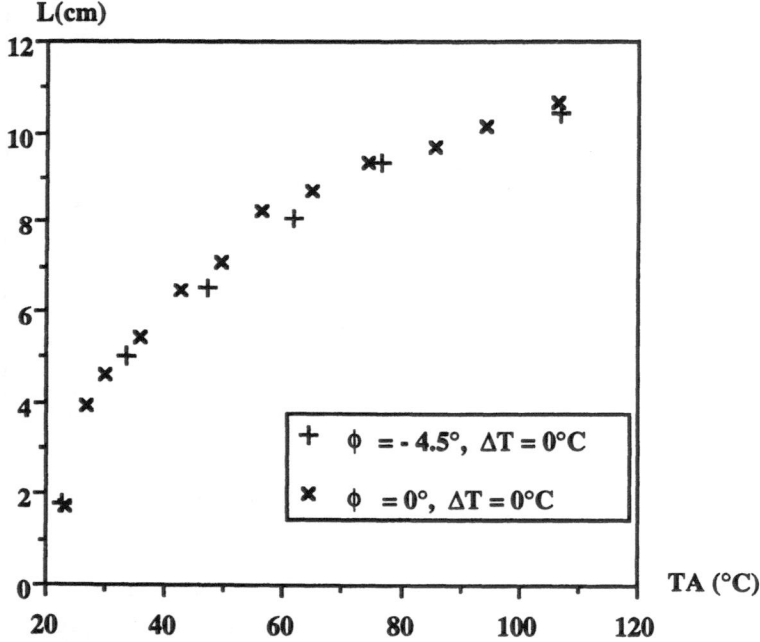

**Fig.5.** Series 3 : Size of the induced roll as a function of TA
Horizontal and inclined ($\phi = -4.5°$) configurations

The bandwidth of the selected wavelengths in Rayleigh-Bénard convection is narrow [5,39], so the pattern cannot be compressed beyond a certain limit. In Fig. 3 we drawn wavenumbers corresponding to Rayleigh-Bénard rolls for series 4. It can be seen that effectively these wavenumbers are included into the range of k observed for the classical Rayleigh-Bénard convection (without lateral heating). A new increase of the induced roll driven by a new lateral heating needs the disappearance of rolls. Due to gear effect a roll pair disappears. Generally the remaining rolls are greater than the natural ones. Then there is a new compression and so on.

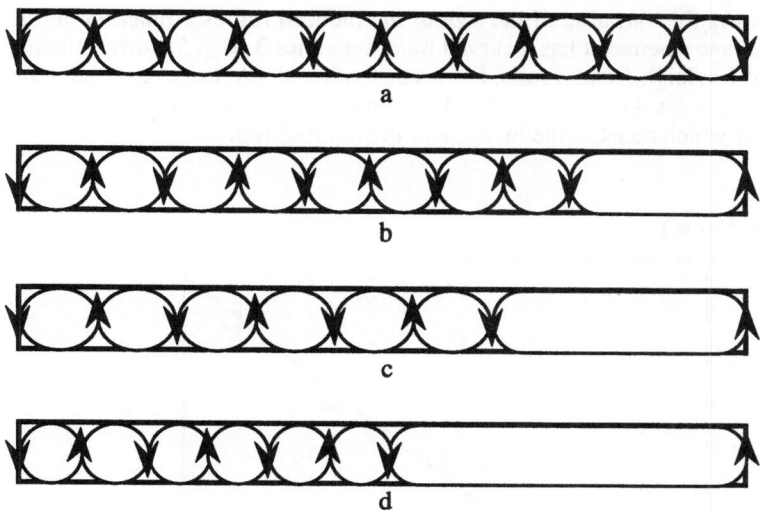

**Fig.6.** Series 4 : Rolls observed in the vertical median plane (Drawn from photos)
a : TA = 23.3 °C, b : TA = 32.5 °C, c : TA = 43.2 °C, d : TA = 61.6 °C

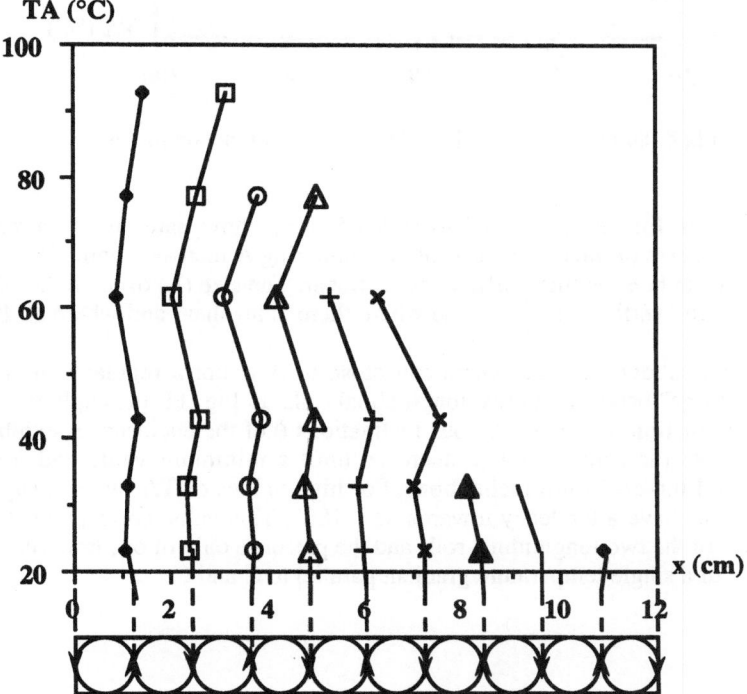

**Fig.7.** Series 4 : Behaviour of the roll pattern for different TA

Figure 8 shows the variation of the size of the induced roll as a function of TA. It is a linear increase whereas it has a curved trend for series 3 (Fig. 5) corresponding to the absence of Rayleigh-Bénard rolls. For a same value of TA, the size of the induced roll is smaller in series 4 than in series 3. It is due to the presence of Rayleigh-Bénard rolls in series 4 which resist to the increase of the induced roll.

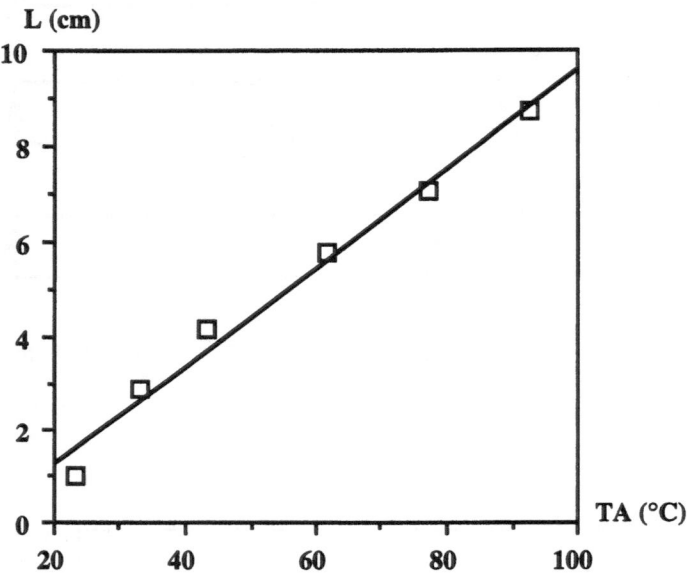

**Fig.8.** Series 4 : Size of the induced roll as a function of TA

The pictures in Fig. 9 correspond to series 5, they show patterns observed for various TA. The size of the transversal roll neighbouring A increases with TA.

In Fig. 10 it can be seen that this increase is linear. The size (L) of the induced roll is measured in the vertical median plane where there is no shear and where all flows are downward.

As mentioned above, the transverse roll close to A is not a regular cylinder, it "drives in a wedge" between the two longitudinal rolls. In Fig. 11, the angle $\alpha$ of this "wedge" as a function of TA for various inclinations $\phi$ of the enclosure is exhibited. The curves have the same trend (a decrease until a minimum value and a slow increase) for all the enclosure inclinations. For high values of TA, we can suppose that these curves have a tendency towards $\alpha = 180°$. This value corresponds to the disappearence of the two longitudinal rolls and the presence only of one transverse roll as in the case of a single temperature gradient parallel to C and C'.

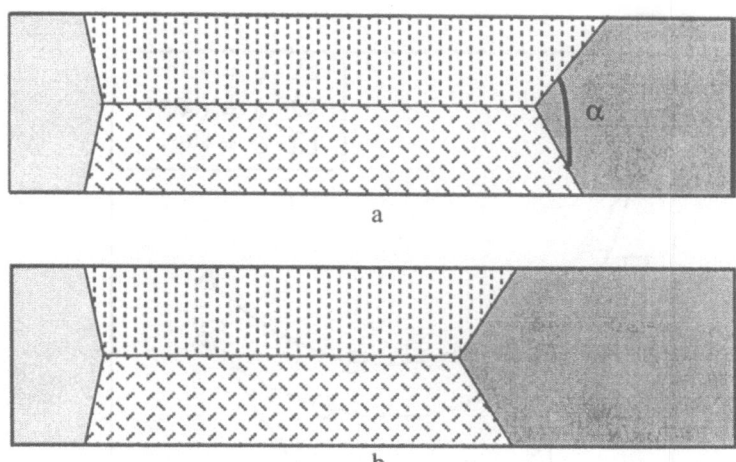

**Fig.9.** Series 5 : Rolls observed in the horizontal median plane (Drawn from photos)
($\phi = +3°$) a : TA = 34.7 °C, b : TA = 44.6 °C

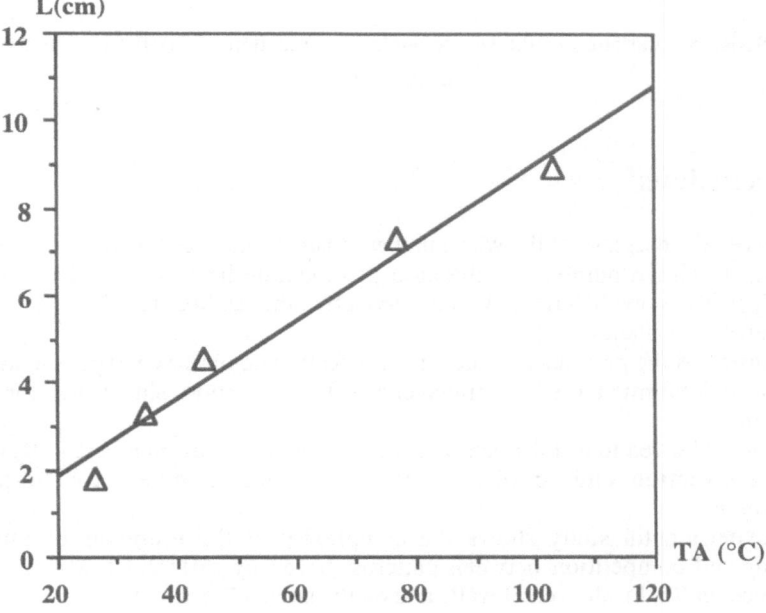

**Fig.10.** Series 5 : Size of the induced roll as a function of TA
($\phi = +3°$)

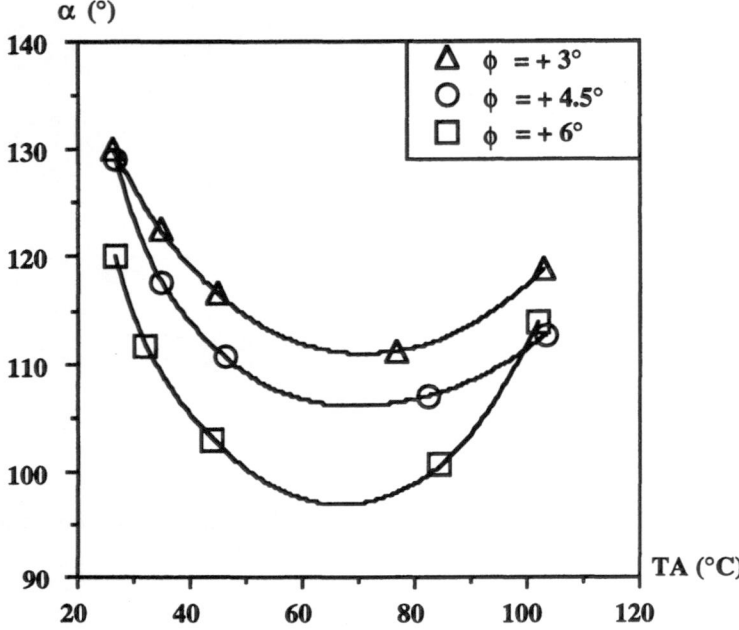

**Fig.11.** Series 5 : Variation of the wedge angle α as function of TA for various angles of inclination φ

# 4 Conclusions

We observed the decrease of the wavenumber of supercritical convective motions with increasing Rayleigh number for the case of moderate heat conductors horizontal boundaries . The same behaviour was observed by other authors [1, 32-37] for perfect heat conductors boundaries.

We confirmed the presence, predicted analytically and observed experimentally by Masuoka and Shimizu [39], of transverse rolls near short side walls for small inclinations.

This study allowed to test the bandwidth of the selected wavenumbers in Rayleigh-Bénard convection and to observe the stable wavenumbers for horizontal configuration.

This experimental study shows the complexity of the problems of interface behaviour and competition between patterns driven by different heatings and its dependence on Ra, Pr, the vessel walls and on the angle of inclination.

Further investigations are under way to precise the influence of aspect ratios, great angles of inclinations, to link structures to heat transfer characteristics (temperature field, heat flux, Nusselt number, etc....) and to analyse the shear effect at the interface of two rolls with opposite flows.

115

# References

[1] E. L. Koschmieder, Beitr. Phys. Atmos. **39**, pp. 1-11 (1966).
[2] K. Stork and U. Müller, J. Fluid Mech. **54**, pp. 599-611 (1972).
[3] S. H. Davis, J. Fluid Mech. **30**, pp. 465-478 (1967).
[4] G. S. Charlson and R. L. Sani, Int. J. Heat Mass Transfer. **13**, pp. 1479-1496 (1970).
[5] F. H. Busse, Rep. Prog. Phys. **41**, pp. 1929-1967 (1978).
[6] R. Krishnamurti, J. Fluid Mech. **42**, pp. 295-307 (1970).
[7] F. H. Busse and J. A. Whitehead, J. Fluid Mech. **47**, pp. 305-320 (1971).
[8] J. P. Gollub, A. R. McCarriar, and J. F. Steinman, J. Fluid Mech. **125**, pp. 259-281 (1982).
[9] P. Kolodner, R. W. Walden, A. Passner, and C. M. Surko, J. Fluid Mech. **163**, pp. 195-226 (1986).
[10] K. R. Kirchartz and H. Oertel Jr, J. Fluid Mech.**192**, pp. 249-286 (1988).
[11] V. Croquette, Contemp. Phys. **30**, pp. 113-133 (1989).
[12] J. W. Elder, J. Fluid Mech. **23**, pp. 99-111 (1965).
[13] J. Imberger, J. Fluid Mech. **65**, pp. 247-260 (1974).
[14] J. C. Patterson and J. Imberger, J. Fluid Mech. **100**, pp. 65-86 (1980).
[15] P. G. Simpkins and K. S. Chen, J. Fluid Mech. **166**, pp. 21-39 (1986).
[16] S. W. Armfield, in "3rd Int. Symp. on Computational Fluid Dynamics", pp. 305- 310. North-Holland, Amsterdam (1989).
[17] P. G. Daniels and P. Wang, Int. J. Heat Mass Transfer. **37**, pp. 375-386 (1994).
[18] P. Wang and P. G. Daniels, Int. J. Heat Mass Transfer. **37**, pp. 387-399 (1994).
[19] J. E. Drummond and S. E. Korpela, J. Fluid Mech. **182**, pp. 543-564 (1987).
[20] D. A. Nield, Int. J. Heat and Fluid Flow. **15**, pp. 157-162 (1994).
[21] W. Schinkel, "Natural convection in air-filled enclosures", Dutch Efficiency Bureau, Pijnacker, (1980).
[22] Douglas W. Ruth, K. G. T. Hollands and G. D. Raithby, J. Fluid Mech. **96**, pp. 461-479 (1980).
[23] S. J. M. Linthorst, W. M. M. Schinkel, and C. J. Hoogendoorn, J. Heat Transfer. **103**, pp. 535-539 (1981).
[24] S. M. Elsherbiny, K. G. T. Hollands, and G. D. Raithby, J. Heat transfer. **104**, pp. 515-520 (1982).
[25] H. Ozoe, K. Fujii, N. Lior, and S. W. Churchill, Int. J. Heat Mass Transfer. **26**, pp. 1427-1438 (1983).
[26] H. Inaba, Int. J. Heat Mass Transfer. **27**, pp. 1127-1139 (1984).
[27] R. J. Goldstein and Q. J. Wang, Int. J. Heat Mass Transfer. **27**, pp. 1445-1453 (1984).
[28] H. Q. Yang, K. T. Yang, and J. R. Lloyd, Int. J. Heat Mass Transfer. **30**, pp. 1637-1644 (1987).
[29] F. J. Hamady, J. R. Lloyd, H. Q. Yang, and K. T. Yang, Int. J. Heat Mass Transfer. **32**, pp. 1697-1708 (1989).
[30] J. A. King and D.D. Reible, Int. J. Heat Mass Transfer. **34**, pp. 1901-1904 (1991).
[31] F. H. Busse and R. M. Clever, J. of Engin. Math. **26**, pp. 1-19 (1992).
[32] A. J. Leontiev, and A. G. Kirdyashkin, Int. J. Heat Mass Transfer. **11**, pp. 1461-1466 (1968).

[33] G. E. Willis, J. W. Deardorff, and R. C. J. Somerville, J. Fluid Mech. **54**, pp. 351-368 (1972).

[34] R. Farhadieh, and R. S. Tankin, J. Fluid Mech. **66**, pp. 739-752 (1974).

[35] E. L. Koschmieder, and S. G. Pallas, Int. J. Heat Mass Transfer. **17**, pp. 991-1002 (1974).

[36] K. Bühler, K. R. Kirchartz, and H. Oertel, Acta Mech. **31**, pp. 155-171 (1979).

[37] G. Z. Gershuni and E. M. Zhukhovitskii, "Convective Stability of incompressible Fluids", Translated from the Russian by D. Louvish, Keter Publications, Jerusalem (1976).

[38] T. Masuoka, and G. Shimizu, Heat transf. Japanese Research. **16**, pp. 82-91 (1987).

[39] I. Catton, J. Heat transfer. **110**, pp. 1154-1165 (1988).

# Surface Deflection in Bénard-Marangoni Convection

P. Cerisier[1] and G. Lebon[2]

[1] IUSTI, UMR-CNRS139, University of Provence, F-13397, Marseille Cedex 20,
France
[2] Institute of Physics,B5, Liège University, Sart Tilman,B-4000 Liège, Belgium
and Department of Mechanics, Louvain University,B-1348 Louvain-la-Neuve,
Belgium

**Abstract.**This paper is a review of some recent experimental and theoretical works
about the deflection of the upper surface in thin layers heated from below and subject
to Bénard-Marangoni instability.

**Keywords.**Surface deflection. Bénard-Marangoni convection.

## 1 Introduction - Origin of Surface Deflection

Bénard-Marangoni convection refers to the motion observed in a thin horizontal liquid
layer submitted to temperature gradients. The upper surface of the layer is free but it
is submitted to a temperature-dependent surface tensio; this effect coupled to buoyancy
is responsible for the onset of convection.

The air layer above the liquid is either at rest or in motion : its thickness can vary
from a value of the order of 1mm to the open atmosphere; it is an important
parameter as it governs the heat transfer at the liquid-air interface.

Bénard in 1901 [1], was the first to study the hexagonal pattern displayed by this
phenomenon. He observed a steady regime of vertical prismatic convective cells with
hexagonal section. The hexagonal pattern fills up the whole liquid layer and the flow
is upwards at the centre of each cell and downwards at the outer periphery. Bénard
noticed also a deflection of the free surface, with a depression over the central part of
the convective cell, where the liquid is rising and warm, and an elevation of the
surface level over the cell edges, where the liquid is descending and cold. This
deflection will be called, in the following, a concave relief. Quoting Bénard : "*La
tension superficielle à elle seule, provoque déjà une depression au centre des cellules et
un excès de pression sur les lignes de faîte qui séparent les cuvettes concaves les unes
des autres* "(thesis p. 32).

However, this observation did seemingly not receive a deep attention, and Bénard
himself did not investigate further about this problem. A few years later, Lord
Rayleigh in 1916 [2], studied theoretically the problem of Bénard's instability : his
basic hypothesis was to assume that only the buoyancy forces are responsible for the
onset of motion. However, Rayleigh's theoretical predictions were not in agreement
with experimental results. Low and Brunt [3], in 1925, noticed that the critical
temperature gradients corresponding to onset of instability were at least ten times less
than Rayleigh's results. Later, Vernotte [4] repeated Bénard's experiment and found a

critical Rayleigh number between 5 and 10, instead of the value 657 calculated by Rayleigh, for two stress-free surfaces. Jeffreys, [5], using Rayleigh's theory, predicted that the relief would be convex (i.e. a dome over the central region).

A few years later, Block [6], experimentally, and Pearson [7], theoretically, proved that the variation of surface tension with temperature is an imùportant factor of instability omitted in Rayleigh analysis. Considering only surface tension forces, Scriven and Sterling [8] showed that convective motions induce concave deflections of the free surface. However Pearson's model was to crude to interpret all the features of Block and Bénard experiments.

Truly, when a liquid layer has a free upper surface, buoyancy forces (Rayleigh-Bénard mechanism) and surface tension forces (Marangoni effect) are both destabilizing, provided the liquid has a surface tension $\xi(T)$ varying with temperature, which is the case for most liquids. This is the main conclusion of a theoretical approach performed by Nield [9] who examined in detail both the coupling of buoyancy and surface-tension effects.

As shown by Jeffreys [5], and Scriven and Sterling [8], both buoyancy and surface tension produce deflection of the upper surface. When these two destabilizing factors act together, they have opposite actions on the surface deformation, but it is easy to foresee the nature of the deflection [8] :

- for a shallow liquid layer, surface tension forces are dominant with respect to buoyancy forces : the deformation is concave.

- for a thick layer, buoyancy forces are playing the major role and the relief is convex.

- when the two effects are of the same order of magnitude, deflections may balance each other and the surface may remain flat. In practice, as will be shown below, the forces do never balance exactly each other. The surface is not strictly flat, but displays a "mixture" of both deformations and exhibits a hybrid relief.

From the physical point of view, it is easy to understand why the two destabilizing phenomena have opposite effects on the deflection of the upper surface. Assume first that surface tension effects are negligible. Under these conditions, the pressure due to a warm (light) column of liquid in the axial part must balance that due to a cold (heavy) column in the periphery : the warm central part will be higher than the cold edges. Moreover this effect will be reinforced, because, due to fluid motion, the liquid is upwards along the central axis and downwards in the periphery. Now for a liquid with a negative surface tension temperature coefficient $\gamma = -\dfrac{\partial \xi}{\partial T} > 0$, a cold point at the surface has a greater surface tension than a warm point. The former attracts the liquid more than the latter, and as a consequence, the cold region will be higher than the warm region.

Very simple models have been proposed to calculate the height $\delta_m$ between the centre and the edge of a cell. Hershey [10], and Anand and Balwinski [11] obtained a rough estimation of $\delta_m$ , but they took only surface tension forces into account. For that reason, their model is unable to describe the influence of the depth layer d .

A decizing work was achieved by Kayser and Berg [12] who studied the deflection of a free surface heated from below but with one restriction : the heating source is not a surface as in Bénar-Marangoni convection but a straight wire. They showed that the relief is concave over the warm rising flow when the liquid depth d is small. On the opposite, it is convex for large values of d. For intermediate depths; the relief is hybrid. The mathematical model of Keyser and Berg provides results in qualitative agreement with experiments; in addition, the role of several parameters like the thermal expansion coefficient ($\alpha$), the surface tension ($\xi$), the temperature coefficient of surface tension ($\gamma$), depth d and thermal power on the sign and on the amplitude of the deflection was quantified. Other in teresting experimental and theoretical studies about surface deflection in Bénard-Marangoni convection were performed in the eighties [13-15].

The present paper is organized as follows. In sect.2, we describe recent experiments about surface deformation in Bénard-Marangoni convection. In sect.3, a linear theory is proposed : the fluid layer is open to the atmosphere and obeys Boussinesq's approximation. In sect.4, recent theoretical improvements are described. Sect.5 is devoted to the problem of forced deformations of the surface while prospectives are proposed in sect.6.

## 2 Experimental Set-up

### 2.1 Apparatus
The apparatus is sketched in Fig.1. It consists of a cylindical vessel (diameter : 14 cm) with a flat and horizontal bottom plate made of copper; the lateral walls are made of glass. The fluid under consideration is silicone oil Rhodorsil 47V100.

**Fig. 1** - Experimental set-up
A : Outer vessel - B : vessel - C : oil to study - D : outer guard ring

An outer guard ring containing the same oil prevents thermal losses through the lateral walls. The whole set rests on a thermostated heating device. The upper surface of the liquid is in contact with the open atmosphere. The temperatures of the upper and the lower boundaries are measured by using two thermocouples. The end of each of them is soldered to a small horizontal metallic disk to ensure a good thermal contact with the liquid. The accurateness of the temperature measurements is about 0.2°C. The thickness of the oil layer is determined by a mechanical procedure with a precision of the order of ±50 μm.

### 2.2 Optical Methods
The maximum surface deformation is of the order of a few μm so that, optical methods are specially well adapted. Many techniques are available : Moiré method (Moiré by reflection or Moiré of slopes) [16], Schlieren technique [12,17,18] or

interferometer measurements [1,11,19]. Cerisier *et al.* [13], used the Poggendorf method coupled to two different interferometric techniques.

## 2.2.1 Poggendorf Method

It is schematized in Fig.2. A laser beam is reflected on the deformed surface of the liquid. The relief profile is determined from the measurement of the deviation X of the spot on the screen S, as a function of the position x of the laser on the free surface (x is the distance between the considered point from the cell center along a perpendicular direction to a cell side).

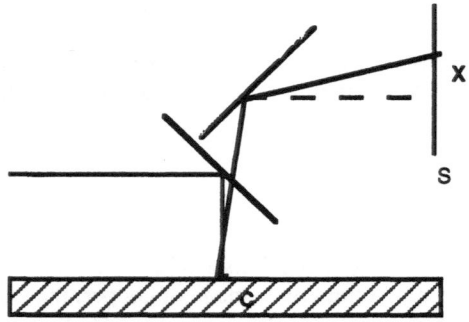

**Fig. 2** - Poggendorf method
S : screen - X : beam deviation - C : Liquid layer

## 2.2.2 Interferometer Methods

Among the interferometer methods, three of them are relevant in regards to thermoconvective instablities. The first one was used by Bénard [1] himself : interferences are created between the deformed free surface B and a horizontal reference glass plane ( see Fig.3).

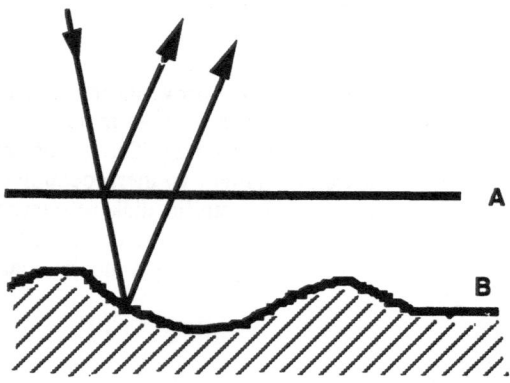

**Fig. 3** - Interferometer method
A : glass plane - B : liquid surface

The second technique is that of Michelson adapted to the study of a liquid surface (see Fig.4). The two mirrors used habitually are replaced, on the one hand by the deformed surface and, on the other hand by a horizontal and motionless surface of the same liquid (using the same liquid increases the contrast of the fringes).Figs. 5 and 6 exhibit examples of fringes displaying the surface relief.

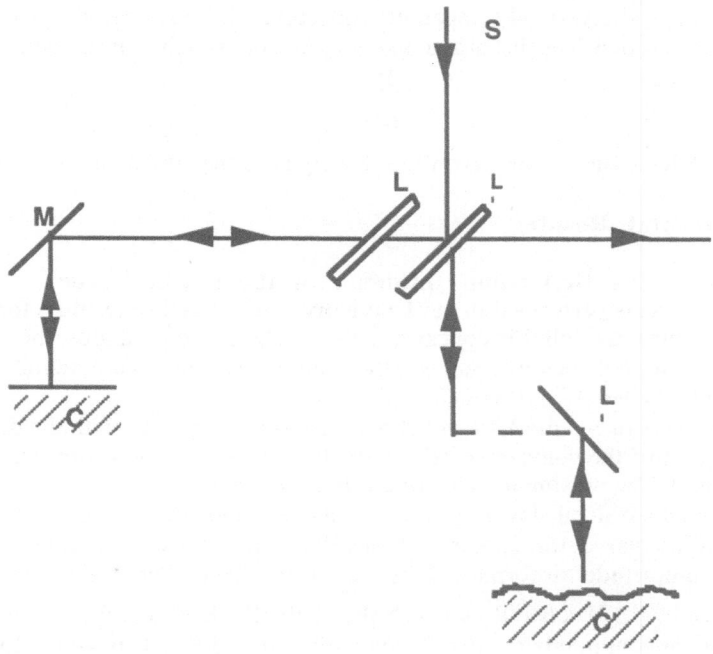

**Fig. 4** - Michelson method adapted to a liquid surface.

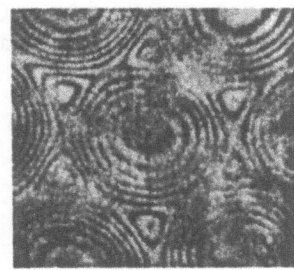 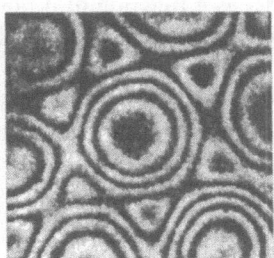

**Fig. 5** - Michelson method          **Fig. 6** - Air layer method
          Convex relief                          Concave relief

## 2.2.3  Foucault Method

The previous techniques provide the value of the amplitude of the deformation but not its sign. As is well known, the Foucault method produczs a shadow on the relief. From the direction of the shadow, it is possible to determine the sign of the deflection, negative for concave deformation and positive for convex deformation. Experiments are performed for layer depths between $d = 1mm$ and $d = 7mm$. For each value of d, the vertical temperature difference DT between the two limiting surfaces is also varied. The distance to the threshold of instability is defined by

$$e = \frac{Ra}{Ra_c} - 1 \qquad (1)$$

where Ra and $Ra^c$ denote respectively the Rayleigh and the critical Rayleigh numbers.

## 3  Experimental Results

### 3.1 Shape of the Deflection. Influence of the Depth Layer

All the experiments performed in our laboratory confirm earlier results : for shallow pools (d < 2 mm), the relief is concave, the six vortices are pyramids, the edges are similar to saddles between two valleys (the central parts) and two pyramids (Fig. 5). The maximum value of $\delta_m$ is equal to $2\mu m$.

For deep layers (d > 3 mm), the relief is inversed : it is convex with the dome over the central part of the convective cell while the vortices are the lower points of the surface (Fig.6). The maximum value of $\delta_m$ is about $4\mu m$.

For intermediate depth layers (1.8 mm < d < 2.3 mm) the relief is hybrid, with concave and convex deformations. It looks like a volcano with a round crest. The maximum amplitude deformation is about 0.3 $\mu m$. The transition zone is characterized by $0.171 < Ra^c/R_0 < 0.175$ ($R_0$ is the Rayleigh number corresponding to a zero Marangoni number ). Fig. 7 summarizes the surface behaviour close to the threshold : the region below the straight line of equation

$$\frac{Ra}{R0} + \frac{Ma}{M0} = 1 \qquad (2)$$

represents stable state, the region above the line refers to unstable convective states, Mo is the Marangoni number corresponding to a zero Rayleigh number.

Variations of $\delta_m$ as a function of d, for values of various $\epsilon$, defined by eq.(1) are represented in Fig.8. It is observed that, whithin the limit of experimental errors and a fixed value of $\epsilon$ :

- for a convex deformation, $\delta_m$ is a linear function of d,

- for a concave relief, $\delta_m$ varies as $d^{-1}$.

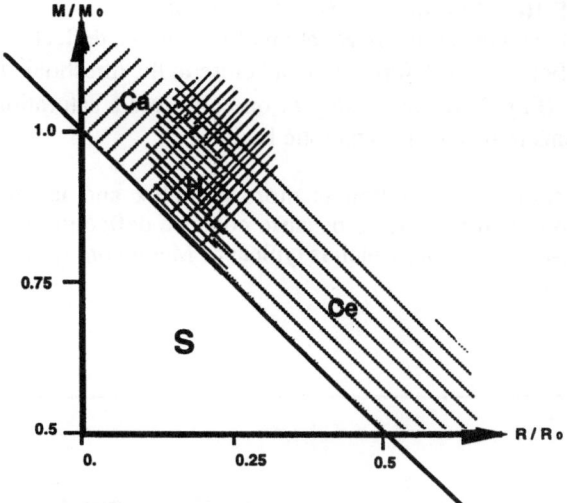

**Fig.7** - Relief shape close to the threshold
Ca : concave - Ce : convex - H : hybrid
S : stable region

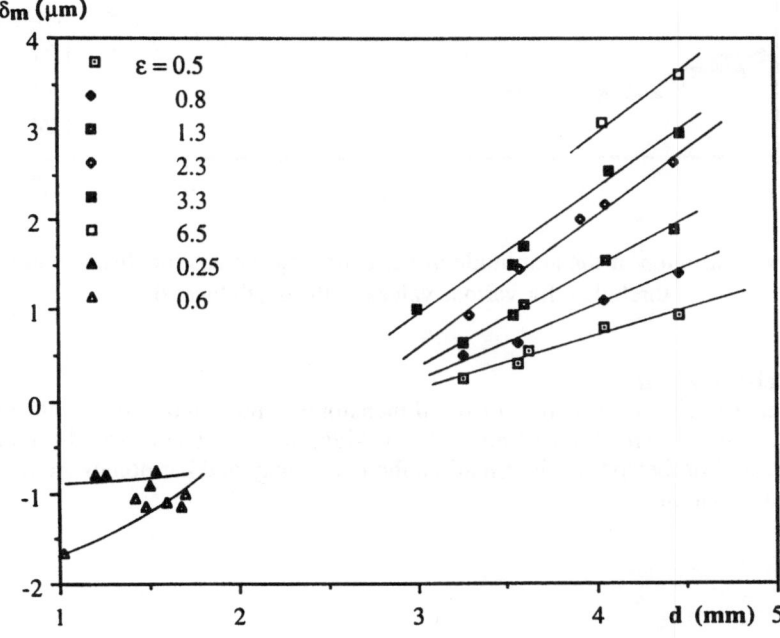

**Fig. 8** - Variation of the amplitude of the deflection $\delta_m$ versus the depth layer d for various values of the distance to the threshold $\varepsilon$.

## 3.2 Influence of the distance to the threshold

We now fix the depth and examine the variation of the surface deflection as a function of the parameter $\varepsilon$. For convex deformation and close to the threshold ($\varepsilon\approx1$), $\delta_m$ is a linear function of $\varepsilon$ (Fig.9). By increasing $\varepsilon$, one observes a saturation : $\delta_m$ varies very slowly and seems to reach an asymptotic limit.

For concave deflection, the situation is more complex and is presently under investigation. The problem is delicate because concave deflection is related to the presence of an inverse bifurcation, which is typical of Marangoni instability [20].

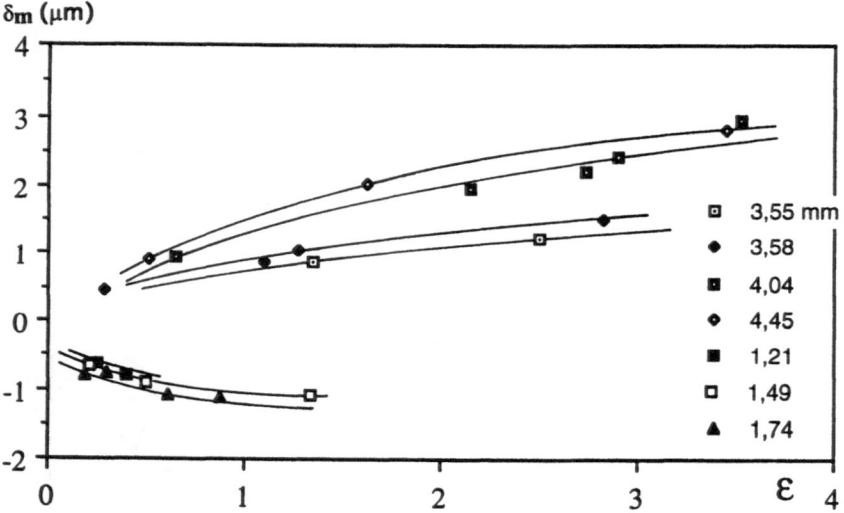

**Fig. 9** - Variation of the amplitude of deflection $\delta_m$ versus the distance to the threhold e for various values of the depth layer d.

## 3.3 Relief profile

The relief profile, as a function of the dimensionless horizontal coordinate x/L, is shown in Fig.10. The Fourier analysis provides a sinus shape for the concave deformation. For the convex deformation, there are many odd harmonics are found in the profile equation.

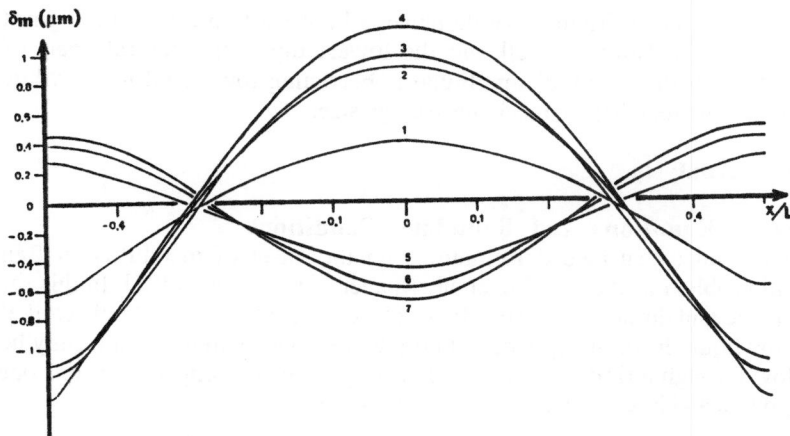

**Fig.10** - Profile of the surface relief for various values of the distance to the threshold. L is the width of the hexagon. Curves (1) to (7) correspond to the following values of ε : 0.3, 1.8, 2.4, 3.26, 0.05, 0.18 and 1.0. The depth layer is 4.04 mm for (1) to (4) and 1.75 mm for (5) to (7)

### 3.4 Surface deformation and structural defects
The most current defects observed in hexagonal patterns are the pentagon(P5)-heptagon(P7) structures. Isolated P5 "flowers" made up of four P5 or two P5-P6* pairs have also been observed; notation P6* means a regular hexagonal planform whereas P6 means an irregular one [21].

The height $\delta_m$ of these defects has been measured for various ε and small d-values in the case of convex deformations [22]. The main results can be summarized as follows :

- Whatever ε or d, $\delta_m$ of a P5 or a P7 cell is smaller than that of any P6* pattern.
- Deformation of a P5-structure is always less important than that of a P6 cell, even if they belong to the same pair configuration.
- $\delta_m$ is generally larger for a P6 cell than for a P6*.
- Whatever the cell (P5,P6 or P7), the larger the cell, the larger the deformation.
- Whatever the nature of the cell (P5 or P7), the more regular the shape, the larger the deflection.
- For any cell, the "thermal relief" strongly parallels the surface relief, which means that the isotherms can be idenfified with the contour lines.

### 3.5 Surface deformation during the birth of a cell
The polygonal pattern is generally not frozen : cells and defects move and transform. Some cells disappear while an equivalent number of cells are born to keep the mean wavelength constant. The creation of a convective cell is very similar to the birth of a living cell. Afterwards, the cell deforms and elongates to take generally the form of an octogon. This is the "mother" cell. During this period, the relief also elongates, the

dome becomes a crest. Then a saddle appears between two tops. The highest top corresponds to the "mother" cell and the lowest top is the central part of the "daughter." Then the "mother" amplitude is becoming lower and lower while the "daughter" size increases up to reach the average size.

## 4 Linear Analysis

### 4.1 Balance Equations and Boundary Conditions
The role played by surface deformations on the onset of instability in Bénard-Marangoni problem is examined in another paper of this booklet [23]. In this section we summarize and discuss some specific results obtained by Pérez-Garcia *et al.* [24].

A shallow liquid layer of depth d, and infinite horizontal extent, is uniformly heated from below through a rigid good heat conducting plate. The upper surface is open to the atmosphere and is deformed as schematized on Fig 11.

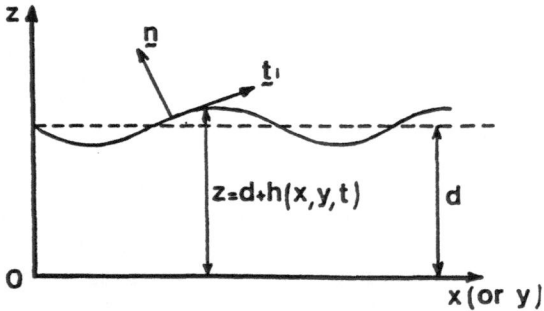

**Fig. 11** - Deflection of the free upper surface

The Boussinesq approximation and the following notation are used : $\rho$ is the density, $v$ (u,v,w), $\theta$ and p are the velocity, temperature and hydrostatic pressure perturbations respectively, g is the acceleration of gravity, $\alpha$ and $\kappa$ are the coefficients of thermal expansion and thermal diffusivity, $V$ the kinematic viscosity and h the coefficient of heat transfer across the upper surface. The variables are made dimensionless by using respectively as units of length, time, velocity, temperature and pressure the values $d/\pi$, $d^2/\kappa\pi^2$, $\pi\kappa/d$, $\Delta T/\pi$, $\kappa V\pi^2/d^2$. The $\pi$ factor is introduced in accordance with the method proposed by Nield [9] to solve a similar set of equations.

This procedure leads to the adimensional numbers:

$$C1=Cr\pi, B1=B_0/\pi^2, L1=L/\pi, M1=Ma/\pi^2, R1=Ra/\pi^4$$

with

$$Cr=\mu\kappa/\xi d, B_0=\rho gd^2/\xi^2, L=h\kappa/d, Ma=\gamma\Delta Td/\rho V\kappa, Ra=\alpha gd^3\Delta T/V\kappa$$

where Cr, Bo, L, Ma and Ra are the crispation, Bond, Biot, Marangoni and Rayleigh numbers respectively..

The balance equations as well as the boundary conditions at the bottom are the same as in the no-deflection case [25]. Since the deflection is small within normal conditions, it can itself be considered as a perturbation, so that the boundary conditions on the free deformable surface can be written at the mean height z=d [8].

We now search for steady solutions in terms of normal modes at marginal stability; they are conveniently cast into the form

$$\{w,\theta,h\} = \{W(z), \Theta(z), E\} \exp\left[-i(k_x x + k_y y)\right] \tag{3}$$

wherein $\mathbf{k}(k_x, k_y)$ denotes the horizontal wavenumber. The classical linearized equations for the amplitudes $W(z)$ and $\Theta(z)$ are

$$\left(D^2 - k^2\right)^2 W - k^2 R_1 \Theta = 0 \tag{4}$$

$$\left(D^2 - k^2\right)Q + W = 0 \tag{5}$$

wherein D stands for $D= \partial/\partial z$ in eq. (3) . The quantity E stands for the amplitude of the deflection.

The corresponding boundary conditions are

$$W(0) = DW(0) = \Theta(0) = 0 \tag{6}$$

$$D\,\Theta(\pi) = -\,L_1\,\Theta(\pi) \tag{7}$$

$$E = \left(k^2 M_1\right)^{-1} D^2(\pi) + \Theta(\pi) \tag{8}$$

$$C_1\left(D^2 - 3k^2\right)DW(\pi) = \left(k^4 + B_1 k^2\right)E \tag{9}$$

The system of balance equations and boundary conditions (4-9) is solved by Nield's method based on a Fourier expansion of solutions [9] .

## 4.2  Results

The main result (Fig.12) is a change in the sign of the deflection at $Ra^c/R_0=0.174$ as recalled in sect.3, this observation is in good agreement with experimental results.

Pérez-Garcia et al. [24] studied also the influence of the various parameters Cr, L and $B_0$. For instance the influence of the crispation number on the relief is reported in Fig. 12. Whatever Cr, the deformationE is zero for $Ra^c/R_0=0.174$. The deformation is increasing with Cr.

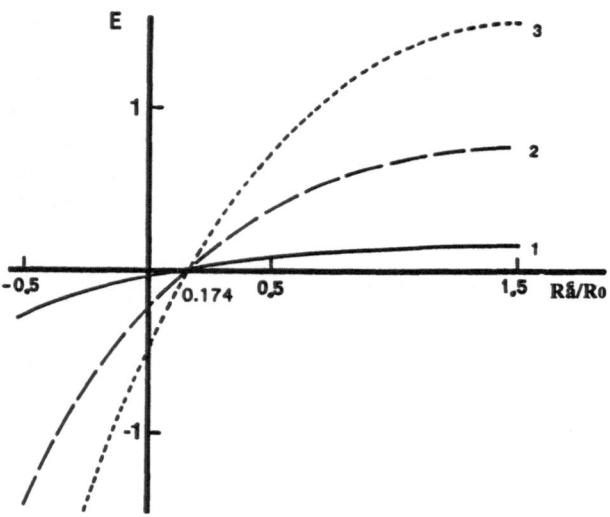

**Fig. 12** - Amplitude of deflection (in arbitrary units)
as a function of $Ra^c /R_0$ for three crispation numbers.
$1 : Cr=1. 10^{-5}$ ; $2 : Cr = 5. 10^{-5}$ ; $3 : Cr = 1. 10^{-4}$

### 4.3 Improvements

The work of Pérez-Garcia et al. [24] has been criticized by Garcia-Sanz and Jimenez-Fernandez [27] because the first authors assumed that the velocity is equal to zero..

Garcia and Jimenez have applied a method due to Joseph [28] which allows for obtaining analytical expressions for both the velocity and the surface deflection amplitude in terms of the fluid parameters and the linear critical wave number. An expression providing the sign of the surface displacement was derived. Instead of calculating the amplitudes from a nonlinear analysis, the latter are determined, from the boundary condition at the upper surface developed at second order in term of a parameter c, called the control parameter and defined by

$$c^2 = N_u - 1 \tag{10}$$

where Nu is the Nusselt number, defined as

$$Nu = \frac{dT(0)}{dz} \frac{d}{\Delta T} \tag{11}$$

dT(0)/dz is the derivative at z=0.

The behaviour of the surface deflection as a function of the layer depth remains in agreement with the predictions of Pérez-Garcia et al. [14] and the experimental results of Cerisier et al. [13]. In particular, the exact thickness at which the transition concave-convex occurs, has been found in good agreement with experimental results.

# 5 Surface Relief with an Imposed Horizontal Periodic Spatial Temperature

Loulergue [29] has studied the surface profile of a liquid layer open to the atmosphere, heated by a spatially periodic imposed temperature gradient, with the prospect to determine the sign and the amplitude of the deflection.

## 5.1 Experimental study

The free surface of a liquid (L) is locally heated by irradiation with parallel and equidistant fringes (Fig.13). This is produced by a one dimensional grid (G) of variable step q, lighted by a tungsten lamp giving a spectrum in the range of about 1-2 μm. An optical device focuses uniformly the image on the free surface of the layer Absorption of the infrared ligth at the upper part of the liquid produces a one dimensional temperature distribution. The surface profile is analysed by shadowgraph technique (ST).

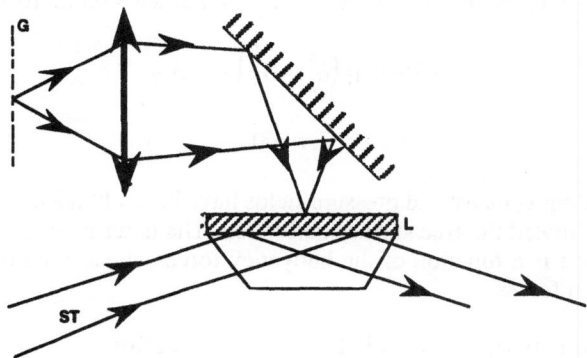

**Fig. 13** - Experimental set-up for forced
Bénard-Marangoni convection
G: grid - ST: ligth beam for shadowgraph technique - L: liquid layer

## 5.2 Theoretical modelling

A two-dimensional model is proposed with u and w, the horizontal and vertical velocity components respectively. The imposed temperature perturbation is supposed to be a periodic function of x, independent of z :

$$T(x) = T_0 + dT \sin kx \tag{12}$$

$T_0$ is the initial temperature of the fluid and k the wave number of the heating. This periodic temperature distribution produces similar variations of density and surface tension. The boundary conditions at z=0 and z=d are given by

$$z=0 : \qquad\qquad u = w = 0 \qquad\qquad\qquad (13)$$

$$z=d : \qquad\qquad w = 0, \quad Dw = 0 \qquad\qquad (14)$$

The continuity of the normal stresses at the upper boundary can be written as :

$$z=d : \qquad\qquad p_a = [p - 2\eta Dw] + \xi \left[ \frac{\partial^2 \delta}{\partial x^2} \right] \qquad\qquad (15)$$

where $p_a$ and $p(x,z)$ are respectively the pressure in the atmosphere and in the fluid, $\eta$ is the dynamic viscosity, $\xi$ the surface deformation and $\delta$ the surface deflection. The continuity of the tangentiel stresses is expressed by

$$z=d : \qquad\qquad \eta\, Du = \partial\xi/\partial x \qquad\qquad (16)$$

In the stationary state, the linearized 2-D Navier-Stokes and continuity equations are :

$$-\nabla p + \eta \left( \partial^2_{xx} + \partial^2_{zz} \right) v = \rho\, g \qquad\qquad (17)$$

$$\nabla.v = 0 \qquad\qquad (18)$$

The corresponding velocity and pressure fields have been obtained by Loulerge [29] who has also calculated the free surface deflection. The latter is the difference of two terms, the first one is a function of the buoyancy forces, the second one depends on the surface tension forces.

## 5.3 Comparison with classical Bénard convection

Some results for silicone oil are reported in Fig. 14 which gives the shape (concave or convex) of the deflection in terms of the layer depth d and the imposed wave-length q of the heating.

Curve OC divides the plane d-q into two regions. The first one, (B), is characterized by a convex deflection. It corresponds to convection mainly driven by buoyancy forces ; it is the part of the plane corresponding to large d and small q-values. The second region (S) represents the domain where surface tension forces are dominant and where the deformation is concave. It is essentially the region wherein q is large and d small. The line OC represents the balance between both driving forces and surface tension force : in this case the upper surface of the layer is flat.

**d (mm)**

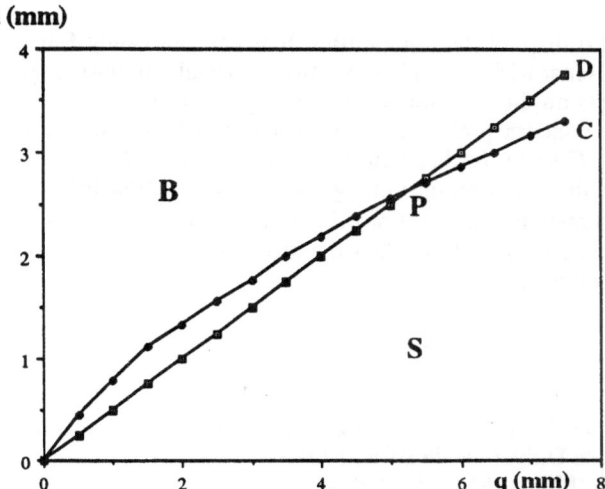

**Fig. 14** - Deflection shape in the plane imposed wave length of the heating (q) and depth layer (d).
B : zone of predominant buoyancy effect (convex relief) ; S : zone of predominant surface tension effect (concave relief) ; OC : balance between both effects (flat surface)

We have also drawn the straight line OD whose slope is 0.5, corresponding to q = 2d. The point P resulting from the intersection of OC and OD gives, according to Loulergue [29], the depth of a fluid layer wherein only the classical Bénard convection should be present. This result is in qualitative agreement with Cerisier *et al.* [13,14] experiments but it could be improved by considering another relation between q and d. Indeed q = 2d describes a roll pattern, close to the threshold, and is not appropriate for a hexagonal cell. It should be interesting to extend Loulergue's analysis in the case of hexagonal patterns, far from the treshold.

Loulergue [29] introduced also a new characteristic length, the so-called "thermal capillary length" $L_T$, to determine the sign of the deformation; the quantity $L_T$ is defined by

$$L_T = \left( \frac{\partial \xi / \partial T}{g \partial \rho / \partial T} \right)^{1/2} \tag{19}$$

For $d < L_T$, the surface is concave and for $d > L_T$, it is convex. For $d = L_T$ there is no deformation.

Finally Loulergue [28] examined also the influence of the thermal properties of the fluid on the deformation of the surface. It is shown that $\xi$, $\rho$ and $\eta$ do not influence the sign of the concavity, but they are important for determining its amplitude. It is obvious that the larger is $\xi$, the larger is the deflection.

# 6 Conclusion

Our objective is to display the competition between buoyancy forces and surface tension forces in Bénard-Marangoni convection ; special emphasis is put on the role of these two effects on the deformation of the upper surface. The results reported in the present work concern preliminary experimental observations and (simplified) linear modellings. Clearly, further non-linear analyses are necessary to improve the quality of the results concerning, among others, the relation between $\delta_m$ and the distance $\varepsilon$ to the threshold, the exact profile of the relief, the deflection shape over an irregular cell, the transient relief when the structure is in a transitory regime, for example when a cell is born or dies.

# References

[1] H. Bénard, Rev. Gen. Sci. Appl. **11**,1261, (1900); Ann. Chim. Phys. **23**, 62 (1901); Thèse Université de Paris, 1901.

[2] Lord Rayleigh, Philos. Mag. **32**, 529 (1916).

[3] A.R. Low and D. Brunt, Nature, **115**, 299 (1925).

[4] P. Vernotte, C. R. Acad. Sci., Paris, **202**,119,(1936); **202**, 733 (1936).

[5] H. Jeffreys, Quart. J. Mech. **4**, 283 (1951).

[6] M. J. Block, Nature **178**, 650 (1956).

[7] J. R. A. Pearson, J. Fluid Mech. **4**, 489 (1958).

[8] E. Scriven and C.V. Sternling, J. Fluid Mech.**19**, 321 (1964).

[9] D. A. Nield, J. Fluid Mech. **19**, 1941 (1964).

[10] A.V. Hershey, Phys. Rev. **56**, 204 (1939).

[11] J.N. Anand and H.J. Balwinski, J. Colloid Interface Sci. **31**,196 (1969).

[12] W.V. Kayser and J.C. Berg, J. Fluid Mech. **57**, 739 (1973).

[13] P. Cerisier, C. Jamond, J. Pantaloni and J.C. Charmet, J. Physique, **45**, 405 (1984).

[14 ] C. Pérez-Garcia, J. Pantaloni, R. Occelli and P. Cerisier, J. Physique, **46**, 2047 (1985).

[15] J. Jiménez-Fernàndez and J. Garcia-Sanz, J. Physique,50, 521 (1991); Phys. Let. A, 141, 161 (1989)

[16] P. Cerisier, J. Pantaloni and J.C. Charmet, Congrès Société Française de Physique, Clermont-Ferrand (1981)

[17] H. Owder Ozbelge, E.N. Ligthfoot and E.E. Miller, J. Phys.E, **14**, 1381 (1981).

[18] R.H. Sellin, J. Sci. Instrum., **40**, 355 (1963); Nature, **217**, 536 (1981).

[19] W.B. Spangenberg and W. R. Rowland, Phys. Fluids, **4**, 743 (1961).

[20] F.H. Busse, Rep. Prog. Phys. **41**, 1929 (1978).

[21] J. Pantaloni and P. Cerisier, Structure defects, in *Cellular Structures in Instabilities*, Proceedings Gif sur Yvette, France, *Lectures Notes in Physics*, Edited by E. Weisfreid and S. Zaleski, vol. 210,197, Springer, Berlin (1983)

[22] P. Cerisier and J. Pantaloni, Physico. Chem. Hydr. **10**, 249 (1988).

[23] G. Lebon , P. Cerisier and O.Dupont, this volume (1995)

[24] C. Pérez-Garcia, J. Pantaloni, R. Occelli and P. Cerisier, J. Physique, **46**, 2047 (1985).

[25] C. Pérez-Garcia in *Stability of Thermodynamics Systems* Proceedings, Barcelona,Spain, *Lectures Notes in Physics*, edited by J. Casas-Vazquez and G. Lebon, vol. 164, 94, Springer, Berlin, 1982.

[26] D. A. Nield, J. Fluid Mech. **81**, 513 (1977).

[27] C. Garcia-Sanz and J. Jimenez-Fernadez, Proceedings of the VII European Symposium on Materials and Fluid Sciences in Microgravity, ESA SP-295 (1990)

[28] D.D. Joseph, *Stability of fluid motions*, II,Springer Tracts in Natural Philosophy, 28, XIII, Springer Verlag, 1976.

[29] J.C. Loulergue, *Competition between Marangoni and Archimedian forces to determine the surface profile of a liquid heated open to air*, Lectures Notes in Physics, 210, 358, 1984 ; ibid, private communication 1988.

**Acknowledgements**

The authors wish to thank the EC program "Human Capital and Mobility under contract ERB CHRCX-CT-94-0481 for financial support. Partial financial support from SSTC, Belgian Science Policy Programming, under contract PAI 29 is also acknowledged.

# Cooling of Small Electronic Devices by Boiling Under Microgravity

J. Straub[1], J. Winter[1], G. Picker[1] and M. Zell[2]

[1]Institute A for Thermodynamics, TU Munich, Arcisstr. 21, 80290 München, Germany

[2]DASA GmbH, Dornier, 88039 Friedrichshafen, Germany

**Abstract.** The boiling heat transfer on a miniature heater has been measured under microgravity conditions during the IML 2 mission and under earth gravity after the mission in 1994. These experiments are simulations for the direct cooling of small electronic devices by boiling heat transfer, which becomes very important due to high thermal loads of modern electronic components.
The boiling process was studied in the liquid FREON 11, by simultaneously measuring the heat-flux and the temperature of the heater. The boiling process was observed in two perpendicular directions.
Several boiling modes, like cavitation and sparkling boiling, as well as the cavitational thermocapillary supported flow-mode, thermocapillary jet boiling and saturated boiling have been observed during the experiments, depending on the subcooled liquid state and the overall heat flux. Under microgravity surface tension driven convection plays an important role in the boiling heat transfer on a miniature heater as used in our experiments. The observations under earth gravity show significant differences especially in terms of bubble size, departure diameter and induced flow-pattern. Nevertheless a remarkable influence of the gravity on the heat transfer could not be measured. That confirms our earlier observation that the evaporation in the liquid wedge between the solid heater and the bubble interface, the capillary force, the surface tension and the wetting condition play the most important role in the boiling process, as recently proposed in the micro wedge model.

**Keywords.** Boiling heat transfer, saturated boiling, subcooled boiling, evaporation, condensation, interface phenomena, thermocapillary flow, Marangoni convection, microgravity.

# 1 Introduction

During the IML 2 Spacelab mission of NASA in July 1994 we performed a pool boiling experiment in microgravity using a small thermistor with a bead of 0.26 mm in diameter simultaneously as a heating element and as a resistance thermometer. The experiment was accomplished in the BDPU (Bubble, Drop and Particle Unit), which is a Spacelab multiuser facility for fluid physics experiments in space, operated by the European Space Agency, ESA.

According to the original planning of the mission this boiling experiment was not foreseen. However, our experiment „Study of vapor bubble growth in a supersaturated liquid" [1] scheduled for this mission worked perfectly, so during the course of the mission we proposed a new experiment using the same experimental equipment. Thanks to the replanning efforts of the NASA team from the MSFC Operation Support Center in Huntsville, AL, we received additional time for a boiling experiment. We could modify the experiment sequence in flight by changing software parameters from ground and control the experiment by telecommanding. After disintegration from the Spacelab we repeated the experiment on ground with the flight hardware, the identical fluid, and the same experimental sequence, in order to have comparable boiling data both under microgravity as well as under earth gravity.

The scientific objectives of these boiling experiments are listed as following:

a) To study the mechanisms of boiling on a small spot heater of nearly spherical geometry at microgravity and to compare the results with 1 g reference data.

b) To observe the bubble dynamics as growth, departure and bubble size, in order to determine the influence of gravity on the dynamics as on the boiling process and heat transfer.

c) To investigate the heat transfer coefficient for a very small hemispherical micro heater at 1 g and under microgravity, to find out what are the most important driving forces for the heat transfer, if the buoyancy force is reduced to a neglectible value.

d) To study the feasibility of boiling heat transfer, using this technique for the cooling of high powered microelectronic components directly immersed into a liquid and its application in space and on earth.

The performance and the speed of computers can be considerable enhanced with the miniaturization of the electronic devices. However, this causes serious problems, because most of the electric power disperses into heat. To protect the electronics from any critical overheating, an efficient method of cooling must be applied. One of the most promising methods is boiling heat transfer. Several papers [2-6] report on integrated circuit chips or simulated micro heaters that are directly immersed into a liquid. These experiments are performed under earth conditions, where the buoyancy supports the vapor transport away from the heater.

In fact, we know from former boiling experiments with various geometries under microgravity [7] that the influence of gravity is much less as predicted by most of the applied correlations for boiling heat transfer if they are extrapolated to lower gravity levels. However, no investigations are known, how the boiling process behaves on such small heaters as electronic chips in the microgravity environment. The question arises, if the heating element is immediately covered with vapor, leading to film boiling and corresponding with it to high temperatures ? Therefore it is justified to ask the question, if the boiling process is still a suitable mechanism for an efficient cooling of small electronic elements under microgravity ? The solution of this question is important for the use of modern electronic equipment in space applications.

But, of course, as mentioned before our main interest is to study the fundamentals of the boiling mechanisms. Because of our former finding from earlier boiling experiments under microgravity conditions was *[7, 10]*, that most of the existing equations for boiling heat transfer, valid for earth conditions, cannot be applied if they are extrapolated to lower or even higher gravity values. According to this the suspicion rises that the basic physical principles of boiling are not well understood and represented within most of the present boiling models, where the buoyancy plays the dominant role.

# 2 Experimental Equipment

## 2.1 BDPU[1] Facility

During the IML 2 Spacelab mission on STS 65 our experiment was conducted in the ESA multi-user facility BDPU, designed by Alenia, Italy. The flexibility of this facility is ensured in that way, that for various experiments specific test containers are built, which meet the special scientific requirements. The BDPU provides the power supply and the possibility of optical observation with background illumination and a point diffraction interferometer. Each observation path is recorded by a video camera or an extra 16 mm film camera. Furthermore the BDPU provides the interface to the Spacelab and the Shuttle system for the video and data link direct to the Operation Support Center at the NASA-MSFC in Huntsville. There, the principal investigators and their teams could observe the experimental runs during the mission in real time video and all important experimental data were directly available on computer screens. The experiments in the BDPU were controlled by telecommanding from ground and the essential experimental parameters were adjusted depending on the observed data. This system ensures that a maximum of scientific results was obtained.

## 2.2 Test Container

The test container was designed and built for the requirements of our experiment *[1]* by the companies Dornier, Germany, in cooperation with Ferrari Engineering, Italy. The container was set in the facility and electrically connected by one of the astronauts. It has the size of $45x15x30$ cm$^3$. Because of safety reasons our container had a triple containment. It had to be ensured that in case of a leakage of the liquid cell no gas of the evaporating liquid was released into the Spacelab environment. This test container was a precision engineered piece of equipment, because all parts necessary for the experiment sequences had to be arranged in the narrow space of the container. The liquid cell itself had an internal dimension of about $50 \times 50 \times 50$ mm$^3$ and was spherical inside. It had four windows of sapphire with 40 mm optical diameter for two perpendicular observation directions. To change the pressure of the liquid in the cell and to compensate the vapor volume in order to keep the pressure constant during boiling a bellows was used with a counter pressure imposed by nitrogen. A small gas compressor supplied a small pressure vessel with the necessary compressed nitrogen, in such a way, that an experimental run could always start at high pressure of about 6.5 bars , and by release of nitrogen, the pressure was reduced

---

[1] Bubble, Drop and Particle Unit

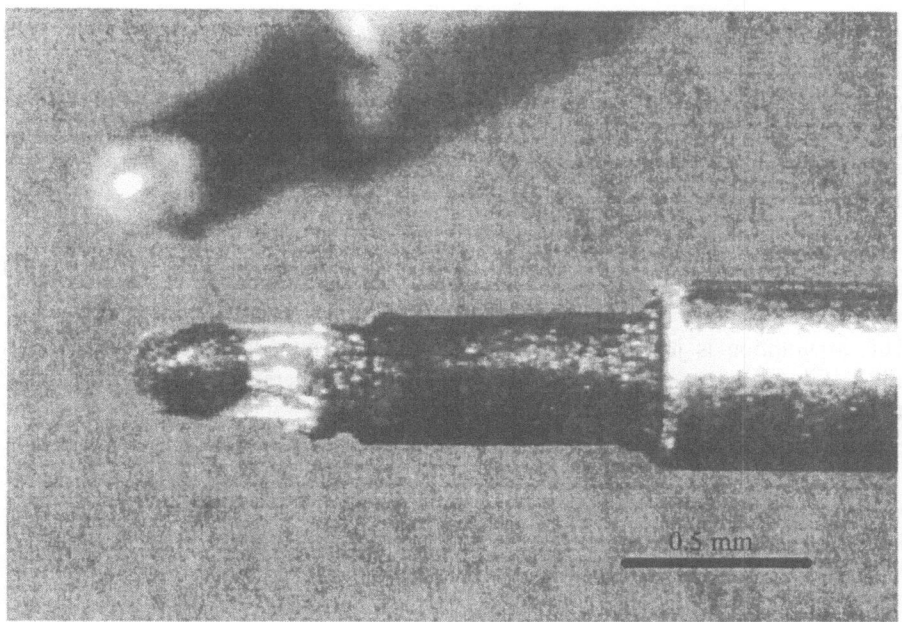

Fig. 1:   Photo of the thermistor used as heater.

to the saturation pressure of the experiment-fluid. Due to safety regulations the upper limit of the pressure was 100 psi ($\approx$ 6.9 bars).

Two Peltier elements on the lateral walls of the fluid cell were used both for heating or cooling. The temperature could be controlled within $\pm$ 0.05 K. The Peltier elements dissipated the heat to the avionics air duct through two heat pipes. The test liquid in the fluid cell was FREON R11. Eight identical thermistors were directly placed in the liquid to measure the liquid temperature close to the one thermistor which was used as heater. The thermistor beads had a total size of 0.26 mm in diameter (fig. 1) and was especially selected with respect to their resistance characteristics. The thermistor top was formed as a hemisphere on the end of a cylinder of the same diameter. They were coated by a lead glass layer of about 0.05 mm thickness. The thermistor, which was used as a heater, was controlled at constant temperature independent from the heating power, which depended on the heat flux through the glass coating and the heat transfer coefficient to the liquid. That means, that the temperature on the outer surface of the thermistor $T_w$ was not identical to the controlled inner temperature $T_i$ and therefore had to be calculated in each case separately. The steady state condition was immediately reached after the heat flux or after a pressure step was changed, because the capacity of the heater was very small. The heat flux $\dot{Q}$, determined by the measured voltage over and the current through the thermistor, is given by:

$$\dot{Q} = \frac{T_i - T_l}{R_\lambda + R_\alpha} \tag{1}$$

where $T_i$ is the controlled inner temperature of the thermistor, $T_l$ is the bulk liquid temperature recorded in the cell, $R_\lambda$ is the thermal conductivity resistance of the thermistor coated with glass and can be determined, if the geometry is assumed spherical:

$$R_\lambda = \frac{(r_2 - r_1)}{\lambda_g 4\pi\, r_1 r_2} \cdot \tag{2}$$

This assumption is justified because in the case of pure heat conduction to the liquid at lower heater temperature a nearly spherical temperature field was observed in the interferometer.

From a 50 times magnification of the thermistor we determined the outer diameter $d_2$ = 0.26 mm and the inner diameter $d_1$ = 0.15 mm. The thermal conductivity of the glass coating $\lambda_g$ is still kept as a trade secret by the supplier (Thermometrics). We assumed a value of $\lambda_g$ = 1.05 W/mK for lead glass. In a later stage, we will conduct a material probe to achieve a accurate value for $\lambda_g$. The heat transfer resistance $R_\alpha$ to the liquid is given by

$$R_\alpha = \frac{1}{\alpha\, 4\pi\, r_2^{\,2}} \cdot \tag{3}$$

With eq. (1) the total thermal resistance $R_\lambda + R_\alpha$ of the thermistor is determined, and respectively with eq. (3) the heat transfer coefficient $\alpha$. The wall temperature on the outside of the thermistor $T_w$ can be determined as:

$$T_w = T_i - \dot{Q} R_\lambda \cdot \tag{4}$$

The heat loss to the cylindrical not heated part of the thermistor is calculated as a fin and is proportional to $T_i$ - $T_l$. The heat losses are considered in the calculation of the heat flux.

## 3 Experimental Procedure

In the introduction it was mentioned that our test container was not specially designed to conduct boiling experiments, however, the software was flexible enough that one thermistor could continuously be heated and controlled at a constant temperature level. To ensure the best use of the valuable microgravity time, the following procedure was executed:

The temperature of the fluid cell was set to the desired value, and while stirring the liquid in the test cell was heated up to the set temperature. After the stirrer was switched off, the movement of the liquid was calmed down within 2 minutes. The thermistor was set to a certain temperature level between 200 and 140 °C in steps of 20 K. The pressure was set to the highest value of about 6.5 bar and, during the run of the boiling process, reduced in about 12 pressure steps down to the saturated pressure corresponding to the liquid temperature. According to the heat transfer to the liquid the power adjusted itself in a way that the inner thermistor temperature Ti was kept constant. The current and voltage was measured to determine the electrical power to the heater.

At each pressure step the boiling process was maintained about 30 sec and more, this time was sufficient for steady state conditions, which are immediately reached after the pressure was changed. All measurement data was recorded with a frequency of 1 Hz and linked down to the control center of MSFC in Huntsville. After the saturation pressure was reached and the saturation measurement was finished, the pressure was increased again to the maximum value, the liquid stirred and the experimental run was repeated with another thermistor temperature. Sometimes at low thermistor temperatures $T_i$ of about 150 or 140 °C and dependent on the liquid temperature no boiling was observed, the heat to the liquid was transported by pure conduction. However, at a certain pressure reduction or increase of the thermistor temperature boiling starts. From that observation we study the incipience of boiling. After about 5 runs at one liquid temperature the liquid was heated up to the next level. The liquid temperatures investigated were 30, 40, 50 and 70 °C and some points at 60°C. By this way we obtained values for a wide range of subcooled and saturated liquid conditions. We repeated the same sequence of parameters for the post flight reference experiments on ground after the mission.

# 4 Results

## 4.1 Nucleation and Incipience of Boiling

At lower heater temperatures $T_w$ < 140 °C incipience of boiling did not occur in space. The heat transfer is managed by pure conduction, and at 1g, if $T_w$ < 135 °C by natural convection. This relative high isobaric superheat up to $T_w$-$T_{sat}$ = 95 K and the reproducibility of the data opens the question of the nucleation process. In fig. 2 we see in the vapor pressure diagram, that the measured superheat data fit to one line for µg and parallel with about 8 K less for 1g. As mentioned, the thermistor was set to a certain temperature of 150 or 140 °C and the pressure was reduced. The thermistor temperature $T_w$ was calculated from the heat flux. As long as boiling did not occur the heat flux was nearly constant during the pressure reduction. When the incipience of boiling occurs because of a further pressure reduction step, the heater temperature immediately decreased by 50 to 60 K and the heat flux increased by a factor of about 4.

We assume that the nucleation occurs as heterogeneous nucleation. In first experimental test the superheat was still higher, therefore the glass of the thermistors was a little corroded at the outer side. To describe the onset of boiling we use the well known theory of nucleation [11, 12] and define the reduction of the activation energy by the ratio of the heterogeneous to the homogeneous activation energy:

$$\Phi = \frac{\Delta G_{het}}{\Delta G_{hom}} \quad \text{with} \quad \Delta G_{hom} = \frac{16\pi \sigma^3}{3\Delta p^2} \ . \tag{5}$$

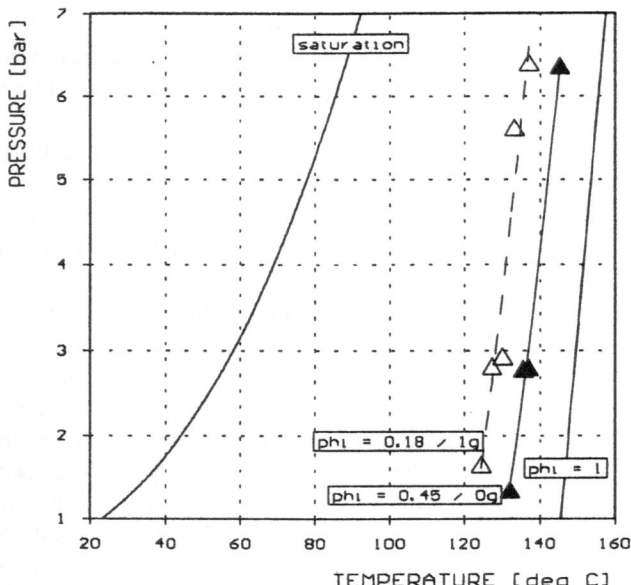

Fig. 2:   Heterogeneous nucleation in μg and at 1g.
$\phi$ = 1 homogeneous nucleation for $I = 1$ mm³sec⁻¹,
$\phi_{\mu g} = 0.48$, $\phi_{1g} = 0.2$ in the pressure, temperature diagram with saturation line

The nucleation probability rate is given by:

$$I = N \cdot B \cdot \exp\left(-\frac{16\pi\sigma^3}{3\Delta p^2 \, kT}\Phi\right)$$ (6)

Where $N$ is the number of molecules per volume, and $B$ the kinetic factor which was determined to $2 \cdot 10^{12}$ 1/sec. The activation energy $\Delta G$ follows from the unstable equilibrium condition for a bubble in supersaturated liquid, $\sigma$ is the surface tension at the superheated temperature $T$, and $\Delta p = p_{sat} - p_l$ is the pressure difference at that temperature, and determines the penetration into the metastable state. In our case $T = T_w$ and the pressure is isothermally reduced and $\Delta p$ increased step by step, the experiment run follows according to the definition of the theory of nucleation. From the observed superheat we calculated the heterogeneous nucleation ratio $\Phi_{\mu g} = 0.45$ for μg, and with that value the heterogeneous nucleation limit for our experimental condition is plotted in fig. 2, all measured supersaturation data fit this curve.

It is of interest to note that in case of 1g the experiments are conducted with the same heater and liquid, but the overheating temperatures are lower, however, at a higher heat flux due to convection. The heterogeneous nucleation ratio for this case is $\phi_{1g} = 0.2$. This lower value may be caused by the fact, that with convection the temperature at the upper stagnation point of the heater is higher than the average value which is registered, and that at this point the onset of boiling occurs. The curve

$\phi = 1$ determines the homogeneous nucleation limit for the assumption of a nucleation probability $I = 1 \text{ mm}^{-3}\text{sec}^{-1}$.

In this experiment the scatter of the points of incipience of boiling is relatively small compared to other experiments, and it can be described by the heterogeneous nucleation theory. The reason is, that the heated area is small and the nucleation sites are limited to a certain amount.

## 4.2 Boiling Sequence

The experiments are carried out at constant bulk liquid temperatures $T_l = 30, 40, 50,$ 70 °C and at various thermistor bead temperatures, controlled to a certain constant value $T_i$.

During one experimental sequence we reduced the pressure stepwise and kept it by means of the compensation bellows and counter pressure at a constant level for 30 sec. This time was sufficient to reach steady-state condition, because after 1 sec the heat flux and the temperature were at a constant level. By this procedure the subcooling changed stepwise from high values down to zero at saturation. In fig. 3a and b an example of such an experimental sequence is shown at a certain pressure versa temperature, respectively heat flux plot for the liquid temperature of 30 °C at various thermistor temperatures $T_i$. The resulting heater surface temperature $T_w$ and the heat flux density are monitored during the sequence and correlated to the liquid and heater state. The curve represents the saturation curve of R11, and the bars on the 30 °C line indicate the pressures, measurements were conducted at.

It is interesting to note that at high subcooling the heater surface superheat $\Delta T_{sat} = T_w - T_{sat}$ is relatively low but is considerably increasing with the approach to the

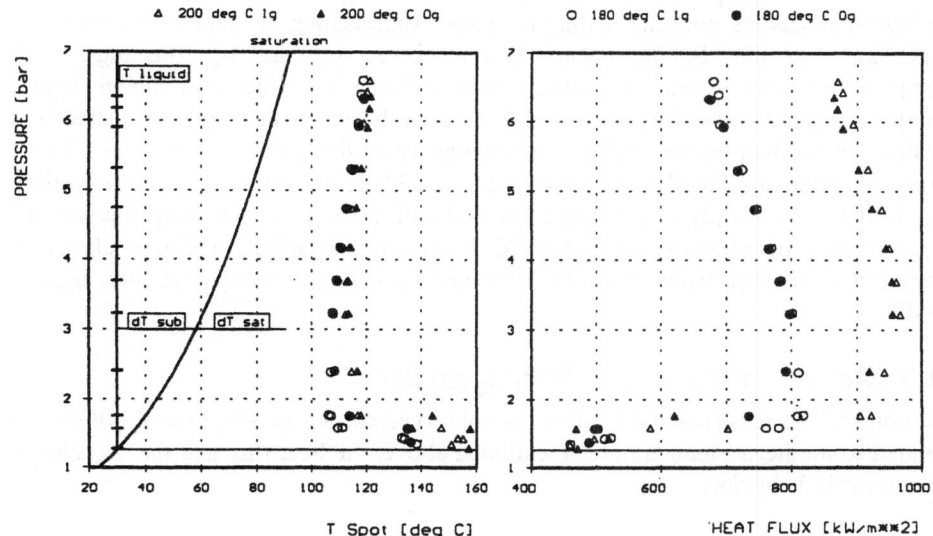

Fig. 3: Boiling sequence at liquid temperature of 30 °C:
a) the heater temperature $T_w$ as function of pressure    b) the heat flux density

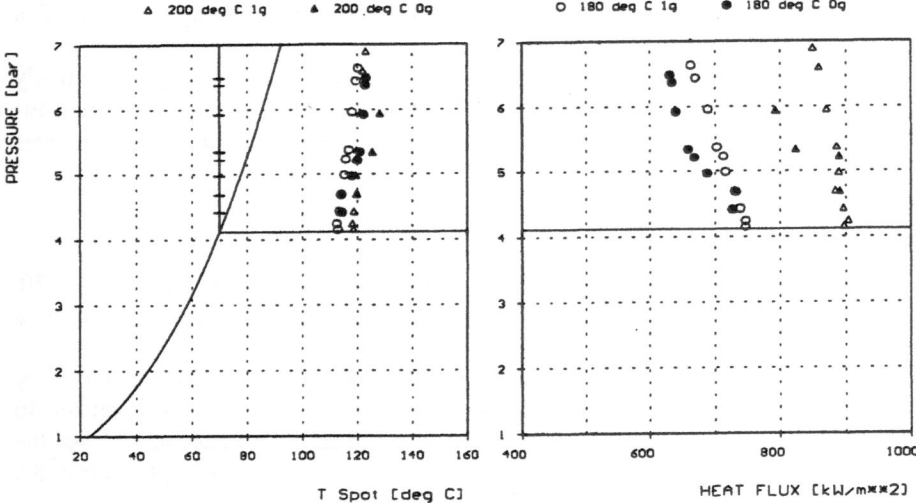

Fig. 4: Boiling sequence at 70 °C:
a) the heater temperature $T_w$ as function of pressure   b) the heat flux density

saturation state. In this figure the 1g reference data are plotted as open symbols whereas the full symbols are the µg data.. At higher subcooling there is nearly no difference between the µg and 1g data, approaching saturation the heater surface temperature at 1g is a little less than at µg. However, the temperature curves have the same tendency. This indicates that this behavior is principally not caused by the absence of buoyancy under microgravity. This is characteristic for the small size of the heater and the liquid conditions. For this case of high subcooling the heat transfer is like in subcooled nucleate boiling on earth. Approaching the saturated conditions transition and film boiling occurs in both cases, µg and 1g. The high wall temperatures near saturation decrease with reduced heat flux and higher liquid temperatures. The heat flux, transferred from the heater to the liquid, is increasing with reducing the pressure and has a maximum value for $T_l$ = 30 °C at about 3.3 bar corresponding to a liquid subcooling of $\Delta T_{sub}$ = 30 K. At further pressure reduction the heat flux is sharply decreasing combined with a simultaneous sharp increase of $T_w$. At a bulk liquid temperature of 70 °C nearly normal nucleate boiling is observed during the sequence from subcooled to saturation with increasing heat flux, fig. 4a and b.

## 4.3 Modes of Boiling in Microgravity

During such an experimental sequence we observed various boiling modes, which are related to the heater temperatures mentioned above, the heat flux and the associated heat transfer behavior.

### 4.3.1 Cavitation Boiling

At high liquid subcooling and at liquid temperatures of 30 °C, 40 °C and 50 °C, we assume the formation of a very thin, not visible micro layer of vapor at the heater surface which grows and collapse very fast and pump by that process the hot liquid into the bulk, acting like a membrane pump, known as microconvection (Forster and Zuber [8]). At the heater a thin liquid layer of the order of the critical nucleation radius is superheated, nucleation occurs and immediately the superheated layer evaporates, increases its volume and pushes the liquid away. The larger vapor surface gets in contact with the subcooled liquid and due to the great difference of $\Delta T_{sub}$ condensation occurs immediately like at a cavitation process. The inertia of the succeeding liquid increases locally the pressure and with it the local subcooling and accelerates the condensation. Our camera with only 30 frames per second was much too slow to follow this process visually. We observed only a hot liquid plume arising from the heater, first only on one side and at lower pressures on both sides of the heater.

### 4.3.2 Sparkling Boiling

At lower pressures and certain fluid states the cavitation boiling mode changed to the sparkling mode. Small, tiny bubbles are formed at the heater and they jump away in all directions from the heater surface. Their velocity decreases with increasing distance to the heater and they condense on their way through the subcooled liquid. They reach about a spherical space around the heater with a radius of about 10 mm to 20 mm. At this boiling mode high heat flux values are obtained.

### 4.3.3 Cavitation Mode Supported by Thermocapillary Flow

At lower liquid subcooling between 20 and 10 K transition boiling occurs at bulk liquid temperatures of 30 to 50 °C. Two to three bubbles of the size of the thermistor and larger occur, they grow and condense, however not completely, and produce a thermocapillary flow around them. Thereby the heat flux is little increasing at an unchanged heater temperature.

Depending on the liquid and thermistor temperature this mode happens during one to two pressure steps. At the next step - the subcooling is below 10 K - vapor bubbles around the heater form one single vapor bubble at the top of the thermistor. The bubble has first a size between 1 to 2 mm and sticks stationary at the top of the thermistor.

### 4.3.4 Thermocapillary Jet Mode

This single bubble produces a very strong thermocapillary flow forming a turbulent trailing jet stream behind the bubble, which looks like a turbulent Kármán Vortex Street, fig. 5 . This jet flow presses the bubble at the thermistor. By the formation of this single bubble the heat flux is spontaneously reduced by 20 to 30% and the heater surface temperature is spontaneously increased up by 30 K depending on the liquid and thermistor temperature. The size of this bubble keeps constant, it does not grow

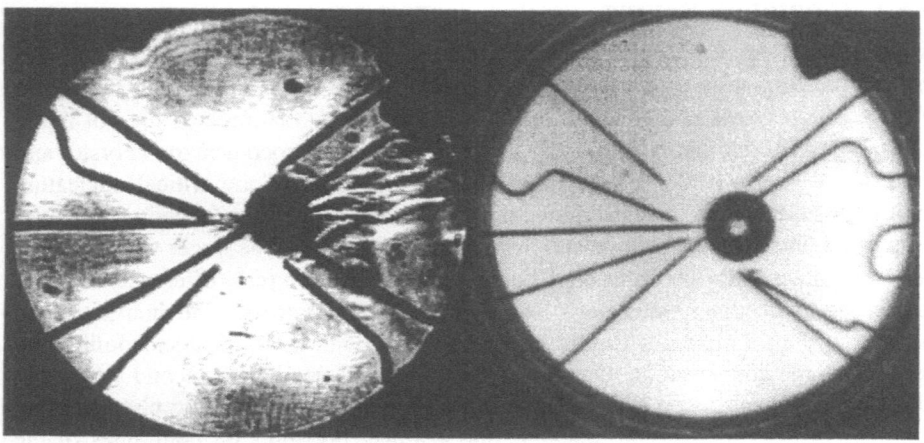

Fig. 5:    Photo of the thermocapillary jet        Fig. 6: Photo of a sticking bubble (9 mm dia.)

any more. All heat, produced at the thermistor, is transported by the thermocapillary flow. It is not quite sure, how the bubble sticks at the thermistor. It seems that the thermistor itself is full or partly covered by a very thin, not visible vapor film, which feeds the large bubble with energy and vapor.

At the following pressure step at subcooling of about 6 K the single bubble is growing up to about 9 mm diameter, but is still stationary sticking at the thermistor by the thermocapillary jet flow (fig. 6). Now the vapor layer around the thermistor becomes visible, with a flow around it, feeding the large bubble. The heat flux is again reduced by about 25 % and the temperature increased.

By the next pressure step - 2 to 3 K subcooled - the bubble is growing with increasing heat flux. Due to the larger surface more heat can be transported by thermocapillary convection. At a certain size the bubble departs from the thermistor and stands stationary in some distance from it, thereby the heat flux is reduced. A flow from the small bubble of the thermistor feeds the great bubble which is still producing a trailing jet stream. This situation, however, is unstable after a while. The bubble migrates away, comes back, stands still and departs finally. A new large bubble forms up at the thermistor and departs. The heat flux is oscillating with the same frequency of about 0.03 Hz. With that the heat flux oscillates with amplitudes of more than 100 kW/m². This behavior is especially observed at liquid temperatures of 40 and 50 °C.

## 4.3.5 Saturation Mode

When the saturation pressure is reached, the bubbles depart in any direction with a size of 1 to 2 mm and the heat flux oscillates with an amplitude of about 40 kW/m². This is typical transition boiling at saturation. The heater seems to be covered partly with a vapor film, which periodically is contracted by the surface tension to a single bubble, by this effect the bubbles are pushed away from the surface. At lower heat fluxes and at higher liquid temperature these effects are systematically reduced. At the liquid temperature of 70 °C normal nucleate boiling is observed.

From these observations a map of the boiling modes will be established, when the final evaluation of the data is completed.

## 4.4 Heat Transfer

### 4.4.1 Saturated Liquid

The heat transfer for saturated liquid conditions is shown in fig. 7 and fig. 8. As usual the heat flux is plotted versus the temperature difference $\Delta T_{sat}$. At the liquid temperature of 30 and 40 °C (fig. 7 )we observe transition boiling with increasing heat flux at decreasing $\Delta T_{sat}$, which is corresponding to the observed bubble dynamics as described before. In microgravity the heat flux is strong oscillating, indicated as bars, caused by the larger vapor volumes which form a bubble before it departs. At the liquid temperature of 50 °C transition boiling is observed with a maximum heat flux followed by nucleate boiling at lower $\Delta T_{sat}$. The heat flux is no more oscillating. The maximum heat flux in microgravity is 12 % lower as at 1g condition. According to the hydrodynamic theory of the peak heat flux this reduction should be more than 30 %, for small heater size [7]. At 60 °C and 70 °C, a normal nucleate boiling curve is achieved (fig. 8 ) with strong increasing heat flux with a small increase of $\Delta T_{sat}$, about 100 kW/m² / 1 K. The 1g data plotted as open symbols lay on the same curve.

It is surprising to note that there is only a small difference in the heat flux values between the µg and the 1g reference data. This observation confirms our former statement that the influence of the gravity on nucleate boiling heat transfer is small [7, 10].

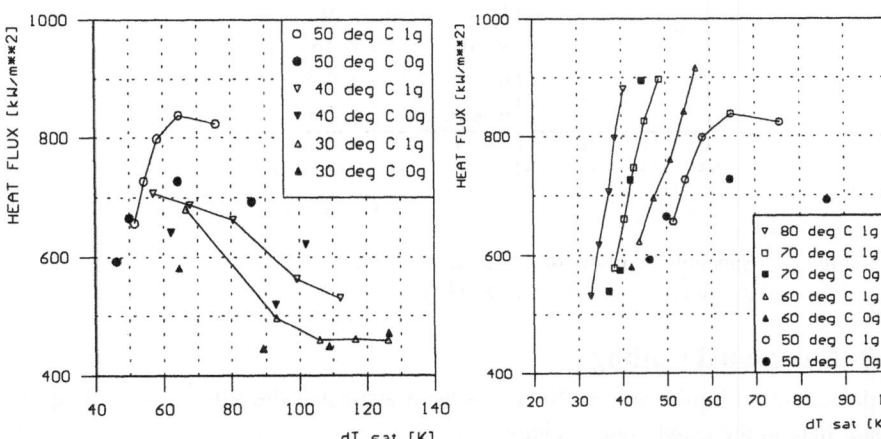

Fig. 7: Saturation boiling for various liquid temperatures ranging from 30-50°C µg: full symbols, 1g: open symbols

Fig. 8: Saturation boiling for various liquid temperatures ranging from 50-80°C µg: full symbols, 1g: open symbols

In any case the influence on the heat transfer is much less as equations developed for boiling predicts. For instance using the well known equation from Rohsenow [9] at the same fluid states and at same heater conditions the influence of gravity for the heat flux ratio at µg related to the value at 1g can be derived for nucleate boiling as:

$$\frac{\dot{q}_{\mu g}}{\dot{q}_{1g}} = \left(\frac{a}{g}\right)^{\frac{1}{2}} \tag{7}$$

If we assume that in the Spacelab the gravity ratio $a/g$ is about $10^{-4}$ and lower, the heat flux should be reduced to 1 % and less of the value observed at 1g conditions. As demonstrated here, this is not the case, even for the very small heater geometry. These results confirm our findings on various geometries before [7] and our statement, published recently, [10], that the surface tension and the wetting conditions are most important for the boiling process itself. The surface tension and the mechanisms caused by it can totally replace the role of the buoyancy and much more it is the most important force likewise at boiling on earth.

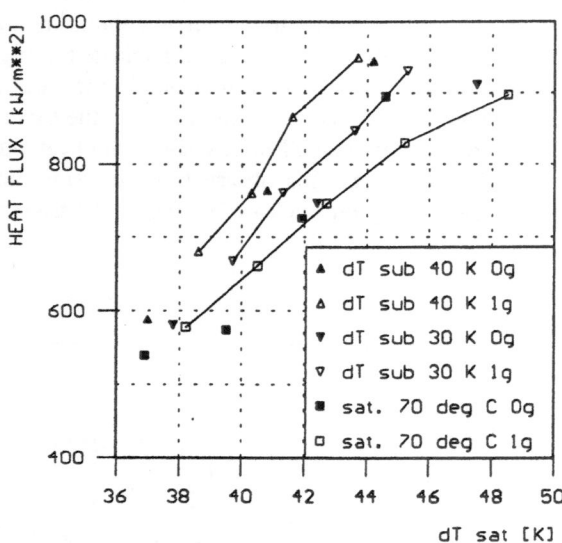

Fig. 9:  Subcooled boiling for the liquid temperature of 70 °C
µg: full symbols, 1g: open symbols

## 4.4.2 Subcooled Boiling

For the saturated liquid state at 70 °C we have evaluated the subcooled boiling heat transfer data in the usual manner along
the isobar line of p = 4.11 bar (corresponding to 70 °C saturation state), fig 9 .
With increasing liquid subcooling ΔTsat is reduced as usually observed, and as seen before, the 1g reference data did not show any systematic deviation.

The experimental work will be continued under earth gravity as well as under microgravity. Saturated boiling on a miniature heater will be studied using the Spanish Drop Tower (INTA). A wide range parametric study about boiling on wires is foreseen for a getaway special experiment (MAUS) scheduled for April 1996.

The configuration of the IML2 experiments is currently refurbished, improved and prepared for a reflight on the Spacelab LMS mission, scheduled for June 1996. In contrast to IML2 hardware the fluid cell will be equipped with several dedicated boiling heaters to study the influence of the geometry on the heat transfer. It is planned to measure the thermocapillary flow and the temperature field around the bubble in a quantitative manner by using Particle Image Velocimetry and Interferometry.

## Nomenclature

| | | | | |
|---|---|---|---|---|
| $a$ | acceleration | | $\Delta$ | difference |
| $B$ | kinetic factor | | $\phi$ | heterogeneous nucleation ratio |
| $g$ | earth gravity | | $\mu g$ | microgravity |
| $\Delta G$ | activation energy | | | |
| $I$ | nucleation probability | | **Subscripts** | |
| $k$ | Boltzmann constant | | | |
| $m$ | mass | | $g$ | glass |
| $N$ | number of molecules | | $het$ | heterogeneous nucleation |
| $p$ | pressure | | $hom$ | homogeneous nucleation |
| $\dot{Q}$ | heat flux | | $i$ | center |
| $r$ | radius | | $l$ | liquid |
| $R$ | thermal resistance | | $sat$ | saturation |
| $T$ | temperature | | $sub$ | subcooled |
| | | | $w$ | wall |
| **Greek Symbols** | | | $\alpha$ | heat transfer |
| | | | $\lambda$ | heat conduction |
| $\alpha$ | heat transfer coefficient | | $\mu g$ | microgravity |
| $\lambda$ | thermal conductivity | | | |
| $\sigma$ | surface tension | | | |

## Literature

[1] Straub, J., Winter, J., Picker, G, Zell, M.; *Study of Vapor Bubble Growth in a Supersaturated Liquid;*. Proc. of the National Heat Transfer Conference, Portland, Oregon, August 1995

[2] Kazuyoshi Fushinobu, Takao Nagasaki, Tadashi Saitoh, Atsushi Ui and Kunio Hijikata; *Boiling Heat Transfer Characteristics from very small heaters on a substrate;* Heat Transfer 1994 Proc of the 10th Int. Heat Conf. Brighton U.K., Vol. 5, pp. 51-56, Ed. by J. Hewitt

[3] Baker, E.; *Liquid Immersion Cooling of Small Electronic Devices;* Microelectronics and Reliability, 1973, Vol. 12, pp. 163-173

# 5 Conclusion

These boiling experiments at saturated and subcooled liquid conditions performed at a very small spot heater under microgravity and under earth gravity conditions lead to the following conclusions:

- Even for the small heater geometry we received very high values of the heat flux at μg and at 1g up to 900 kW/m², comparable higher heat flux values as on wires.

- A remarkable influence of the gravity on the overall heat transfer could not be observed. That confirms our earlier observation and our resulting statement that the evaporation in the liquid wedge between the solid heater and the bubble interface, the capillary force, the surface tension and the wetting condition play the most important role for the boiling process, as recently proposed in the micro wedge model.

- Furthermore, we must conclude that likewise the critical heat flux is not caused by a hydrodynamic instability alone, it is much more a complex problem of the flow limitation in the micro wedge, the dry out at the three phase line, and of wetting conditions. It can be regarded as a „three phase thermo-hydrodynamic instability".

- We could observe several different modes of boiling in the subcooling state of the liquid, which are characterized by different kinds of vapor formation and of bubble dynamics. These modes depend on the liquid state, on the subcooling, and on the heat flux respectively the surface temperature of the heater.

- These different regimes of bubble dynamics change immediately with the change of the subcooling state. The transition from one mode to the other is stimulated by thermo-hydrodynamic instabilities in the boundary layer of the heater, which needs a more detailed investigation.

- For the first time in boiling the thermocapillary jet mode was observed for moderate subcooling states, and it is interesting to note that the high heat flux produced up to 600 kW/m² can be transported by this flow alone. This is possible by the large surface area of a big bubble forming a trailing jet stream.

- Furthermore it is important to note that even in microgravity, boiling is a very efficient process of heat transfer, and that it controls over a very wide range of heat dissipation its temperature itself by increasing or decreasing the heat flux. Therefore the boiling process allows the application for direct cooling of high powered electronic microelements in all space systems. For such an application the liquid condition can be optimized as seen from our experimental results.

- Furtheron it must be emphasized that in boiling no general statements can be made if not a wide range of parameters, which influence the process, are investigated.

The evaluation of the data is still going on. It is intended to perform a detailed quantitative correlation between measured heat-flux and fluid state with the observed bubbles sizes and flow-patterns.

[4] Bar-Cohen, A.; *Fundamentals of Nucleate Pool Boiling of Highly-Wetting Dielectric Liquids;* ASI Proceedings, Cooling of Electronic Systems, Izmir, Turkey, 1993, pp. 415-455, Ed. by S. Kakac

[5] Bergles, A. E. & Bar-Cohen, A; *Immersion Cooling of Digital Computers;* ASI Proceedings, Cooling of Electronic Systems, Izmir, Turkey, 1993, pp. 539-621, Ed. by S. Kakac

[6] Nagasaki, T., Hijikata, K., Fushinobu, K., & Saitoh, T.; *Boiling Heat Transfer from a Small Heating Element;* 3, to be presented at ASME Winter Annual Meeting, New Orleans

[7] Straub, J., Zell, M., and Vogel, B.; *Pool Boiling in a Reduced Gravity Field;* Proc. Ninth Int. Heat Transfer Conf., Jerusalem, Israel, 1990, Ed. G. Hetsrony, Vol. 1, pp. 91-112, Hemisphere, New York 1990

[8] Forster, H. K., Zuber, N.; *Dynamics of Vapor Bubbles and Boiling Heat Transfer;* AICHE J., vol. 1, 1955, pp. 531 - 534

[9] Rohsenow, W. H.; *A Method of Correlating Heat Transfer Data for Surface Boiling of Liquids;* Trans. ASME Ser. C, J. Heat Transfer 74, pp. 969-976

[10] Straub, J.; *The Role of Surface Tension for Two-Phase Heat and Mass Transfer in the Absence of Gravity;* Experimental Thermal and Fluid Science 1994, 9, pp. 253-273

[11] Skripov, V. P., *Metastable Liquids,* New York, John Wiley & Sons, 1974

[12] Thormählen, I., *Grenze der Überhitzbarkeit von Flüssigkeiten - Keimbildung und Keimaktivierung,* Fortschritt-Berichte VDI, Reihe 3, Nr. 104, VDI-Verlag, 1985

# 6 ACKNOWLEDGMENT

We like to express our appreciation to all, who made this research possible. Especially we are grateful to the Space Shuttle crew of STS 65, the mission manager, the mission scientist and the team of NASA Operation Support Center in Huntsville, AL, to ESA and ESA/ESTEC for providing the BDPU facility and the test container, the individuals of the industrial companies designing and building the facility and the test container of Alenia, Laben, Ferrari and Dornier. Furthermore we thank all individuals and teams supporting us during the mission and we gratefully acknowledge the support of the Germany Space Agency DARA.

# Modelling of Transient Boiling in Microgravity

K.Sefiane[1] and A.Steinchen[1][2]

[1] Laboratoire de Thermodynamique CP 512 Fac. des Sciences St Jerome Bd Escadrille
Normandie Niemen
13397 Marseille cedex 20, FRANCE
[2] IUSTI Univ. Provence Marseille
Fac des Sciences St Jerome Bd Escadrille Normandie Niemen
13397 Marseille , FRANCE

## Abstract

Under steady conditions of pool boiling, the observed maximum heat flux is described
by the so-called Hydrodynamical theory stated by Zuber in 1959. This theory is in a
reasonable agreement with experiments under normal gravity, however under vanishing
gravity, the experimental maximum heat flux is larger than the one predicted by the
Zuber formula. The hydrodynamical theory breaks down.

In unsteady boiling, under very low subcooling and in microgravity conditions, a
peak of heat transfer coefficient corresponding to transitory nucleate boiling, is
observed after a short time of bubbles formation immediately followed by a dryout
and thus by a strong decrease of heat transfer coefficient.

The principal aim of this work is to modelize the maximum heat transfer coefficient
observed under microgravity [4] for low subcoolings. The model could be extended to
represent boiling phenomenon under microgravity for wider range of conditions.

## 1 Introduction

Heat transfer to boiling liquids with phase change is of a great practical significance in
a wide number of processes. The outstanding feature is the existence of definite
nucleation sites where bubbles form on the heat transfer surface. The heat passes from
the heating surface into the liquid wherever the two touch, and the liquid becomes
superheated everywhere. The greatest superheat occurs in a thin zone near the heat
transfer surface, with increasing surface temperature difference. Nucleate boiling is
initiated and the rate of heat transfer is determined by convection of the liquid due to
formation, growth and motion of the vapor bubbles and to natural convection.

## 2 Maximum heat flux :

The most successful approach to the prediction of the maximum heat flux is predicted by the so-called Hydrodynamical theory. This theory is based upon the observation that the transferred heat is contained in the heat of the vapor's evaporation.

Furthermore it is observed that upon approaching $q_{Max}$ the vapor bubbles coalesce into vapor columns.These columns are spaced regularly by distance b, and their diameter ,D , is about b/2. The maximum heat flux is then computed:

$$\dot{q}_{Max} = \frac{\Pi}{4} \rho_g \, h_{fg} V_g$$

Vg represents the vapor speed in the columns at the occurrence of the maximum heat flux. It is known that when two immiscible fluids are in relative motion, their contact surface may become unstable under the action of inertia and surface tension forces (Helmholtz instability); in particular, with case of gas flowing over a stagnant liquid with speed Vg (Fig. 1 ). Also the vapor-liquid interface of a vapor column may become unstable. This leads to the break-up of vapor columns in separate bubbles, to fulfill the classical theory we have to consider :

$$\delta = \frac{2R}{L_C} \qquad \text{where} : L_C = \sqrt{\frac{\sigma}{\Delta \rho g}}$$

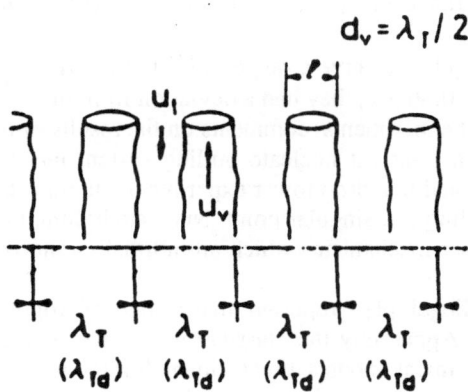

Fig. 1 : Kelvin-Helmholtz instability

The dimensionless parameter $\delta$ is deduced from the squaare root of the Bond number and it compares lengths related to the solid and gaseous phase, namely the diameter of a solid ( R ) to the capillary length of a bubble ( $L_C$ ). Its value helps to select the various boiling mechanisms.

The Zuber formula for the maximum heat flux according to the hydrodynamical model is:

$$\dot{q}_{Max} \approx \left[\sigma\left(\rho_L - \rho_g\right)g\right]^{1/4}$$

This last formula , based on the classical theory and it is of a good agreement with experimental results under normal gravity.

## 3 Validity of the theory in microgravity

The N. Zuber [1] theory (formula) has a limit for its use which is often encountered in microgravity. In 1973 J.H. Lienhard and K. Dhir [2] derived the following conclusions :

a - for $\delta$ greater or equal to 2.2 the ( $g^{1/4}$) power law is reasonable.

b - for $0.2 < \delta < 2.2$ the ($g^{1/8}$) power law is used.

c - for $\delta$ less than 0.2 other mechanisms may intervene to explain the existence of a maximum heat flux, such as Marangoni effect and recoil effect.

d - for $\delta$ less than 0.02 there is a jump from nucleate boiling to film boiling the recoil effect may be of considerable importance as proposed by Steinchen et al..

In 1967 R. Siegel made an analysis of the then available experiments. The Maximum Heat flux varies as ($g^{1/4}$) for numerous results. In contrast in other cases the experimental values are higher than those predicted by the ($g^{1/4}$) law. As a result he concluded that in the range of [ 0.1 g to 1g ], the ($g^{1/4}$) law is a reasonable approximation.

C. Usiskin and R. Siegel [ 3 ] give a range of $\delta$ [ 0.2 , 0.5], for distilled water under low gravity ( less than $< 0.06$ g ) , they had a deviation from the ($g^{1/4}$ ) law.

H. Merte [4] in recent experiments comments on the results of measurements of heat transfer coefficient in a transient nucleate boiling system under a short time where reduced gravity is obtained in a drop tower experiment. He reported " In microgravity in many cases pool boiling is a singular point. Nucleate boiling may be expected to be quite different. Can other mechanisms which are normally ignored become significant to nucleate boiling? ".

C. Usiskin and R. Siegel [3] reported that under microgravity bubbles remain attached to the heater. Apparently the classical theory has a limit to its use - to our mind - the hydrodynamical models break down for reduced gravity environment. Seemingly, under microgravity conditions, another heat transfer mechanism must be taken into account to describe the maximum heat flux.

It is clear that the agreement between the experimental results and N. Zuber law is quite real. Nevertheless, at a very low gravity level we notice a deviation in the results.

The hypothesis is that under low gravity the bubbles remain attached to the heating surface. These bubbles will grow without forming ascending vapor jets. The wetting of the heater by the liquid and its displacement by the vapor will be of a major importance. In recent experimental observations H. Merte [4] comments on a plot (Fig. 2 ) of heat transfer coefficient versus time "(Fig. 2) shows a drop test. Nucleation occurs at t = 3.29 sec and propagates relatively slowly as indicated by the slow increase in « h » until about t = 4.15 sec at which time the dynamic growth of the vapor bubble takes place followed by the gradual dryout of the heater surface.".

# 4 Pool boiling under reduced gravity

## 4.1 Heat fluxes under microgravity

Under micro gravity according to A. Steinchen et al [5] there are three remaining heat fluxes that might contribute to the heat transfer process:
The evaporation heat flux.
Marangoni heat flux.
The Vapor recoil flux, observed for rapidly evaporating liquids and analyzed by H.J. Palmer [6].

In addition to these three fluxes we must add the heat flux due to conduction that prevails between the bubbles. As previously mentioned the wetting of the heater by the liquid and its displacement by vapor, will be of major importance and will be detailed further in this paper. In the suggested model, the maximum heat transfer coefficient observed in transient boiling (see Fig. 2) will be mainly due to this phenomenon

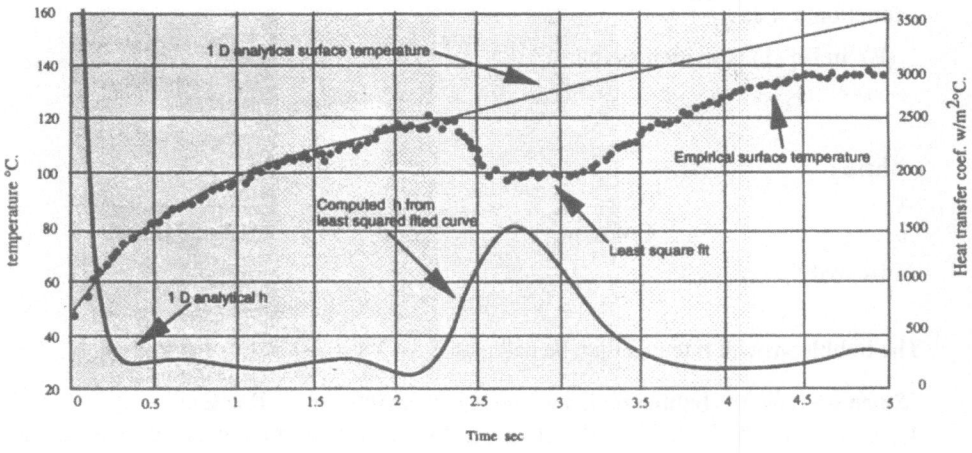

Fig. 2 : Drop test a/g ≅ 10⁻⁵ . Boiling and dryout. $\dot{q}$ =7.7 W/cm²
$\Delta T_{subΔ}$ =0.5 °C. Tsat = 78.0 °C (from H. Merte ).

## 4.2 Nucleation and bubbles growth parameters

The process of bubble formation is known as nucleation. In many ways it is analogous to crystal nucleation. The surface properties are obviously important in all these nucleation phenomena. The surface used for heating is not smooth. It contains pits and cracks, these will constitute nucleation sites (Fig. 3 ).

Fig. 3 : Nucleation process.

As a first approximation we can use the law of the vapor bubble growth in the form proposed by H. Beer [7] :

$$R = E \tau^n \tag{1}$$

The equation (1) has been found from experimental results and are plotted as a function of ( $Ja \sqrt{a_L}$ ).

(E) in Eq (1) is written as follows :

$$E = b \, Ja \, \sqrt{a_L}$$

where :

| | | | | | |
|---|---|---|---|---|---|
| $\tau$ | : | Time. | Ja | : | JACOB number. |
| b | : | Constant. | $a_L$ | : | Liquid thermal |

diffusivity.

The bubble growth rate can then be calculated as $Vs = \dfrac{dR}{d\tau}$ .

Since we have the bubble radius which can be written as $R = E \tau^n$

Then one can obtain the growth velocity as the derivative of the previous expression

$$Vs = \frac{dR}{d\tau} = nE \, \tau^{(n-1)}$$

## 4.3 Heat fluxes calculation

### 4.3.1 The evaporation heat flux

The amount of heat required to evaporate the bubble vapor can be expressed as follows :

$$\dot{q} = 2 v_s \rho_g h_{fg} \tag{2}$$

## 4.3.2 The recoil heat flux

The total recoil heat flux is a result of the remixing effect of the overheated evaporating liquid layer of surface S= $\pi R^2$ attached to the bubble which is in turn due to the vapor recoil instability ( momentum transfer from the evaporative surface to the overheated liquid ), also called Hickman instability. The layer is then expelled and a violent hydrodynamic motion is transmitted to the neighboring bulk liquid, and K. Hickman experiments [ 8] are clearly describing this phenomenon. The recoil force depends on evaporation rate and thus on overheating of the microlayer. It is a constant for every value of evaporation rate :

$$\dot{q}_{rec} = \Pi R^2 \rho_L \, C_{PL} (T_w - T_s) U_{rec} \tag{3}$$

Using this equation and the one suggested for the bubble radius we can compute the recoil heat flux.

## 4.3.3 Marangoni heat flux :

The Marangoni effect is the convective motion around the bubble due to the surface tension difference caused by the temperature gradient (lower surface tension in the warmer region inducing a surface motion from the region of lower surface tension to the region with higher surface tension and driving the adjacent hot liquid towards cooler regions). It occurs with a velocity :

$$U_\sigma = L_\sigma \, \frac{grad\sigma}{\eta}$$

where :

$L\sigma$ : is the characteristic length for the Marangoni convection related to the size of the bubble ( curvature ).

$$\dot{q}_M = 2\Pi R \delta_n \rho_L \, C_{PL} \, (T_w - T_m) L_\sigma \, \frac{grad\sigma}{\eta} \tag{5}$$

| | | |
|---|---|---|
| $\delta$ | : | Boundary layer thickness. |
| $\sigma$ | : | Surface tension. |
| $C_{PL}$ | : | Liquid heat capacity. |
| $\eta$ | : | Dynamic viscosity. |
| $\rho_L$ | : | Liquid density. |

### 4.3.4 Conduction heat flux :

The conduction heat flux prevails mainly during two periods. Before nucleation and during nucleation through the surface not covered by bubbles. The usual expression is used

$$\dot{q}_{cond} = \lambda_L ( T_W - T_S ) \qquad\qquad (6)$$

where :

$\lambda_L$ : The liquid thermal conductivity.

# 5 Presentation of the model

In recent experiments on nucleate boiling under reduced gravity, performed in a drop tower H. Merte et al [ 4 ] concluded "Unless some mechanism exits for the removal of the vapor from the immediate vicinity of the heater surface, pool boiling will remain inherently transient process.". The model is developed for transient nucleate boiling under microgravity conditions.

Under vanishing gravity, the buoyancy force vanishes. Consequently bubbles grow and remain attached to the heated surface. The heat fluxes observed in the bulk around the bubble are mainly recoil flux, Marangoni flux and conduction heat flux which are dependent on the outside bubble surface ( Fig. 4 ). This latter is time dependent the heat flux is then time dependent, and a transitory state of nucleate boiling is observed followed by the formation of a film. The flux of evaporation carries heat from the heater through the microlayers, as well as Marangoni and recoil fluxes around the bubbles carry heat from the warmer bubbles base thus from the heater. The conduction flux between the bubbles also contributes to removal of heat from the heater through the liquid thermal boundary layer on the fraction of surface covered by liquid. Hence, under microgavity the heat flux in the boiling phenomenon is expected to behave in the transitory state as detailed in subsequent paragraph.

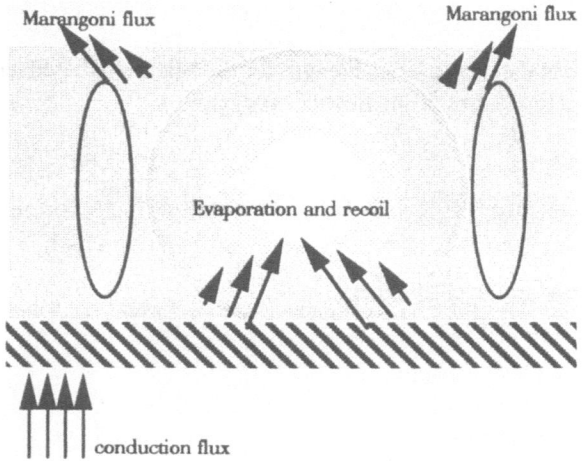

Fig. 4 : Heat fluxes around a single bubble.

- heat conduction through the liquid without nucleation.
- nucleation occurs and an observed increasing in heat flux coefficient; the spreading of the nucleated bubbles over the surface increases heat flux due mainly to Recoil, Marangoni and Evaporation heat fluxes.
- growing bubbles begin to coalesce and reduce the area through which the main heat flux is transferred. Then a decrease in the heat flux coefficient is expected.
- bubbles continue to coalesce until they overlap the whole area of the heater where a film boiling is observed.

## 5.1 Heat transfer calculation

### 5.1.1 Introduction

The surface through which the main heat fluxes are transferred is proportional to the outside surface of the bubble and between the bubbles. This area depends on time and on the temperature gradient $T_W - T_S$.

This surface is reduced when bubbles coalesce. As previously detailed, the total heat flux which is increasing with nucleation will decrease with the bubbles coalescence which inhibits heat transfer.

To compute the heat flux, we assume that the heat of evaporation for bubble growth is provided from the wall at the bubble base .

At the same time, there is a contribution from heat convection around the bubble due to the Marangoni effect and to the recoil flux, where as the conduction flux prevails through the area between the bubbles.

According to this model we may have two cases :

1 - the Hickman instability will not occur before the bubbles coalesce and the recoil heat flux will not contribute to the total heat flux .

2 - the Hickman instability will occur before the bubbles coalesce, and the recoil heat flux will contribute to the total heat flux. In this case the maximum heat flux is expected to be greater (Fig. 5 ).

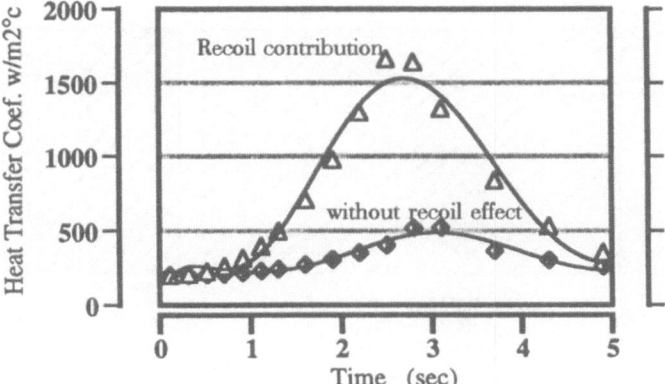

Fig. 5 : The heat transfer coefficient range.

## 5.2 Bubbles Growth Dynamics

### 5.2.1 Description and analysis

When boiling occurs on a heater in a liquid, spherical entities are observed (bubbles). The bubbling is induced at some points of the heating surface named nuclei, from which bubbles grow in all available directions of space.

Bubbles progressively cover the whole surface therefore, the bubbles formation kinetics can be described by the appearance (nucleation) and the growth of spherical entities which progressively cover the whole surface in a coalescence process.

To study this phenomenon, a statistical description of bubbles morphology and their formation which is controlled by nucleation followed by radial propagation is performed. The covered surface fraction, $\alpha$, can then be defined as the ratio of the surface covered by bubbles, to the total surface of the heater (Fig. 6 ).

This covered surface fraction ranges between $(0-1)$. To describe the bubbling phenomenon, the following hypotheses are considered :

1. The bubbles are nucleated by tiny nuclei which already exist on the heating surface and whose initial number per unit surface $N_o$, depends on the heaters materials nature, as well as the temperature.

2. These nuclei can be activated at any time if they are in a still liquid region (i.e not covered by a bubble ). As soon as they are activated they immediately become growing entities. Conversely if a nucleus is overlapped by a bubble before its activation it will never be activated. The probability per unit time of the occurrence of an activation in an infinite surface also named Activation Frequency $q$ , depends only on the temperature at which boiling takes place.

3. An entity grows in all the available directions until it reaches another entity.

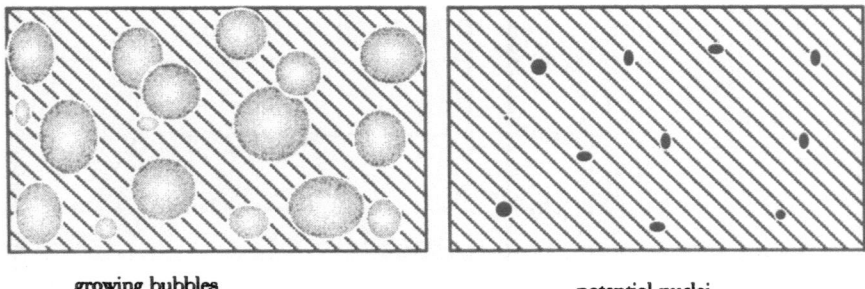

growing bubbles                                          potential nuclei

Fig. 6 : Bubbles nucleation and coalescence kinetics.

At this moment they coalesce. The growth rate $V_s \doteq \dfrac{dR}{d\tau}$ depends on the temperature of boiling. According to this description, the bubbling phenomenon is entirely described by these three parameters : $N_o$; q; $V_s$ (Fig. 6 )

The probability that "n" expanding spheres reach an arbitrary chosen point (A) on the heating surface up to time "$t_m$"obeys the Binomial Distribution

$$P = \binom{M}{n}\left[\left(\frac{V_C}{V_S}\right)^n\left(1-\frac{V_C}{V_S}\right)^{M-n}\right]$$

Where :
M : The total number of nucleation of attempts in the sample.
$V_S$ : The volume of the space-time.
$V_C$ : The volume of the space time cone.

$$V_C = \frac{1}{3}\Pi V_s{}^3 t_m{}^4$$

For an infinite number of attempts ( $M \rightarrow \infty$ ) the binomial law distribution transforms into Poisson Distribution:

$$P(n) = \lim\left[\frac{M!}{n!(M-n!)}\left(\frac{V_c^n B^n}{M^n}\right)\left(1-\frac{V_c B}{M}\right)^M\right]$$

The probability that an arbitrary point (A) is included by one expanding spheres nucleated at time $t_i < t_m$ and none of others nucleated up to time $t_m$ can be obtained by multiplication of the respective components probabilities because of the independence of the component events .

Further the probability that this point (A) will be included by only one growing bubble and not by the remaining others is computed.

One can obtain a general form of the probability of simultaneous inclusions of point (A) at time $t_m$ by (n) bubbles nucleated at times $(t_{i1}, t_{i2},...t_{in})$
This describes the probability of the inclusion of an arbitrary point (A) at time $(t_m)$ by a bubble nucleated at time $t_i$.

The integration of equations describing these probabilities over the range $t_i < t_m < t$ gives the probability that an arbitrary chosen point will be occupied before time 't', by the bubble nucleated at time $t_i$. This *is proportional to the surface fraction* occupied up to time 't ' by bubbles nucleated at time $t_i$.

If $t = \infty$, we obtain the surface fraction of bubbles nucleated at time $t_i$. After integration over the range $0 < t_i < t_m$, the probability that the point will be included by a bubble front at time $t_m$ can be computed.

After integration of the above described probabilities expressions over the range

$0 < t_m < t$, one can obtain the formula describing the time dependence of the surface fraction covered by the bubbles.

$$\alpha(t) = 1 - \exp\left(-\frac{4}{3}\Pi\langle V_S\rangle^3 \int_0^t F(\tau)(t-\tau)^3 d\tau\right) \qquad (7)$$

It is worth noticing that in the polymers field, theories of overall crystallization kinetics have been developed to express the average evolution of the transformation. As it has been already mentioned, crystallization phenomenon is practically analogous to bubbles formation. This analogy relies on the fact that both phenomena are governed by the same factors affecting the evolution of the process. An analogous equation was given to describe the transformation rate. Among the approaches proposing an expression for the surface fraction evolution, M. Avrami's theory [9] is the most popular. Evans initial theory is limited to isothermal crystallization and to special cases of nucleation, and no general demonstration is given. Further, it has been shown that Avrami 's and Evans' theories are formally equivalent when the potential nuclei are randomly distributed.

Finally, it is noteworthy that several years before, Kolmogoroff has obtained the same results as Avrami and Evans, using a statistical approach. The Avrami-Kolmogoroff-Evans theory leads to the following expression also used by E. Piokowska and A. Galeski [10]

$$\alpha(time) = 1 - \exp[-E(time)]$$

In recent works [11] the following parameter has been detailed as follows ( Fig. 7 ) :

$$E(time) = 2\Pi N_\circ \left(\frac{\langle V_s\rangle}{q}\right)\left[1 - \exp(-q\tau) - q\tau + \frac{(q\tau)^2}{2}\right] \qquad (8)$$

where :

$N_\circ$         : The potential initial nuclei on the heating surface.$(m^{-1})$.

$q$          : The nucleation frequency.$(sec^{-1})$.

$\langle V_s\rangle$        : The bubbles average growth rate.$( m.sec^{-1} )$.

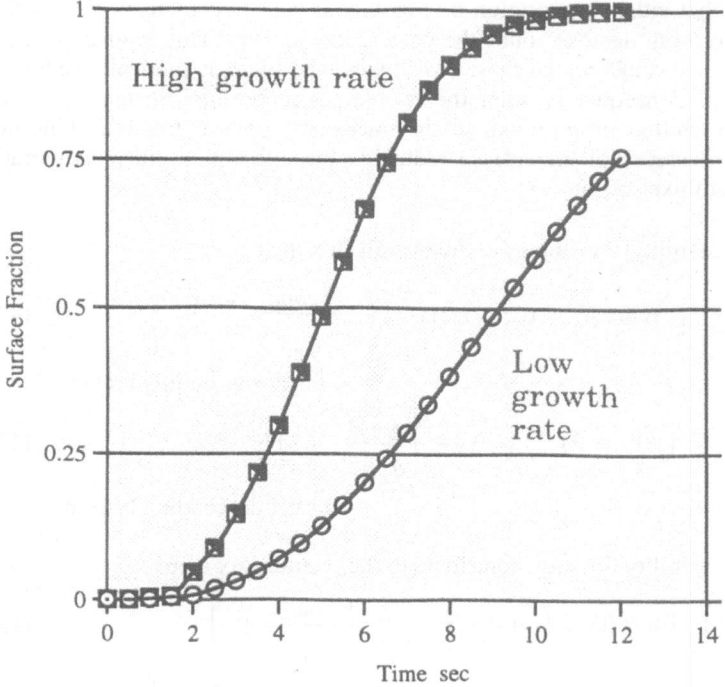

Fig. 7 : Vapor evolution kinetics during boiling process.

## 5.3 Heat Flux Coefficient Expression

In the previous paragraphs, we have computed the different heat fluxes that participate to the total heat flux in pool boiling under reduced gravity. According to the model we have suggested, the maximum heat flux will occur when the bubbles which are covering the heating surface will coalesce and inhibit the heat transfer. Writing the total heat flux as :

$$\dot{q}_{tot} = h \, (Tw - Ts) \tag{9}$$

h        :  Heat transfer coefficient.
Tw,Ts  :  Respectively wall, Saturation temperatures.

From the previously calculated heat fluxes, the other parameters that intervene in the explanation of the pool boiling, such as the surface fraction, covered by the growing bubbles; the coalescence which determines the change in the heat flux. According to all these computed parameters we can write the heat flux as follows

From previous studies [5] the following ratios $\dot{q}_M / \dot{q}_{ev}$ , $\dot{q}_{rec} / \dot{q}_{ev}$ have been approximated to :

$$\dot{q}_M / \dot{q}_{ev} = 0.8 \quad \text{and} \quad \dot{q}_{rec} / \dot{q}_{ev} = 5 \tag{10}$$

We can note that either evaporation or Marangoni heat flux increase during the first period, however both decrease once the coalescence occurs. This is mainly due to the fact that the area through which these two fluxes take place is gradually reduced after the coalescence. Consequently, after the two fluxes represented in the total heat flux expression, are shifted proportional to the uncovered surface fraction. This latter is describing the decrease of the surface reduction through which either evaporation or Marangoni heat fluxes occur.

1 . the recoil instability contributes to the overall flux then :

$$\dot{q}_{tot1} = = \dot{q}_{ev} \left\{ 6.8\alpha(t) + \left( 1 - \alpha(t) \right) \left( \frac{\dot{q}_{cond}}{\dot{q}_{ev}} \right) \right\} \tag{11}$$

for the ascending branch

$$\dot{q}_{tot2} = \dot{q}_{ev} \left\{ \left( 1 - \alpha(t) \right) \left( 6.8 + \frac{\dot{q}_{cond}}{\dot{q}_{ev}} \right) \right\} \tag{12}$$

for the descending branch

2 . the recoil instability does not contribute to the overall flux then:

$$\dot{q}_{tot3} = \dot{q}_{ev} \left\{ 1.8\alpha(t) + \left( 1 - \alpha(t) \right) \left( \frac{\dot{q}_{cond}}{\dot{q}_{ev}} \right) \right\} \tag{13}$$

for the ascending branch.

$$\dot{q}_{tot4} = \dot{q}_{ev} \left\{ \left( 1 - \alpha(t) \right) \left( 1.8 + \frac{\dot{q}_{cond}}{\dot{q}_{ev}} \right) \right\} \tag{14}$$

for the descending branch.

The last expression gives the variation of the heat flux during the nucleation period, as well as during the transition period. Using this last expression for the heat flux, the heat transfer coefficient can be computed during the different steps. Taking into account equations (1),(2),(4),(6),(9),(10) it can be written as follows :

1. the recoil instability contributes to the overall flux then :

$$h_1 = 2n\beta\tau^{n-1}\sqrt{a_L}\,\rho_L c_L \left\{ 6.8\alpha(t) + (1 - a(t)) \frac{\lambda_L}{2\tau^{n-1}n\beta\sqrt{a_L}\rho_L c_L} \right) \right\} \tag{11'}$$

$$h_2 = 2n\beta\tau^{n-1}\sqrt{a_L}\,\rho_L c_L \left\{ \left( 1 - \alpha(t) \right) \left( 6.8 + \frac{\lambda_L}{2\tau^{n-1}n\beta\sqrt{a_L}\rho_L c_L} \right) \right\} \tag{12'}$$

2. the recoil instability does not contribute to the overall flux then :

$$h_3 = 2n\beta\tau^{n-1}\sqrt{a_L}\,\rho_L c_L \left\{ 6.8 + (1 - \alpha(t)) \left( \frac{\lambda_L}{2\tau^{n-1}n\beta\sqrt{a_L}\,\rho_L c_L} \right) \right\} \qquad (13)'$$

$$h_4 = 2n\beta\tau^{n-1}\sqrt{a_L}\,\rho_L c_L \left\{ \left( 1 - \alpha(t) \right)\left( 1.8 + \frac{\lambda_L}{2\tau^{n-1}n\beta\sqrt{a_L}\,\rho_L c_L} \right) \right\}$$

$$(14)'$$

## 5. 4 Numerical results

The numerical application of the presented model is applied to three components (mainly refrigerants) which are frequently used in boiling experiments : R113, R11, R114. In equation (1) the exponent n, varies from n=0.9 at higher growth rate and n=0.2 at lower rates. in the numerical calculation the exponent,n is taken to be n=0.9 for the first stage (before the coalescence occurs), where bubbles are expected to grow rapidly. Whereas it is taken to be 0.2 for the second stage (after the coalescenc, where the bubbles are growing slowly. The constant $\beta$ is found to be $\beta = \dfrac{1}{2}\sqrt{\pi}$ (ref.7).

Th e following results are obtained:

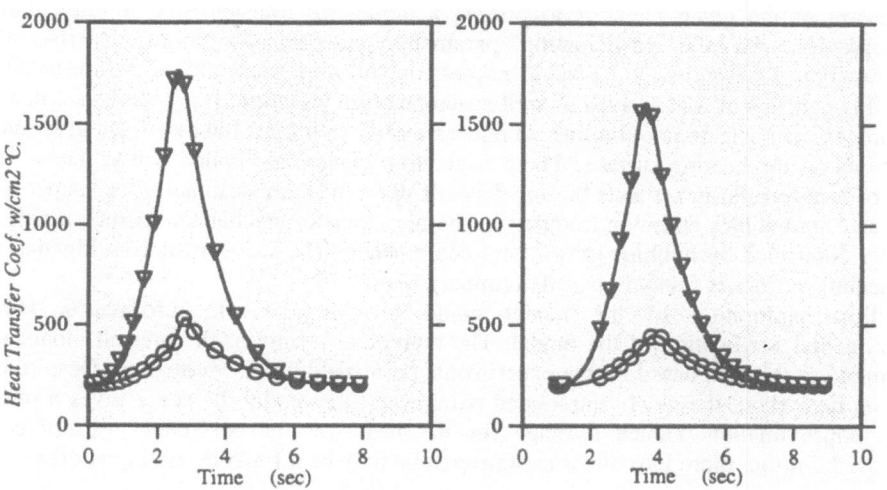

Fig. 8 Boiling curve in microgravity of Forane R11 Fig. 9 Boiling curve in microgravity of forane R114
Ts= 23.8°C, q=7.7w/cm2,$\Delta T_{sub}$= 0°C.        Ts=3.6°C, q=7.7 w/cm2,$\Delta T_{sub}$ = 0°C.

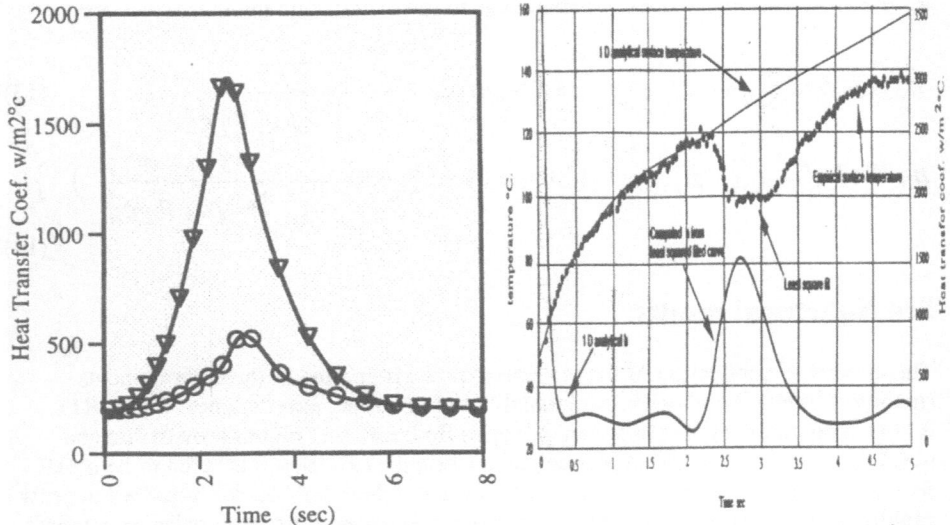

Fig. 10 Boiling in microgravity of forane R113
Ts=48°C, $q$=7.7w/cm$^2$ $\Delta T_{sub}$=0°C.

Fig. 11 Boiling in low gravity a/g=10-5
$q$=7.7 w/cm2, $\Delta T_{sub}$=0.5°C,

# 6 Conclusion

The aim of the given paper is to present a model for transient pool boiling under microgravity. All the participating parameters are introduced to describe the phenomenon. Consequently a general expression that represents the heat flux and the heat flux coefficient was derived as well as temperature variation. It is established that in the process of nucleate pool boiling in microgravity conditions, bubbles form at specific locations on the heating surface. These nucleation centers are called actives sites. The rate of heat transfer in nucleate boiling depends upon the number of actives sites per unit area and the bubbles emission frequency. The heat transfer mechanism in such a case is wholly controlled by bubbles growth and coalescence. The vapor evolution kinetics on the heating surface is found to be a determining factor.

Further exploration of the model could provide valuable information about experimental verification of the model. The numerical results of the present model are compared to the data based on the experiments performed by H. Merte et al. Despite the lacking data about the newly introduced parameters, the model, however gives a pretty good approximation. Hence a range for the heat transfer coefficient variations is proposed. Further more introduces parameters that may be refined in subsequent studies.

## Nomenclature

| | |
|---|---|
| $\rho_L$ | Liquid density. |
| $\rho_g$ | Vapor density. |
| b | The distance between the columns in Helmholtz instability. |

| | |
|---|---|
| $\sigma$ | Surface tension. |
| $V_g$ | Bubble growth rate = dR/dτ. |
| τ,t | Time. |
| R | Bubble' s radius. |
| $Ja = \dfrac{T_W - T_S}{H_{fg}\rho_g}\rho_L C_L$ | Jacob number. |
| $a_L$ | Liquid thermal diffusivity. |
| b | Constant for bubbles' growth parameters. |
| $C_{PL}$ | Liquid heat capacity. |
| $\lambda_L$ | Liquid thermal conductivity. |
| N° | Initial potential nuclei density. |
| q | Nucleation frequency. |
| $\langle V_S \rangle$ | Average growth rate. |
| h | Heat transfer coefficient. |
| $T_s$ | Saturation temperature. |
| α | The surface fraction covered by vapor. |
| $T_w$ | Wall temperature |
| δ | boundary layer thickness. |
| η | Dynamic viscosity. |
| $\dot{q}$ | Heat flux. |

## References.

[1]  N. Zuber, Atomic Energy Commission  Report N° AECU-4439, (1959).

[2]  J.H. Lienhard, K.Dhir, J. Heat Transfer, Trans. ASME,**158**, pp. 97-152,(1973).

[3]  R. Siegel and C.Usiskin , J. HeatTransfer, Trans., ASME., **81**,p.  3, (1959).

[4]  H. Merte, H.S. Lee and J.S. Ervin , Microgravirty Sc. and Technology,**VII**, pp. 173-179, (1994).

[5]  A.Steinchen, A. Sanfeld, L. Tadrist and J. Pantaloni, Microgravity Sc and Technology, **VII/2**, pp. 180-186, (1994).

[6]  H.J. Palmer, J.Fluid Mechanics,**15**, p. 487, (1976).

[7]  H. Beer, Heat Mass Transfer,**2**, pp. 2 311-370 (1969).

[8]  K. Hickman, Ind. and Eng. Chem. , **44**, P. 1892, (1952).

[9]  M. Avrami, J. Chem. Phys. **9**, P. 177, (1941).

[10]  E. Piorkowska and A. Galeski, Journal of  Polymer Physics, **23**, p. 1723, (1985).

[11]  N. Billon and J.M. Haudin, Colloid and Poly. Sc. ,**271**, pp.  343-356,  (1993).

# Surface Dynamics of Surfactant Solutions

J.C. Earnshaw, E. McCoo, C. Nugent, D.J. Sharpe

The Department of Pure and Applied Physics, The Queen's University
of Belfast, Belfast BT7 1NN, Northern Ireland

## 1 Introduction

Waves on liquid surfaces have been studied for many years [1,2]. While the associated aspects of liquid surface dynamics are well established, theoretical studies continue in various directions, including incorporation of processes within or affecting a surface molecular film [3,4] and the recent discovery of certain unsuspected effects of the established corpus of theory [5,6]. Surface light scattering provides a tool to study the surface viscoelasticity governing the capillary waves [7]. The present experimental studies of capillary waves on solutions of ionic surfactants enable us to probe certain of the novel aspects of surface wave behaviours.

In surfactant solutions various processes influence the surface viscoelastic properties. For example, diffusive exchange of surfactant molecules between bulk and surface leads to a specific frequency dependence of the dilatational modulus [8]. Extensions to the model have recently been advanced [3,4]: the Marangoni effect transforms chemical energy, due to temperature or surface concentration gradients in the fluid, into mechanical energy, and under appropriate conditions this energy may exceed the viscous dissipation, generating sustained oscillations of the surface modes [3]. The critical conditions for this to occur can be modified by other processes [4], the effects of different processes occurring at different characteristic wave frequencies.

In particular the presence of a barrier to adsorption at the surface can facilitate the destabilisation of the longitudinal modes, by destroying local equilibrium between surface and bulk [4]. One may imagine free diffusive exchange between the bulk and a fluid layer immediately below the surface, but surfactant molecules must overcome the barrier for adsorption to occur. This barrier reduces the stability of the dilatational waves at high surface wave frequencies.

We present data for solutions of various surfactants, which combine to suggest the presence of an electrostatic barrier to adsorption at the surface.

# 2 Theoretical Background

The random rough surface due to thermal excitation can be Fourier decomposed into a complete set of surface waves. In the present work, the real wave number $q$ of a particular mode is selected experimentally, and the complex propagation frequency $\omega$ $(\omega_0 + i\Gamma)$ is measured.

The frequency relates to the wave number via the surface wave dispersion equation, containing the relevant material properties of the system [9,10]. This involves two surface moduli which affect the capillary waves: $\gamma$ and $\epsilon$. Viscous dissipation within the film can be incorporated into the formalism by expanding $\gamma$ and $\epsilon$ as linear response functions:

$$\gamma = \gamma_0 + i\omega\gamma' \tag{1}$$

$$\epsilon = \epsilon_0 + i\omega\epsilon' \tag{2}$$

where $\gamma_0$ and $\epsilon_0$ are elastic moduli and $\gamma'$ and $\epsilon'$ are specific surface viscosities. The modulus $\gamma$ governs shear transverse to the surface plane, while $\epsilon$ relates to in plane dilatation. $\gamma_0$ is the surface tension and $\epsilon_0$ the dilatational elastic modulus $(-d\gamma_0/d\ln\Gamma_s$, $\Gamma_s$ being the surface excess of surfactant).

The dispersion equation has two roots, corresponding to capillary and dilatational waves respectively: $\omega_C$ and $\omega_D$ for brevity. We are concerned only with the low $q$ regime in which both modes propagate as damped oscillations [1,10],

$$\omega_C \approx \sqrt{\frac{\gamma q^3}{\rho} + i\frac{2\eta q^2}{\rho}} \tag{3}$$

$$\omega_D \approx \frac{1}{2}(\sqrt{3}+i)\left(\frac{\epsilon^2 q^4}{\eta\rho}\right)^{\frac{1}{3}} \tag{4}$$

being useful indicators of the dependences of $\omega$ upon $q$ and the system properties for capillary and dilatational waves respectively. The principal effects of $\gamma'$ and $\epsilon'$ are to increase the damping of the capillary [11] and dilatational [10] waves.

The two surface modes are coupled. The best known effect is a resonance between the two surface modes [12], occurring when the real frequencies of the modes coincide (at $\epsilon_0/\gamma_0 \approx 0.16$). Here $\Gamma_C$ rises to roughly twice its value for a clean surface, before falling to a plateau.

If the damping of the two surface modes can be brought together, the real frequencies will separate, leading to the appearance of manifest signs of mode mixing. In mode mixing that mode which at low $q$ is capillary wave-like crosses over to dilatational wave-like, and vice versa. At intermediate $q$ neither mode can be identified as capillary or dilatational: they are manifestly mixed modes. Theory predicts two routes to mode mixing for waves on a liquid surface [5,6], one involving $\gamma' > 0$, the other $\epsilon' < 0$. The detailed dispersion behaviours differ significantly for the two cases. This is easiest seen by plotting $\omega(q)$ in the complex plane, after normalization by the approximate *capillary wave* frequency

(3). Capillary wave behaviour falls about $(1,1)$ in the complex plane; the dilatational waves are wider-ranging. Fig. 1 shows the variations expected: full lines correspond to conditions such that mode mixing occurs, while the chain and dashed lines are successively further removed from such conditions. Plotting experimental data in this way should enable identification of the route to mode mixing, as the two routes appear quite different.

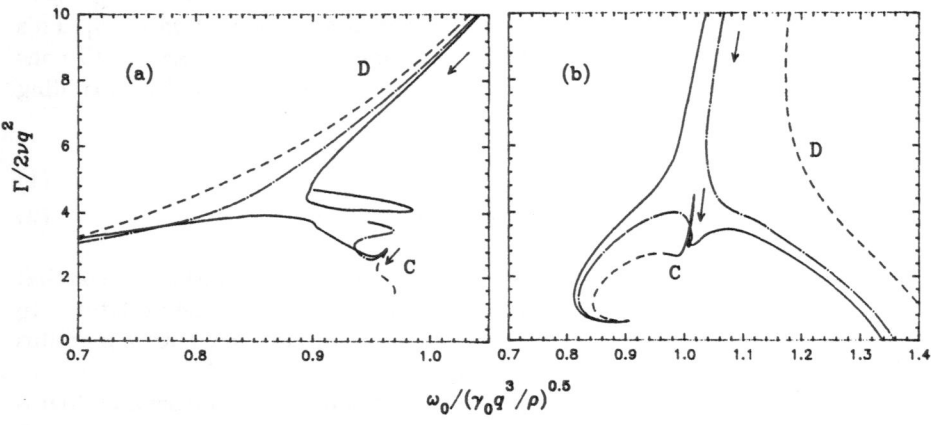

**Fig. 1.** Theoretical capillary (C) and dilatational (D) wave dispersion ($100 < q < 4000\,\mathrm{cm}^{-1}$, $q$ increasing as arrows) for two cases; data normalised by the approximate *capillary* wave values. (a) Effects of $\gamma'$: $\gamma_0 = 65$ mN/m, $\epsilon_0 = 15$ mN/m, $\epsilon' = 0$, and $\gamma' = 0$ (dashed), $2 \times 10^{-5}$ (chain), and $3 \times 10^{-5}$ mN s/m (solid). (b) Effects of $\epsilon'$: $\gamma_0 = 70$ mN/m, $\gamma' = 0$ mN s/m, $\epsilon' = -2 \times 10^{-5}$ mN s/m and $\epsilon_0 = 15$ (dashed), 13 (chain), and 12.5 mN/m (solid).

## 3 Experimental

Our surface scattering set-up has been described elsewhere [11]. We measure correlation functions which are the Fourier transforms of the power spectra of thermally excited capillary waves of the experimental $q$. Complex frequencies ($\omega_0 + i\Gamma$) of the waves can be determined [13]. A second analysis [14], based on the theoretical power spectrum of thermally excited capillary waves on a fluid surface [9] evaluated as a function of $\gamma_0$, $\gamma'$, $\epsilon_0$ and $\epsilon'$, estimates these surface properties directly. This direct analysis is appropriate for pure fluids and insoluble monolayers [14,15]. It should also apply to adsorbed monolayers involving only processes (e.g. diffusive interchange) such as can be fully incorporated into the formalism. However, if processes occur in surfactant solutions which are not encompassed by the conventional theoretical framework then the functional form fitted to the data will be inappropriate.

# 4 Results and Discussion

For both anionic and cationic surfactants, fitted values of $\epsilon'$ are systematically negative. A representative data set [16] for cetyltrimethylammonium bromide (CTAB) is shown in Fig 2. For $c \leq 0.09$ mM, $\epsilon'$ varies relatively smoothly with $\ln(c)$, increasing from about $-2 \times 10^{-5}$ mN.s/m to zero at the resonance between the capillary and dilatational waves ($c \sim 0.085$ mM). At and above 0.1 mM, the average $\epsilon'$ appears essentially constant at about $-6 \times 10^{-5}$ mN.s/m. Such systematically negative values of $\epsilon'$ have never been observed for surface light scattering from insoluble amphiphile films [15,17], for which the occasional negative value was always compatible with zero within errors and always lay in regimes where $\epsilon'$ was of rather small magnitude. We believe that the systematically negative values of $\epsilon'$ for the present *soluble* surfactants represent a real effect.

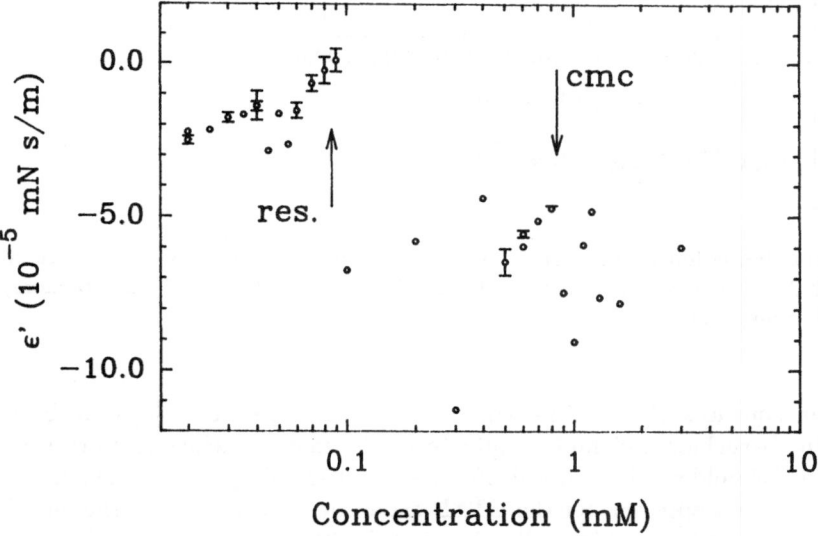

**Fig. 2.** The dilatational surface viscosity as a function of the CTAB concentration. The arrows indicate the surface wave resonance and the cmc.

The dilatational surface viscosity, $\epsilon'$, directly reflects dissipative effects within the system which increase the damping of the dilatational waves [12]. Negative values of $\epsilon'$ must thus correspond to *reduced* damping of the dilatational surface mode from the value appropriate to $\epsilon' = 0$.

We consider the present negative values of $\epsilon'$ as *effective* values only. They are determined by fitting observed correlation functions with the Fourier transform of a particular power spectrum of thermally excited capillary waves. If processes which cannot be accounted for by constitutive equations for the surface moduli occur, then the present direct analysis will involve force-fitting an inappropriate form to the data. Indeed Fig. 3 shows such a fit to a correlation function observed for a complex mixed solution, primarily comprising triethanolamine stearate. Such poor fits are never observed for pure fluids or insoluble monolayers. In such circumstances some or all of the fitted parameters may not be determined correctly. We believe that $\epsilon' < 0$ is such a case: some process or processes affecting the damping of the dilatational waves is not properly incorporated in the spectrum and so only *effective* values of $\epsilon'$ are determined.

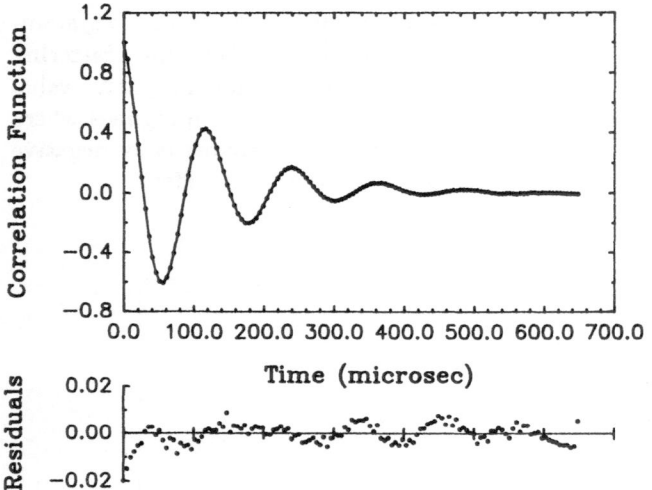

**Fig. 3.** A correlation function observed for an ionic surfactant system, with the best fit using the FT of the accepted power spectrum. The residuals are clearly non-random, indicating the inadequacy of the spectrum used.

The consequences of $\epsilon' < 0$ for the diffusive exchange model [8] have been explored by Earnshaw and McLaughlin [18]. The main conclusions were that both $\epsilon_0$ and $\epsilon'$ should exhibit marked and rather unusual frequency dependences. While this was not apparent for the CTAB experiments, the data for the mixed solution are in very reasonable qualitative accord with these predictions (Fig. 4), indicating the consistency of the behaviour of the negative $\epsilon'$.

The negative values of $\epsilon'$ have consequences for the observed capillary wave dispersion on solutions of ionic amphiphiles. In particular, the data for 0.04 mM CTAB constitute the first experimental observations of mode mixing for liquid surface waves [19].

Detailed analysis of the data has established the nature of the route to mixing. The dispersion behaviour is quite different for the two known routes to mode mixing (Fig. 1). Fig. 5 shows the measured frequency and damping values for a 0.04 mM CTAB solution, appropriately normalised. Given that we can only detect transverse or capillary waves, the signal of manifest mode mixing induced by $\epsilon' < 0$ will be normalised $(\omega_0, \Gamma)$ data which start near $(1,1)$ for low $q$, move

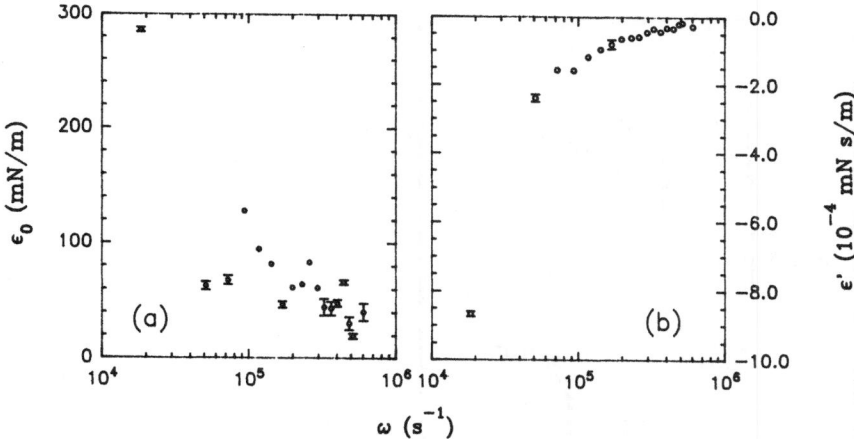

**Fig. 4.** The frequency dependence of the dilatational modulus $\epsilon$ for a mixed surfactant solution (see text for details).

down and then, above some critical $q$ value, jump discontinuously to $\Gamma > 1$ before moving down and left (Fig. 1b). This essentially describes Fig. 5 exactly. The expected behaviour for mode mixing induced by $\gamma'$ is quite different (Fig. 1a). The 0.04 mM CTAB data are thus consistent with the observation of mode mixing, and in particular are of the form predicted for $\epsilon' < 0$. As the CTAB concentration decreases from $\sim 0.1$ to 0.04 mM the data pass through a sequence exactly as sketched in Fig. 1b as the system approaches the conditions necessary for the appearance of manifest mode mixing [16].

In summary, the observed capillary wave dispersion for low concentration solutions of ionic surfactants is entirely consistent with theoretical predictions based on the accepted surface wave dispersion equation incorporating negative effective values of the dilatational surface viscosity.

Recent theoretical developments [3,4] suggest that in surfactant solutions various processes may compete with diffusional exchange between surface and bulk. This competition may result in the reduction of the damping of the surface modes, particularly the dilatational waves, the effects of different processes appearing at different wave frequencies. For example, the effects of an adsorption barrier are expected to be most apparent at frequencies $\sim 10^5$ s$^{-1}$ [4], close to the value at which the perturbations in the present dispersion data occur [16]. Recent experiments on solutions of decanoic acid acting as a non-ionic surfactant reveal none of the peculiarities observed for *ionic* amphiphiles. This strongly suggests a role for electrostatic effects in the underlying mechanism.

The considerations of dilatational wave stability [3,4] are framed in terms of the Marangoni elasticity number ($M$) and frequency ($\omega_0$). In the absence of appreciable volume-to-surface transport $M$ relates directly to $\epsilon_0$ [3]:

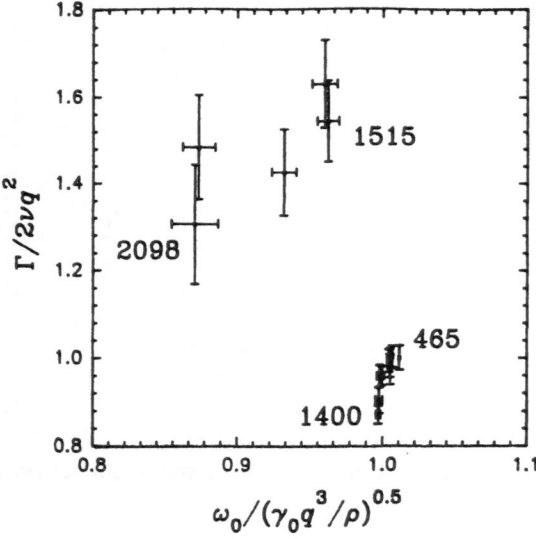

**Fig. 5.** The 0.04 mM CTAB dispersion data normalised and plotted in the complex plane. The surface wave number varies from 465 to 2098 cm$^{-1}$: the numbers indicate $q$ values corresponding to selected points.

$$M = -\frac{\epsilon_0 \gamma_0 q}{\rho g \eta D} \sqrt{1 + \frac{i \omega \rho}{\eta q^2}}$$

where $D$ is the diffusion coefficient of the solute.

Fig. 6 shows the trajectory traced out by the experimental data for 0.04 mM CTAB in the $M - \omega_0$ plane, together with the locus of neutral stability of the dilatational mode [4]. While this locus depends somewhat on the parameters of the theory, it appears that the trajectory of the experimental data does approach it, and that over the frequency range studied some degree of reduced stability of the dilatational waves should be apparent. A further phenomenon is required to cause the observed gross consequences of this upon the capillary wave dispersion: the two surface modes must be relatively close to resonance. Under these conditions the mode coupling is so strong that changes in the dilatational wave stability can have marked consequences for the capillary waves.

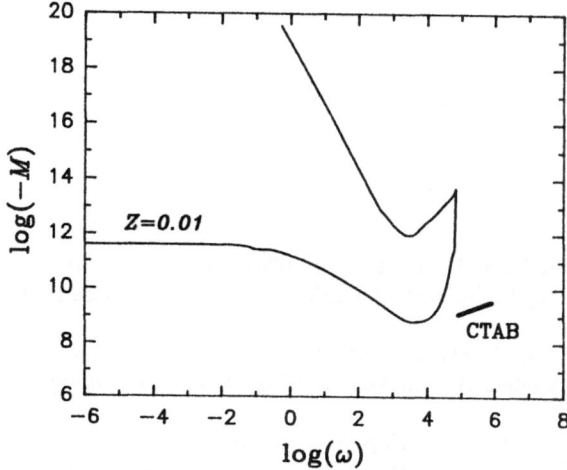

**Fig. 6.** The locus of marginal stability for the longitudinal waves, adapted from Fig. 4 of [4]. The heavy line indicates the trajectory followed experimentally (see text). $Z$ is the ratio of the time scales for adsorption and diffusive exchange.

## 5 Conclusions

The primary conclusion is that there exists some process (or processes) at the surfaces of solutions of ionic surfactants which reduces the damping of the dilatational waves. This is apparent in both the observed complex capillary wave frequencies and the negative effective values of the dilatational surface viscosity. The latter are interpreted here as reflecting a reduction in the damping of the corresponding surface mode compared to its value when $\epsilon' = 0$.

While there have been predictions of impaired stability of dilatational waves on surfactant solutions [3,4], the underlying mechanism in the present cases is not yet clear. However, the perturbation of the observed capillary wave propagation from accepted trends seems to occur at wave frequencies at which effects of adsorption/desorption kinetics would be significant. As noted above, none of the effects described for ionic surfactants are observed for non-ionic amphiphiles, suggesting an electrostatic origin.

174

# Acknowledgements

This work has been supported by the AFRC and SERC.

# References

[1]     Lamb H. *Hydrodynamics*; New York: Dover, 1945; pp 626–628.
[2]     Levich B. *Physicochemical Hydrodynamics*; Englewood Cliffs: Prentice-Hall, 1962; Chap. 11.
[3]     Chu X.-L. and Velarde, M.G. *Physicochem. Hydrodynamics* 1988, *10*, 727.
[4]     Hennenberg, M., Chu, X.-L., Sanfeld A. and Velarde M.G. *J. Colloid Interface Sci.* 1992, *150*, 7.
[5]     Earnshaw J.C. and McLaughlin, A.C. *Proc. Roy. Soc. London A* 1991, *433*, 663.
[6]     Earnshaw J.C. and McLaughlin, A.C. *Proc. Roy. Soc. London A* 1993, *440*, 519.
[7]     Langevin, D. *Light Scattering by Liquid Surfaces and Complementary Techniques*; Dekker: New York, 1992.
[8]     Lucassen, J. and van den Tempel, M. *Chem. Eng. Sci.* 1972, *27*, 1283.
[9]     Langevin D. *J. Colloid Interface Sci.* 1981, *80*, 412.
[10]    Kramer L. *J. Chem. Phys.* 1971, *55*, 2097.
[11]    Earnshaw, J.C. and McGivern, R.C. *J. Phys. D* 1987, *20*, 82.
[12]    Lucassen-Reynders E.H. and Lucassen J., *Adv. Colloid Interf. Sci.* 1969, *2*, 347.
[13]    Earnshaw, J.C. and McGivern, R.C. *J. Colloid Interface Sci.* 1988, *123*, 36.
[14]    Earnshaw, J.C., McGivern, R.C., McLaughlin A.C. and Winch P.J. *Langmuir* 1990, *6*, 649.
[15]    Earnshaw J.C. and Winch P.J. *J. Phys: Condens. Matter* 1990, *2*, 8499.
[16]    Earnshaw, J.C. and McCoo, E. *Langmuir*, Mar. 1995.
[17]    Earnshaw, J.C., McGivern, R.C. and Winch P.J. *J. Phys. (Paris)* 1988, *49*, 1271.
[18]    Earnshaw J.C. and McLaughlin, A.C. *Progr. Colloid Polym. Sci.* 1989, *79*, 155.
[19]    Earnshaw J.C and McCoo, E. *Phys. Rev. Lett.* 1994, *72*, 84.

# The Capillary Pressure Method: a New Tool for Interfacial Tension Measurements

L. Liggieri[1,a], F. Ravera[1], A. Passerone[1], A. Sanfeld[2] and A. Steinchen[3]

[1] ICFAM CNR-Istituto di Chimica Fisica Applicata dei Materiali,
via De Marini 6, 16149 Genova, Italy.

[2] Univ. Paris VII, LISA, Jussieu, 75351 Paris Cedex 05, France.

[3] Univ. Aix-Marseille III, Lab. Thermodynamique, Fac. St. Jerome,
13397 Marseille Cedex 20, France.

## 1. Introduction

The measurement of the interfacial tension of liquid-liquid or liquid-gas systems is an important tool for the study of surface properties. In particular, when a surfactant solute is present, accessing this quantity and its time variation provides important information about the surfactant, the surface and their interactions.

From a theoretical point of view, adsorption kinetics has been recently studied and a new approach has been formulated in order to give a general description of this phenomenon taking into account diffusion, transfer of mass at the surface and presence of adsorption potential barriers [1 ,2 ,3 ]. This theory can be used to treat experimental data and to point out important characteristics of the system under study.

Dynamic interfacial tension measurements represent one of the most powerful tools through which interfaces can be studied during their evolution with time.

Today a large number of experimental approaches exists to collect dynamic interfacial tension data under different experimental conditions. Several methods deal with drops and bubbles continuously formed as, for example, the dynamic maximum bubble pressure [4 ,5 ] or dynamic drop volume: a single interfacial tension value is drawn from each bubble or drop formed. Another class of methods is based on the evaluation of dilational rheologic properties of the surfaces. These methods may use oscillating bubbles [6 ] or Langmuir-Blodget-like balances [7 ,8 ,9 ]. The dynamic interfacial tension can also be measured by methods based on the acquisition of the drop (or bubble) profile [10 ], due to recent improvements in the techniques of image elaboration and in the automatic calculus algorithms.

In the last years, the development of pressure transducers apt to measure very low pressures in liquids offered the possibility to measure surface tensions by the direct measurement of the capillary pressure, opening a new class of experimental methods: the Capillary Pressure methods [11 , 12 ]. At present, these methodologies are applied more and more [13 ,14 , 15 ,16 ] and they seem to be sufficiently reliable to collect accurate data.

One of the improvements offered by these methods is the fact that the measurements are done on static or quasi-static interfaces, so that surfaces are not stressed and the distribution of surfactant inside the bulk is not affected by external disturbances. Moreover, the evolution of the surface tension can be followed on a single surface and with a high sampling rate.

In this paper we treat in particular the Pressure Derivative [11] and the Expanded Drop methods [13]. Applications of this latter to the study of adsorption kinetics at liquid-liquid interfaces are also discussed.

## 2. Experiments

The Capillary Pressure techniques are based on the direct measurement of the pressure difference existing across the curved interface between two immiscible fluids. This is the capillary pressure, P, which is linked to the interfacial tension, $\sigma$, through the Laplace equation which, for a spherical interface with radius R, reads

$$P = \frac{2\sigma}{R} \tag{1}$$

The experimental apparatus (Fig. 1) is composed by a measurement cell in which a drop of a liquid 1 is grown on the tip of a suitably shaped capillary tube inside

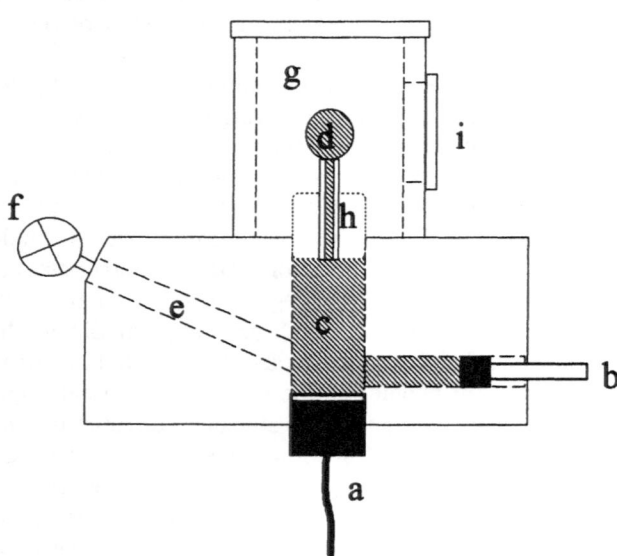

**Fig. 1:** Sketch of the experimental cell. a) pressure transducer; b) injection system; c) liquid 1 reservoir; d) drop of liquid 1; e) gas reservoir; f) gas injection valve; g) liquid 2; h) nozzle; i) optical window.

another liquid 2. The typical nozzle radius ranges between 0.1 and 0.5 mm. The volume of the drop is controlled by an injection system (motor plus piston). The liquid forming the drop is in contact with a pressure transducer (the other liquid phase is usually at constant pressure) which makes it possible to acquire a pressure signal during the experiment. Moreover, if necessary, the drop shape can be monitored through optical windows by a video camera. The whole apparatus is contained in a thermostatic chamber allowing the temperature to be controlled with a precision of 0.1 °C.

This experimental apparatus is used in two different modes: the Pressure Derivative (PD) mode [11] and the Expanded Drop (ED) mode [13] apt to measure respectively the static interfacial tension of pure liquids and the dynamic interfacial tension when at least one of the two liquids is a surfactant solution.

As a matter of fact, these two methods are suitable only for drops with zero Bond Number (in our case spherical caps):

$$Bo = \frac{\Delta \rho g d^2}{\sigma} = 0 \qquad (2)$$

where $\Delta \rho$ is the difference of density across the interface, g the acceleration of gravity and d the typical drop dimension. This condition can be matched either working with liquids of the same density, or by performing experiments in weightless (microgravity: $g \approx 0$) conditions. Indeed, both techniques have been tested onboard spacecrafts. However, as deviation from the sphericity are small, these methods have been proved to be reliable also in normal laboratory conditions, up to Bo values of the order of $10^{-1}$ which can be obtained by working on small drops (R<0.8 mm) and differences of density less than about 0.1 g/cm$^3$. Higher values of Bo up to 0.5 can also be considered by introducing some simple corrections [13].

## 2.1 The Pressure Derivative Method
The Pressure Derivative method is based on the fact that the Laplace Equation states a linear dependence between the capillary pressure and the curvature of the interface. Thus, by a linear fit of Eq. (1) of the capillary pressure vs. curvature data collected during the growth of a drop, the interfacial tension can be calculated as a half of the slope.

In practice, a drop of liquid 1 is grown at constant flow rate inside the liquid 2, while the capillary pressure is recorded. The capillary pressure versus time signal is converted into a set of capillary pressure vs. curvature data, by using the relationship between the injected liquid and the volume of the drop,

$$w(t - t_0) = \frac{2}{3}\pi(R^3 - \alpha^3) + \frac{\pi}{3}\sqrt{R^2 - \alpha^2}(2R^2 + \alpha^2) + V_0\left(\frac{P - b}{P_T + b}\right) \qquad (3)$$

where w is the flow rate ( typically $8 \cdot 10^{-5}$ cm$^3$/s) $\alpha$ is the base radius of the drop (R=$\alpha$ corresponds to a hemispherical drop), t the time and $t_0$ the time of the hemispherical drop (maximum of the pressure signal: see fig. 2). The last term of the right hand side takes into account the expansion of any volume of gas possibly trapped inside the liquid forming the drop, $V_0$ being the volume of such a gas at the maximum capillary pressure $b=2\sigma/\alpha$, P the capillary pressure and $P_T$ the total (atmospheric) pressure.

In order to apply Eq. (3) the parameters w, $t_0$, $\alpha$ and $V_0$ have to be known.

It is worth noting that $\alpha$, the effective base radius of the drop, due to wetting effects, can be slightly different from the nominal radius of the nozzle tip.

The flow velocity, w, is determined with a satisfactory precision by a direct measurement and $t_0$ is evaluated with a sufficient precision by fitting a 3rd degree polynomial to the point of the signal around the maximum. A direct and safe measurement of $\alpha$ and $V_0$ is not possible: however, as only the right values of these parameters can render the linear relation between the capillary pressure and the curvature, imposed by the Laplace Equation, the values of $\alpha$ and $V_0$ are determined, as those which maximise the correlation coefficient in the linear fitting of the capillary pressure vs. curvature data. Further details on this procedure can be found in ref. [11].

Moreover, the PD method uses only pressure variations. Thus no information are needed about the pressure offset, $P_{off}$, which in these kinds of experimental set-ups is always measured together with the capillary pressure. In the PD method, $P_{off}$ is merely a product of the data elaboration (shift of the fitting straight line).

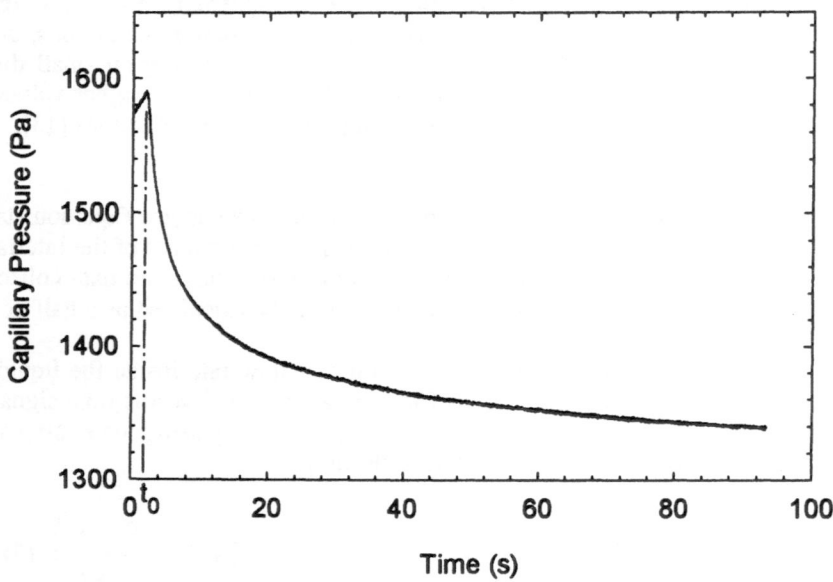

**Fig. 2**: Typical pressure signal recorded using the Pressure Derivative method. At time t (maximum in the signal), the drop is exactly hemispherical.

**Table 1:** Measured values of the interfacial tension of some organic liquid vs. water by the Pressure Derivative method. Reference values from literature are also reported for comparison purpose.

| Organic liquid phase | Interf.Tension (dyne/cm) | Stand. Deviation (dyne/cm) | Ref.values (dyne/cm) |
|---|---|---|---|
| Hexane | 50.3 | 0.2 | 49.4 ÷ 50.9 |
| Paraffin Oil | 51.0 | 0.9 | 49.0 |
| Dioctylphtalate | 29.5 | 0.4 | - |
| Benzene | 34.1 | 0.3 | 33.3 ÷ 34.9 |

Table 1 reports some data of interfacial tension of pure liquids obtained by the PD method. These values are the average of 10-15 experimental runs. The absolute values of $\sigma$ are in good agreement with the literature and the values of the standard deviations show a reproducibility of the measurements ranging between 0.5 and 1.5%.

## 2.2 The Expanded Drop Method

When dealing with surfactant solutions, $\sigma$ cannot be considered a constant parameter in Eq. (1) because it depends on the surface excess concentration of the surfactant or adsorption ($\Gamma$). For example, if the surface is freshly formed, the value of $\sigma$ is close to that of the pure system, while, if the system had the time to reach the thermodynamic equilibrium, the value of $\sigma$ is the equilibrium one and it depends only on the surfactant bulk concentration.

In order to study adsorption kinetics it is useful to start from a configuration with a "fresh" surface and to follow the aging of the surface through the variation of the interfacial tension value.

The Expanded Drop (ED) method was conceived for this kind of measurements.

In the experimental cell described above, filled in the upper part with a surfactant solution, a fresh surface is obtained by a quick and large expansion of the drop. After that, the radius of the drop is kept constant and the capillary pressure is recorded as a function of time. The "dynamic" surface tension is then obtained from the capillary pressure signal by applying the Laplace Equation (1) to get the instantaneous interfacial tension values.

An important point concerns the way to obtain the rapid expansion of the drop for the formation of the fresh surface.

Very good results were obtained by utilising the particular effect [17] that a gas volume $V_0$ inside the liquid 1 has on the growth of the drop. Indeed, under particular conditions, the presence of this gas volume provides an unstable state for the system: when the drop reaches the hemispherical shape a very rapid expansion of such gas volume takes place, causing a corresponding variation of the drop volume.

The mathematical modelling of a growing drop/gas-bubble system [17] shows that the condition for the onset of this instability, in terms of the dimensionless "Bubble Stability number" (BSN), is:

$$BSN = \frac{27\pi P_T \alpha^4}{8\sigma V_0} < 1 \qquad (4)$$

where $P_T$ is the external pressure. The value of the BSN also determines the relative variation in the drop surface area obtained during the expansion: this is the most significant parameter for judging if the surface can be considered to be fresh at the initial time.

By such an effect a dilation of the surface of 50-100 times in 0.1-0.3 s. is easily obtained in the normal experimental conditions, providing surfaces that can be considered to be practically clean.

At variance with the PD, the ED method requires the knowledge of the pressure offset $(P_{off})$. This parameter, necessary to obtain the capillary pressure from the measured pressure signal, can be calculated, by knowing the drop area after the expansion, as described, with other experimental and computational details, in Ref. [13].

A main advantage of this experimental methodology is that adsorption occurs at an interface which is mechanically at equilibrium. Moreover, in a single experiment it is possible to collect a large amount of dynamic interfacial tension data with an error lower than 1% and with a sampling rate of 20-40 Hz. This makes this method very powerful for testing current theories and to develop new ideas about adsorption kinetics.

However, due to the time needed to expand the surface, only adsorption kinetics with characteristic times larger than few seconds can be studied by this method.

## 3. Adsorption Kinetics: Results and Discussion

We present here, as an example, some selected results in order to illustrate the potentiality of the ED method and the theoretical approach used to treat the data acquired in these experiments.

Dynamic interfacial tensions of the system Hexane-TritonX100 aqueous solution were measured at two sub-micellar concentrations: $c=7.7210^{-9}$ mol/cm$^3$ and $c = 2.04\ 10^{-8}$ mol/cm$^3$ at T = 20 ± 0.1 °C.

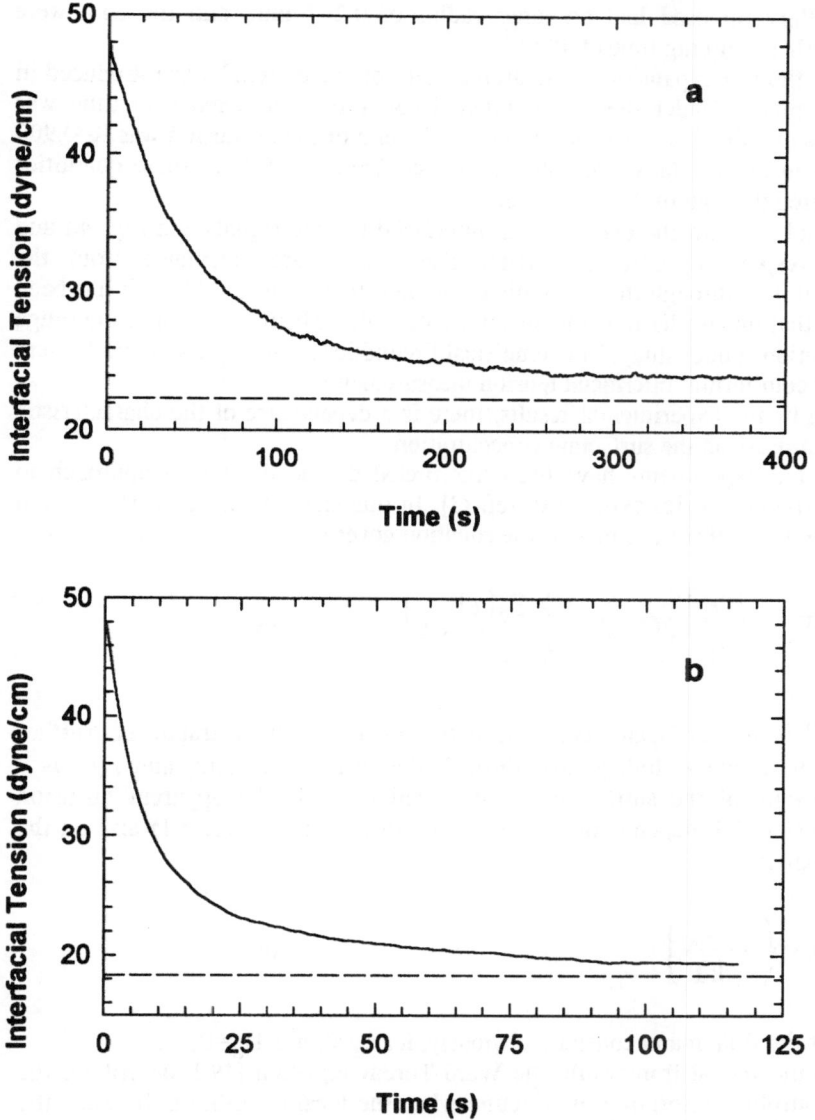

**Figs 3**: Dynamic interfacial tension data (continuous line) and corresponding equilibrium values (dashed line) of hexane vs. aqueous Triton X-100 at different surfactant concentrations. a) $c=7.72 \ 10^{-9} \ mol/cm^3$; $\sigma_{eq}=22.34$ dyne/cm. b)$c=2.04 \ 10^{-8} \ mol/cm^3$; $\sigma_{eq}=18.34$ dyne/cm.

The capillary tip used had an inner radius of 0.247 mm, and the data were acquired with a sampling time of 40 ms.

To obtain the fast expansion of the drop an air volume of 1cm$^3$ was introduced in the hexane phase. Under these experimental conditions, the expansion time was $t_{exp}$= 0.24 s and the drop curvature radius at the end of the expansion was R=0.900 mm, which means a relative variation of the surface area of 24.5 (these quantities were measured through the video image).

Figs 3a and 3b show the experimental interfacial tension signals and Figs 4a and 4b the corresponding adsorption data. These latter are calculated from the experimental data through the Freundlich equation. Indeed, in ref. [18], it has been established that this isotherm is the most reliable to describe this system in the range of concentrations under study. The empirical Freundlich's parameters m and K were obtained by equilibrium interfacial tension measurements .

As shown by the experimental results, there is a dependence of the characteristic time of the process on the surfactant concentration.

The reported experiments have been interpreted by the theoretical approach to mixed adsorption kinetics exposed in ref. [3]. In this study, it has been shown that for a large number of surface models the equation governing adsorption kinetics is

$$\int_0^\Gamma \frac{1}{g(\Gamma)} d\Gamma = \sqrt{\frac{D_a}{\pi}} \left[ 2C^\infty \sqrt{t} - \int_0^t \frac{C_0(\tau)}{\sqrt{t-\tau}} d\tau \right] \qquad (5)$$

where $c^\infty$ is the surfactant bulk concentration, $c_0$ is the extrapolated surface concentration, which is linked to $\Gamma$ through the surface isotherm, and $g(\Gamma)$ is a function specific of the surface model considered. $D_a$ is the apparent diffusion coefficient [1] which depends on the surfactant diffusion coefficient D and on the adsorption barrier $\varepsilon_a$:

$$D_a = D \exp\left( -\frac{2\varepsilon_a}{kT} \right) \qquad (6)$$

where k is the Boltzmann constant; obviously, for $\varepsilon_a$=0, it is $D_a$=D.

In such a theoretical framework, the Ward-Torday equation [19], describing the diffusion controlled adsorption, is re-found when the local equilibrium between the sub-surface and the surface is imposed.

In general, the $\Gamma(t)$ obtained from Eq. (5) differs from the Ward-Tordai equation also when $\varepsilon_a$=0 ($D_a$=D). Thus, the theory foresees the possibility of mixed kinetics also in the absence of potential barriers to adsorption.

However, for $g(\Gamma)$=1 and for $\varepsilon_a$=0, Eq. (5) reduces to the Ward-Tordai equation: in this case the distinction between an adsorption process with $\varepsilon_a$=0 and a diffusion controlled one no longer exists. These latter models are those in which the rate of adsorption does not depend on the surface coverage; the Freundlich model, describing the surfactant solution under study, is just one of them.

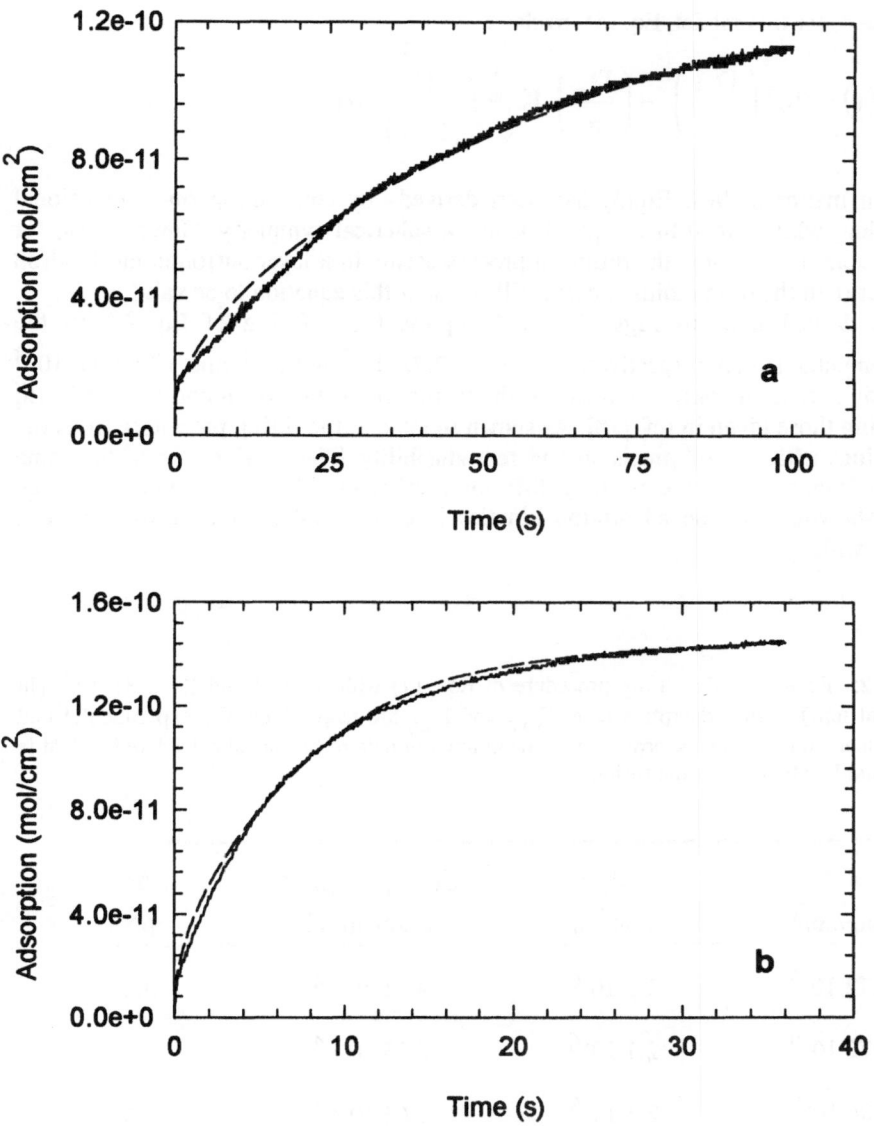

**Figs 4**: Dynamic adsorption data (continuous line) of hexane vs. aqueous Triton X-100 and corresponding fitting curve of Eq. (7) (dashed line) at different surfactant concentration. a) $c = 7.72 \ 10^{-9} \ mol/cm^3$. b) $c = 2.04 \ 10^{-8} \ mol/cm^3$.

Indeed, for this model, Eq. (5) reads:

$$\Gamma(t) = 2c^{\infty}\left(\frac{D_a t}{\pi}\right)^{\frac{1}{2}} - \left(\frac{D_a}{\pi}\right)^{\frac{1}{2}} K^{-\frac{1}{m}} \int_0^t \frac{\Gamma^{\frac{1}{m}}}{(t-\tau)^{\frac{1}{2}}} d\tau \qquad (7)$$

As a matter of fact, Eq.(5) has been derived by considering one-dimensional diffusion, while our diffusion problem has a spherical symmetry. However, as, for the system under study, the diffusion process occurs in a layer surrounding the drop smaller than the drop radius, we can still consider this equation to be valid.

The dashed lines in Figs. 4 and 5 represent the fitting of Eq. (7) to the experimental signal respectively for $c^{\infty}$= 7.72 $10^{-9}$mol/cm$^3$ and $c^{\infty}$=2.04 $10^{-8}$ mol/cm$^3$. $D_a$ is the only parameter of the fitting, the values of m and K used in Eq (7) being those given in ref. [18]. As shown in table 2, the fitting procedure finds the $D_a$ values with a good precision and reproducibility. These values are of the same order of magnitude as the ordinary diffusion coefficient of Triton X-100 in water, D, [20 ] showing that the adsorption kinetics is either controlled by diffusion or it occurs with $\varepsilon_a$=0.

**Table 2**: Results of the fitting procedure of Eq. (7) with m=0.12 and K= 1.31 $10^{-9}$ (in coherent unit) to the adsorption data. $\Gamma_{exp}$ and $\Gamma_{teo}$ are respectively the experimental and theoretical values of the adsorption, n is the number of data of the signal considered and $\Delta t$ is the signal length used for the fitting.

| $c^{\infty}$ (mol/cm$^3$) | $D_a$ (cm$^2$/s) | $\Sigma(\Gamma_{exp}-\Gamma_{teo})^2/n$ (mol/cm$^2$)$^2$ | $\Delta t$ (s) |
|---|---|---|---|
| 7.72 $10^{-9}$ | 1.9 $10^{-6}$ | 4.11 $10^{-24}$ | 100 |
| 7.72 $10^{-9}$ | 2.1 $10^{-6}$ | 5.35 $10^{-24}$ | 100 |
| 2.04 $10^{-8}$ | 2.3 $10^{-6}$ | 1.64 $10^{-23}$ | 36 |
| 2.04 $10^{-8}$ | 2.4 $10^{-6}$ | 1.70 $10^{-23}$ | 36 |
| 2.04 $10^{-8}$ | 2.5 $10^{-6}$ | 2.50 $10^{-23}$ | 36 |
| 2.04 10-8 | 2.3 $10^{-6}$ | 7.76 $10^{-24}$ | 36 |

# References

[1] F. Ravera, L. Liggieri and A. Steinchen , J. Colloid and Interface Sci. , **156**, 109 (1993).

[2] F. Ravera, L. Liggieri, A. Passerone and A. Steinchen, J. Colloid Interface Sci., **163**, 309 (1994).

[3] L. Liggieri, F. Ravera and A. Passerone, submitted to J. Colloid and Interface Sci. (1995).

[4] X.Y. Hua and M.J. Rosen, J. Colloid Interface Sci., **124**, 652 (1988).

[5] K.J. Mysels, Colloids and Surfaces, **43**, 241 (1990).

[6] K.D. Wantke, K. Lunkenheimer and C.J. Hempt, J. Colloid Interface Sci., **159**, 28 (1993).

[7] J. Lucassen and D. Giles , J. Chem. Soc. Faraday Trans.1, **71**, 217 (1975).

[8] R. Miller, G. Loglio, U. Tesei and K.H. Schano, Adv. Colloid Interface Sci., **37**, 73 (1991).

[9] I. Panaiotov , A. Sanfeld, A. Bois and J.F. Baret , J. Colloid Interface Sci., **96**, 315 (1983).

[10] F. Boury, Tz. Ivanova, I. Panajotov, J.E. Proust, A. Bois and J.J. Richou, J. Colloid and Interface Sci., **169**, 380 (1995) .

[11] A. Passerone, L. Liggieri, N. Rando, F. Ravera and E. Ricci, J.Colloid and Int.Sci., **146**,152 (1991)

[12] F. Ravera, L. Liggieri, A. Passerone and A. Steinchen, in "Proceedings of the first European Symposium on Fluids in Space", ESA SP-353,p.213 (1991).

[13] L. Liggieri, F. Ravera and A. Passerone, J. Colloid Interface Sci., **169**, 226 (1995).

[14] C.A. MacLeod and C.J. Radke, J. Colloid Interface Sci., **160**,435 (1994).

[15] R. Nagarajan and D.T. Wasan, J. Colloid Interface Sci., **159**, 164 (1993).

[16] K.D. Wantke , R. Miller and K. Lunkenheimer, Z.Phys. Chem., **261**, 1177 (1980).

[17] L. Liggieri, F. Ravera and A. Passerone, J. Colloid and Interface Sci., **140**, 436 (1990).

[18] L. Liggieri, F. Ravera and A. Passerone, J. Colloid Interface Sci., **169**, 238 (1995).

[19] A.F.H. Ward and L. Tordai, J.Chem.Phys., **14**, 453 (1946).

[20] J. Van Hunsel, G. Bleys and P. Joos, J. Colloid and Interface Sci., **114** , 432 (1986).

# Effects of Evaporation-Condensation on Thermocapillary Convection

G.PETRE[1] and    A.AZOUNI[2]

[1] U.L.B.,Chimie-Physique E.P.(C.P.165) 50 av. F.D.Roosevelt,1050 Brussels,Belgium.
[2] Lab.d'Aerothermique, 4 ter Route des Gardes F92190 Meudon,France.

## 1 Introduction

For some systems consisting of two fluid phases in contact with an interface (liquid/liquid or liquid/gas) the equilibrium surface tension as a function of the temperature presents an extremum. At the temperature of this extremum the Marangoni number becomes zero

$$Ma = -\frac{\partial \sigma}{\partial T} \frac{\Delta T d}{\rho \nu \kappa}$$

This situation presents a great interest for the study of two kinds of systems with a free fluid interface.

1) Systems in which a temperature gradient is created perpendicular to the surface.
2) Systems in which a temperature gradient is established horizontally.

The typical example of the first kind is the "Bénard phenomenon". The temperature gradient is downwards. Usually the density and the surface tension decrease by increasing the temperature. Therefore the less dense fluid is near the bottom, it tends to exchange its place with the denser liquid which is near the surface.
· A local positive fluctuation of the suface temperature creates a small patch of lower a surface tension. The resulting expansion of the patch brings into the surface a fluid with higher temperature which lowers more the surface tension.For some threshold value of the temperature gradient the conductive heat transfer regime becomes unstable and a convective regime sets in.In this case both thermogravitational and thermocapillary forces contribute to destabilise the conductive regime. When the surface tension increases with temperature, the thermocapillary forces counteract the thermogravitational ones and have a stabilising effect. An upward vertical temperature gradient can then have a destabilising effect [1].
The second kind of systems differs from the first one by the absence of a temperature gradient threshold for the induced convection. We are mainly interested in the latter systems.

## 2 Studied systems

The studied systems are aqueous solutions of fatty alcohol. They present a minimum of the equilibrium surface tension ($\sigma_e$) versus temperature. For normal alcohols the temperature ($T_M$) for the minimum of $\sigma_e$ decreases with the increase of the number of carbon atoms from 70°C for n-butanol to 10°C for n-dodecanol.

For each alcohol the concentration must exceed a certain value in order that a minimum exists. From this concentration up to saturation the temperature of the minimum changes only in a weak way with concentration. An example of $\sigma_e$(T) curve is given in fig.1.

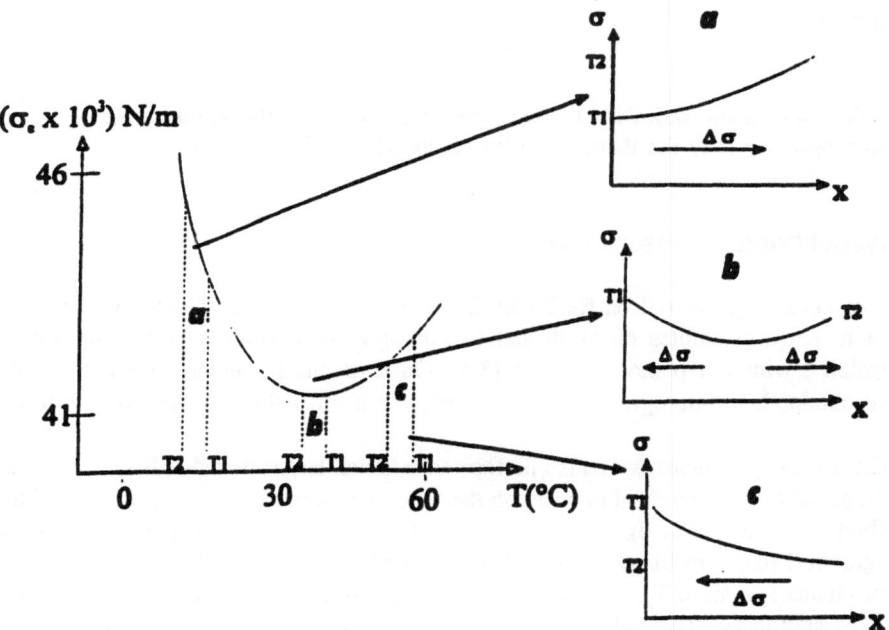

**Figure 1.** Aequous solution of n-heptanol 0.00624 molal

## 3 Expected observations on systems with minimum in $\sigma_e$(T) curves

On fig.1 we indicate the expected sense of the surface movements caused by the indicated surface temperature gradients. Numerical simulations, using a parabolic function for $\sigma$(T) were performed by several authors [2].

When a temperature gradient is established between the lateral walls of a vessel as indicated on fig.2 we expect that:

1) if $T_2 < T_1 < T_M$ thermocapillary and thermogravitational forces cooperate to sustain a clockwise convection roll

2) if $T_1 > T_2 > T_M$ thermocapillary forces tend to induce a counterclockwise convection roll and thermogravitational forces a clockwise one.

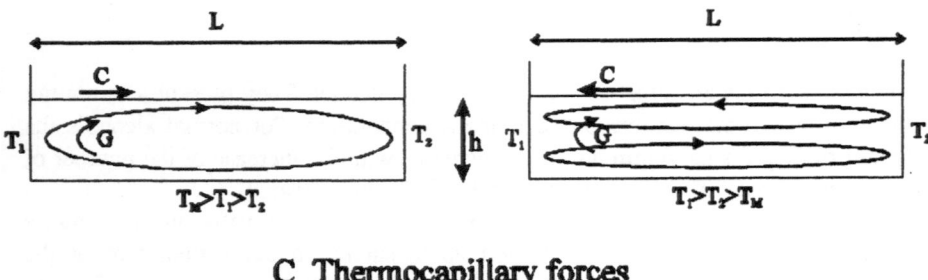

## C Thermocapillary forces
## G Thermogravitational force

**Figure 2**

For appropriate ratios h / L  two superposed rolls should be observed.
In microgravity only the thermocapillary induced roll should exist.

## 4 Experimental observations

a) Vessels as represented on fig.2 with h = 0.01 m,L = 0.03 m (depth 0.01 m) were
used to perform studies on earth and in microgravity ( Texus and D1 mission in
spacelab ) with n-heptanol solutions [3,4]. On earth the predicted superposed rolls
were obtained. In microgravity only the surface tension induced convection appears.

b) On earth experiments were also performed in much larger vessels (diameter = 0.18
m) to avoid any influence of menisci on the observed movements. The principle of the
method is illustrated in fig.3. In the bottom of a Pyrex vessel, two parallel channels
are grooved (0.08  m long). In one of the channels water at $T_1 > T_3$ circulates, in the
other channel water at $T_2 < T_3$. The resulting temperature gradient in the fluid becomes
quickly stationary. The velocity in the surface is detected by small talc particles
observed by means of a profile projector(for details refer to [5]).

189

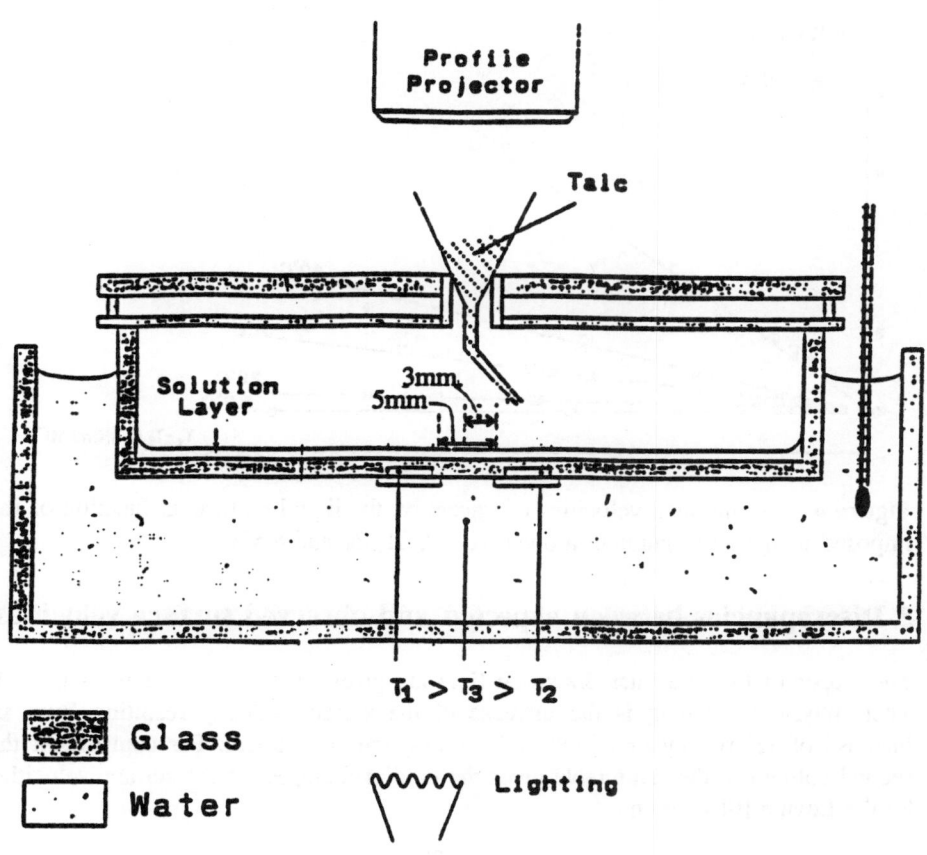

**Figure 3.** Experimental device

The velocities obtained experimentally with a 0.00624 molal aqueous solution of n-heptanol are summarised in fig.4. $v_3$ and $v_5$ are the velocities obtained by measuring the time necessary for a tracer to cover a distance of 0.003 or 0.005 m (cfr fig.3).

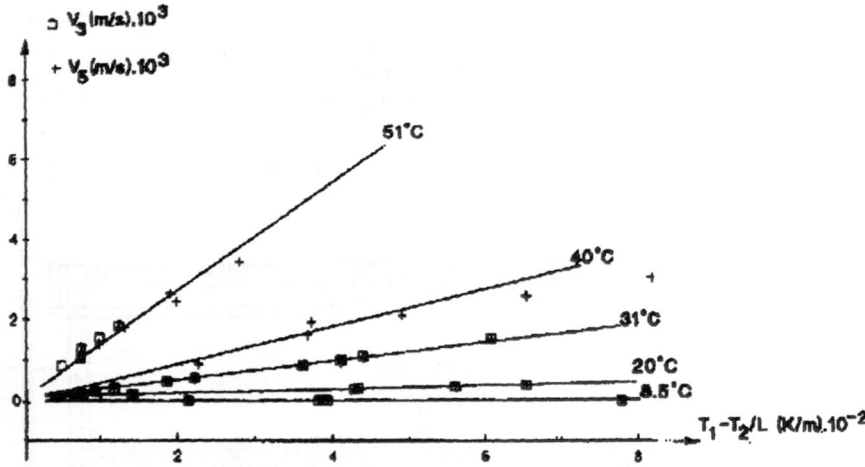

**Figure 4.** The measued velocities of tracers on the liquid surface as function of the imposed temperature gradient around 51, 40, 31, 20 and 8.5°C

## 5 Discrepancies between expected and observed surface velocities

The slopes of the five lines drawn on fig.4 are given in the first column of table 1. Their physical meaning is the increase of the surface velocity resulting from an increase of 1K per meter of the surface temperature gradient. For comparison the second column of the same table gives the predicted slopes for the surface velocities by the Levich [6] relation

$$v_s = \frac{h}{4\mu} \frac{\partial \sigma}{\partial T} \frac{\partial T}{\partial x}$$

in which $\partial T / \partial x = (T_1 - T_2)/12.2 \ 10^{-3}$ and $\partial \sigma / \partial T$ has been obtained from the $\sigma_e(T)$ curves for n-heptanol of ref.[7]. Owing to the great dilution of the alcohol, the viscosity of water was used for $\mu$. Negative values in table 1 corresponds to $\partial \sigma_e / \partial T$ or $\partial \sigma / \partial T < 0$ .

**Table 1**

| $T_3 °C$ | $\alpha = \nu \dfrac{12.2 \ 10^{-3}}{T_1 - T_2} m^2 sec^{-1} K^{-1}$ | $\nu_s \dfrac{12.2 \ 10^{-3}}{T_1 - T_2} m^2 sec^{-1} K^{-1}$ |
|---|---|---|
| 8.5 | $0.09 \ 10^{-6}$ | $-0.89 \ 10^{-4}$ |
| 20 | $0.5 \ 10^{-6}$ | $-0.77 \ 10^{-4}$ |
| 31 | $2.4 \ 10^{-6}$ | $-0.40 \ 10^{-4}$ |
| 40 | $4.6 \ 10^{-6}$ | $\sim 0$ |
| 51 | $13 \ 10^{-6}$ | $+0.65 \ 10^{-4}$ |

Comparison of the values in columns (1) and (2) in table 1 reveals two important discrepancies :

1) the temperature at which σ becomes minimum in the experiments is shifted from 30°C to lower values as compared to the value for the equilibrium surface tension.
2) the observed velocities are much lower than the predicted ones.

A further problem is that, with the used n-alcohol solutions, we have not yet succeeded to observe the "surface opening" expected with a gradient $(T_1-T_2)$ in the range $T_1 < T_M < T_2$ .

## 6 Causes of the discrepancies

Two phenomena can contribute to explain these discrepancies:
a) The actual temperature gradient may be lower than $(T_1-T_2)/\Delta x$, where $T_1$ and $T_2$ are measured in the water of the channels. This was established for a water layer of nearly the same thickness (0.002 m) in the same device. The temperature gradient on the surface measured by means of a bolometer was almost ten times smaller than $(T_1-T_2)/\Delta x$ [5].
b) Alcohol evaporates from the surface region of the highest temperature and condenses on the surface region of lowest temperature. The surface tension increases to values higher than $\sigma_e$ where the alcohol evaporates. The increase of σ by the evaporation of the alcohol was proved by Hommelen [8] and formulated theoretically by Hansen [9]. Similarily the surface tension must be lowered where alcohol condenses.
This could of course result in a shift of the temperature of the minimum surface tension to values lower than that on the $\sigma_e(T)$ curve. The difficulty to observe a "surface opening" with a temperature gradient around the temperature of minimum for σ could moreover be related to the stability problems of stationary convective patterns.

# 7 Conclusions

To get a better knowledge of the actual surface tension gradients governing the studied thermocapillary phenomena we suggest:

1) to verify experimentally the existence of an evaporation-condensation effect in a fatty alcohol solution by means of a device using a two-arms-electrobalance,

2) to study the possibility of "in situ" measurements of dynamic surface tension variations in a thermal Marangoni experiment by means of pressure transducers,

3) to modelize the observed "shift to lower temperatures" of the minimum surface tension versus temperature in dynamic systems,

4) to build a new experimental device to observe the "opening" of the surface when a surface temperature gradient is created around $T_M$ ,

5) to study the stability of induced thermocapillary patterns.

### Acknowledgements

This text presents results of the Belgian programme I.P.A. initiated by the Belgian state, Prime Minister's Office ,Science policy Programming

### References

[1] P.Queeckers,J.C.Dupin and J.CL.Legros, National Heat Transfer Conference 1989. H T D - vol 107,Heat Transfer in Convective Flows.

[2] Villers D, Platten J.K., PCH 6 pp 435-451 (1985).

[3] M.C.Limbourg-Fontaine,G.Petre and J.Cl.Legros, PCH vol 6 , 1985.

[4] J.CL.Legros,M.C.Limbourg and G.Petre pp 209-223 in Physicochemical Hydrodynamics Interfacial Phenomena.Ed.M.G.Velarde,Plenum WPF-DFVLR 1988

[5] G.Petre,M.A.Azouni & K.Tshinyama Applied Scientific Research 50: 97-106,1993

[6] Levich L., Physicochemical Hydrodynamics.Englewood Cliffs:Prentice Hall 1962

[7] R.Vochten and G.Petre J.Coll.Interf.Sci. 42 (1973) 320

[8] J.Hommelen J.Coll.Interf.Sci. 14 (1959) 385

[9] R.Hansen J.Phys.Chem. 64 (1960) 637

# Behaviour of the Liquid Between a Solid Particle and an Approaching Crystallization Front: Forces Balance

Patrick Casses and Aza Azouni
Laboratoire d'Aérothermique du C.N.R.S,
4 ter, route des Gardes, 92190 Meudon, France

**Abstract.** The phenomenon of particles segregation by a solidification front is very important in several areas like metallurgy, materials science, biology...
When a particle is close to a freezing front, it may be trapped in the solid region or may be repulsed by the moving freezing interface. The behaviour of the particle depends essentially on its size and on the distance between the particle and the interface.
In this contribution, we develop a theoretical investigation of the different forces exerted on a solid particle. The results should apply also to gas bubbles and droplets.
The involved forces i.e. the hydrodynamic force which favours capture and the van der Waals forces which induce rejection are determined for various sets of conditions and assumptions :
- ideal case (pure liquid, planar interface, smooth particle),
- realistic cases (curved interface, different surface properties of the particle).
Moreover, the boundary conditions for viscous drag are time dependent and clearly expressed. This represents a first step towards the development of a dynamical model.

**Keywords.** Interface solid-liquid, van der Waals forces, viscous force, capture, repulsion, particle

## 1 Introduction

The study of particles behaviour near a solidification front has several practical applications :
- in metallurgy : elaboration of alloys or elimination of impurities,
- in materials science : elaboration of artificial composites by controlled incorporation of particles,
- in crystal growth : elimination of inclusions (small defects) for better crystal properties,
- in engineering and studies of soil genesis in cold regions (frost effects in soils for construction and agriculture),
- in cryopreservation of biological cells or of food products (fruit juices...),
- in chromatography (segregation of solid particles among others types of particles).

Here, we are concerned with a theoretical model on a water-ice interface interacting with solid spherical particles.

## 2 Mechanisms of repulsion and capture of particles by a solidification front

Let us first give the expressions of the forces in the ideal case (pure liquid, planar interface and smooth particle).

### 2.1 Disjoining force

In a liquid film between two solids appears a pressure $\Pi_h$ which favours or not the compression of the film according to the nature of the considered media. This pressure depends on the film thickness h, it is named disjoining pressure (figure 1).

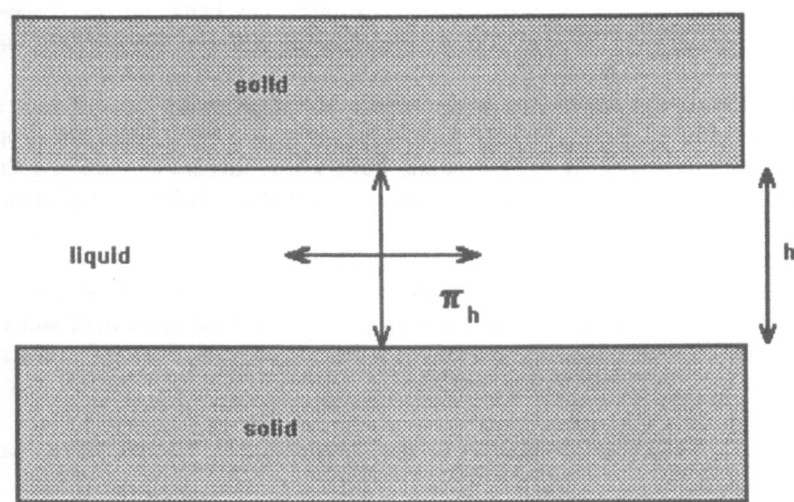

Figure 1
Schematic represention of the disjoining pressure.

The disjoining pressure implies three types of interactions :
- a dispersion interaction ( Van der Waals),
- an electrostatic interaction ( Debye ),
- a structural interaction.
In our case (particle-liquid-solid), the force that produces a disjoining effect on the particle is due to van der Waals interactions.
The non retarded van der Waals interaction energy between two macroscopic bodies, one sphere of radius R and one solid plane (figure 2), can be expressed as [1] :

$$W(h_0) = -\frac{A}{6} \int\limits_{z=h_0}^{z=2R} \frac{(2R-z)z\,dz}{(h_0+z)^3}$$

$$= -\frac{A}{6} \left( \frac{R}{(h_0+2R)} + \frac{R}{h_0} - \ln\left(\frac{h_0+2R}{h_0}\right) \right)$$

(1)

where $h_0$ is the minimum value of the distance between the particle and the solid plane, and A is the Hamaker constant.

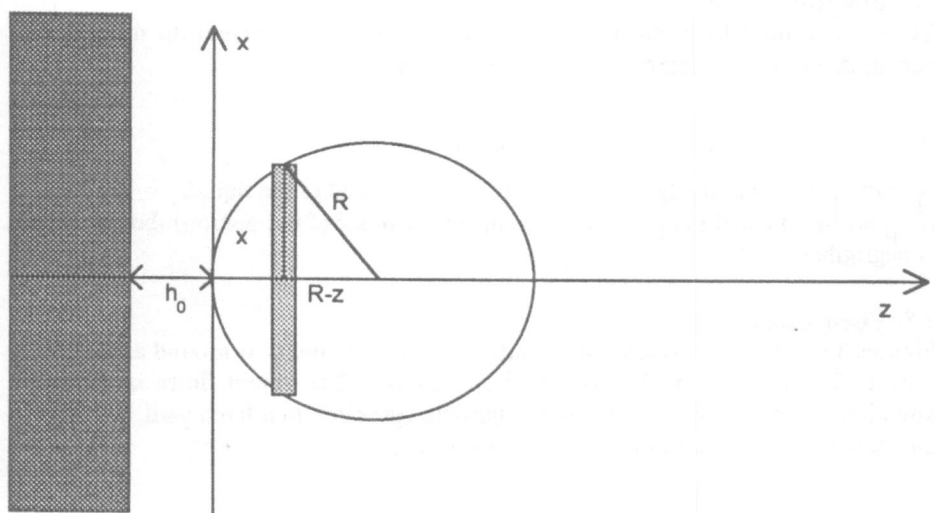

Figure 2
Interaction between a spherical particle and a solid plane

The disjoining force is obtained by differentiating the energy with respect to $h_0$ :

$$F_d = -\frac{\partial W}{\partial h_0} = -\frac{A}{6} \left( \frac{R}{h_0^2} + \frac{R}{(h_0+2R)^2} - \frac{2R}{h_0(h_0+2R)} \right)$$

(2)

With the hypothesis $h_0 \ll R$, the expression of the disjoining force $F_d$ becomes :

$$F_d = -\frac{AR}{6\,h_0^2}$$

(3)

The Hamaker constant A can be calculated in terms of the macroscopic properties of the three media by the Lifshitz theory [1] : $A = f\,(\,n,\,\varepsilon)$ with n the reflective index and $\varepsilon$ the dielectric constant.

In the water-ice-particle system, the Hamaker constant is negative, then the disjoining force $F_d$ is repulsive.

Here are some calculated values of A [2].
A(nylon-water-ice)~ -1.33 $10^{-21}$ J
A(steel-water-ice)~ -6.2 $10^{-21}$ J
A(glass-water-ice)~ -8.2 $10^{-22}$ J
A(carbon-water-ice)~ -1.42 $10^{-21}$ J

## 2.2 Buoyancy force
The buoyant force for a spherical particle Fg may be obtained from the difference of densities. The expression of the buoyancy force is :

$$F_g = -\frac{4}{3}\pi R^3 \left(\rho_p - \rho_l\right) g \qquad \text{with } \rho_p > \rho_l \qquad (4)$$

$\rho_p$ and $\rho_l$ are respectively the density of the particle and of the liquid.
If $\rho_p$ is almost equal to $\rho_l$ or R very small (of the order of the micron) then this force is negligible.

## 2.3 Viscous force
Figures 3 a, 3 b, 3 c represent three configurations of liquid flow around a particle.
Figures 3 a and 3 b are hydrodynamically similar. The liquid flows in the same direction although in the first case the particle gets close to a fixed wall, whereas in the second one, the wall moves towards a still particle.

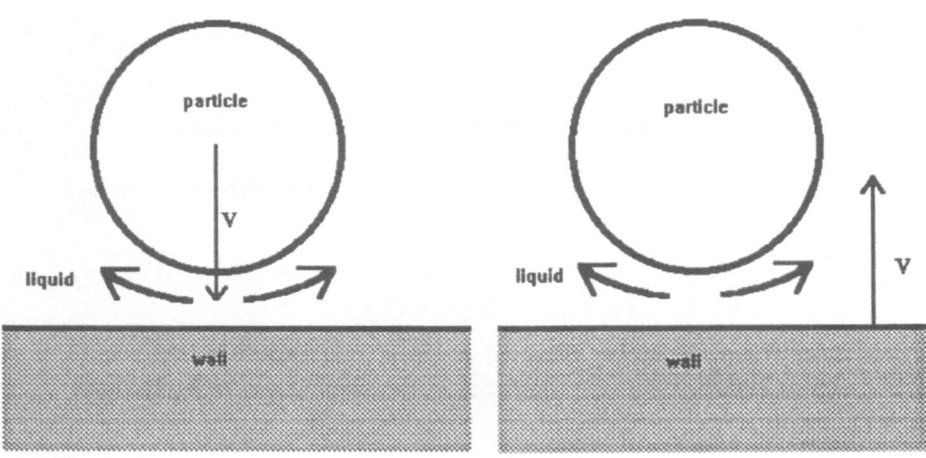

Figure 3 a
a particle approaching a fixed wall

Figure 3 b
a wall approaching a still particle

The third configuration (figure 3 c) is the more interesting as it corresponds to our experimental situation. A moving solid-liquid interface is approaching a particle, both of them move upwards. $v_f$ and $v_p$ are respectively the velocity of the front and of the particle. In this case the flow enters in the gap between the particle and the front as the motion of the liquid is only due to the displacement of the particle (the motion of the liquid caused by the difference of densities between the liquid and the solid is very small near the front). Consequently, the particle is pushed against the solid and the viscous force is then attractive.

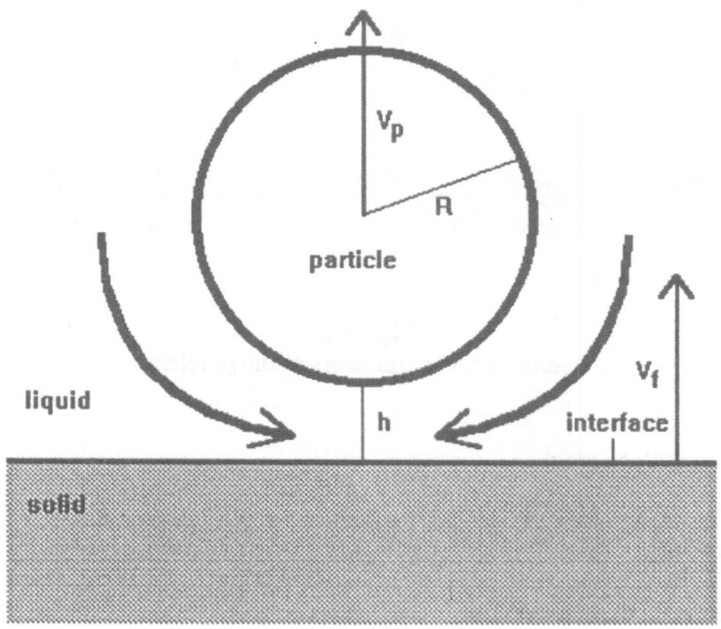

Figure 3 c
a particle and a solidification front moving upwards

Before solving the viscous flow problem described in figure 3 c, let us consider some approximations and impose the boundary conditions.

The motion of the solidification front is very slow ($v_f \sim \mu m/s$), the problem can be considered as a steady-state one, and the temporal term in the momentum equation can be neglected.

As the problem has an axial symmetry, we use the cylindrical coordinates system (figure 4).

The behaviour of the particle is non-brownian (Pe << 1), we use fluid mechanics to solve the problem.

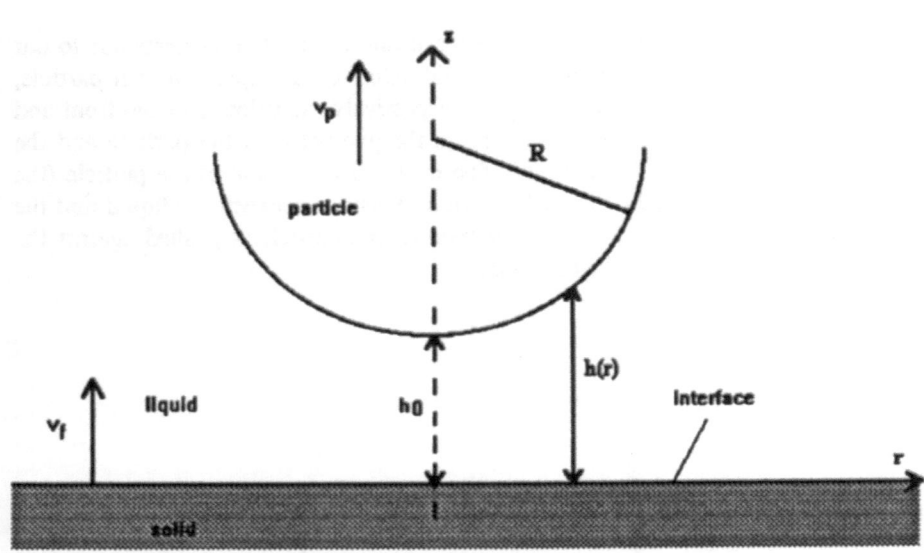

Figure 4
Geometry and parameters of the problem

Boundary conditions are :

$$
\begin{cases}
\text{at } z=0: & v_r = 0 \quad \text{and} \quad v_z = \dfrac{\rho_l - \rho_s}{\rho_l} v_f \\[4mm]
\text{at } z=h: & v_r = 0 \quad \text{and} \quad v_z = v_p
\end{cases}
\tag{5}
$$

The liquid velocity at $z = 0$ is the rejection or the absorption of liquid during solidification.

With the assumptions Re $\ll 1$, stationary flow and $v_z \ll v_r$, Navier-Stokes equation can be written as follows :

$$
\frac{\partial p}{\partial r} = \mu \frac{\partial^2 v_r}{\partial z^2}
\tag{6}
$$

The continuity equation is :

$$
\frac{1}{r} \frac{\partial (r v_r)}{\partial r} + \frac{\partial v_z}{\partial z} = 0
\tag{7}
$$

By using boundary conditions (5) and after integration of (7), we get for (6) :

$$\frac{\partial p}{\partial r} = \frac{6\mu r}{h^3} \beta v_f \tag{8}$$

with $\beta = \frac{v_p}{v_f} \cdot \frac{\rho_l - \rho_s}{\rho_l}$ : a non dimensional parameter derived

from the boundary conditions

For a spherical particle of radius R, the thickness of the liquid film h(r), with the approximation $r \ll R$, is :

$$h(r) \cong h_0 + \frac{r^2}{2R} \tag{9}$$

The viscous force acting on the particle is :

$$F_\mu = \int_0^R 2\pi r(p(r) - p(R))dr = \int_0^R 2\pi r \, dr \int_R^r \frac{\partial p}{\partial \rho} d\rho$$

$$= -6\pi\mu \, \beta \, v_f \left( \frac{R^2}{h_0} - \frac{R^2}{h(R)} - \frac{R^3}{2 \, h^2(R)} \right) \quad \text{with } h(R) = h_0 + \frac{R}{2} \tag{10}$$

As $h_0 \ll R$, we have :

$$F_\mu = -6\pi\mu \, \beta \, v_f \frac{R^2}{h_0} \tag{11}$$

If $v_p = v_f$ and $\rho_s = \rho_l$, the parameter $\beta$ is equal to 1 and we find the formula encountered the most frequently [3,4,5].

Here, some comments are needed. The viscous force was calculated under the assumption that the problem of flow was considered in the framework of continuum fluid mechanics, which implies that the gap between the particle and the front must not be smaller than $10^{-9}$ m [6].


## 3 - Expressions of the forces in the case of a deformed interface

In a realistic situation, when the particle is macroscopic and when there is a difference of conductivity between the particle and the liquid, the interface is not planar. If the particle is less conducting than the liquid the interface is convex, and conversely it is concave [7].

Let us consider a front profile S(r) which has a curvature equal to R/α with 0<α<1. The thickness of the liquid film h(r), with the approximation r << R, is :

$$h(r) \cong h_0 + \frac{r^2}{2R} \pm S(r) \tag{12}$$

with -S(r) for a concave interface and +S(r) for a convex one.

As r << R, the expression of S(r) is $\frac{\alpha r^2}{2R}$ and $h(r) \cong h_0 + (1\pm\alpha)\frac{r^2}{2R}$. $\quad$ (13)

Now, we are able to calculate new expressions for the viscous and the disjoining forces.

The disjoining force can be written as :

$$F_d = \int_0^R 2\pi\, r\Pi(r)dr \qquad \text{with } \Pi(r) = -\frac{A}{6\pi\, h^3}$$

$$F_d = -\frac{A}{3}\int_0^R \frac{rdr}{\left(h_0 + (1\pm\alpha)\dfrac{r^2}{2R}\right)^3} \tag{14}$$

After integration, the result is :

$$F_d = -\frac{A\,R}{6\,(1\pm\alpha)\,h_0^2}\left(1-\left(\frac{h_0}{h(R)}\right)^2\right) \tag{15}$$

For $F_\mu$, we have :

$$F_\mu = \int_0^R 2\pi\, rdr \int_R^r d\rho \left( \beta\, v_f \frac{6\mu\rho}{\left(h_0 + \dfrac{\rho^2}{2R}(1\pm\alpha)\right)^3} \right) \tag{16}$$

$$= -\frac{6\pi\mu\,\beta\,v_f\,R^2}{h_0(1\pm\alpha)^2}\left(1-\frac{h_0}{h(R)}-\frac{R(1\pm\alpha)h_0}{2h^2(R)}\right)$$

With the assumption $h_0 <<$ h(R), expressions (15) and (16) are :

$$F_d = -\frac{A\,R}{6(1\pm\alpha)\,h_0^2} \tag{17}$$

$$F_\mu = -\frac{6\pi\mu\,\beta\,v_f\,R^2}{h_0(1\pm\alpha)^2} \tag{18}$$

We can notice that the two forces evolve in the same way, but for a concave profile, the viscous force increases faster than the disjoining force and conversely for a convex profile.

## 4 - Critical velocity : forces balance

In experiments dealing with particles segregation [8,3,9,10,5,11,12], the critical velocity is the main experimental parameter which can be measured and, therefore it is important to evaluate it theoretically. The critical velocity $v_c$ is the front velocity $v_f$ below which repulsion occurs and above which the particle is trapped.
According to the fundamental relation of dynamics, we have :

$$m_p \frac{d v_p}{d t} = F_\mu + F_d + F_g \tag{19}$$

with $m_p$ the particle mass and $v_p$ its velocity

We have the following relation for the evolution of the distance h between the particle and the front :

$$\frac{dh}{dt} = v_p - v_f \text{ and } m_p \frac{dv_p}{dh}(v_p - v_f) = F_\mu + F_d + F_g \tag{20}$$

We assume that the particle is pushed in a continuous way at a velocity $v_p = v_f$ which imposes to the three forces to balance each other.
The equilibrium condition is the following :

$$F_\mu(h) + F_g + F_d(h) = 0 \tag{21}$$

The front velocity $v_f$ is in this case the critical velocity $v_c$ which can be deduced from the expressions of $F_\mu$, $F_g$ and $F_d$. $v_c$ is considered as a function of the particle radius R, all the others parameters including h being fixed.

We can notice that in all the calculated expressions, we assume that $v_p = v_f$, so the coefficient $\beta$ must be equal to $\rho_s / \rho_l$.

If we consider a smooth particle in interaction with a planar front, we can use expressions (3), (4) and (11) in (21) to find the critical velocity.

$$v_c = - \frac{A}{36\pi \mu \beta R h} - \frac{2}{9} R h \frac{(\rho_p - \rho_l)g}{\mu \beta} \tag{22}$$

In the case of a smooth particle interacting with a curved front, the critical velocity becomes :

$$v_c = -\frac{A(1\pm\alpha)}{36\pi\,\mu\,\beta\,R\,h}\frac{\left(1-\left(\dfrac{h}{h(R)}\right)^2\right)}{\left(1-\dfrac{h}{h(R)}-\dfrac{R(1\pm\alpha)h}{2h^2(R)}\right)}-\frac{2}{9}Rh\frac{(\rho_p-\rho_l)(1\pm\alpha)^2 g}{\mu\,\beta\left(1-\dfrac{h}{h(R)}-\dfrac{R(1\pm\alpha)h}{2h^2(R)}\right)} \qquad (23)$$

With the hypothesis $h \ll h(R)$, we obtain :

$$v_c = -\frac{A(1\pm\alpha)}{36\pi\,\mu\,\beta\,R\,h}-\frac{2}{9}Rh\frac{(\rho_p-\rho_l)(1\pm\alpha)^2 g}{\mu\,\beta} \qquad (24)$$

It is worth noting that for small particles of the order of the micron, the gravity term in expressions (22) and (24) is negligible compared to the others.

Using expressions of the critical velocity (22) and (24), we can know if the particle is captured or not : if $v_f \gg v_c$, there is capture and if $v_f \ll v_c$, there is repulsion.

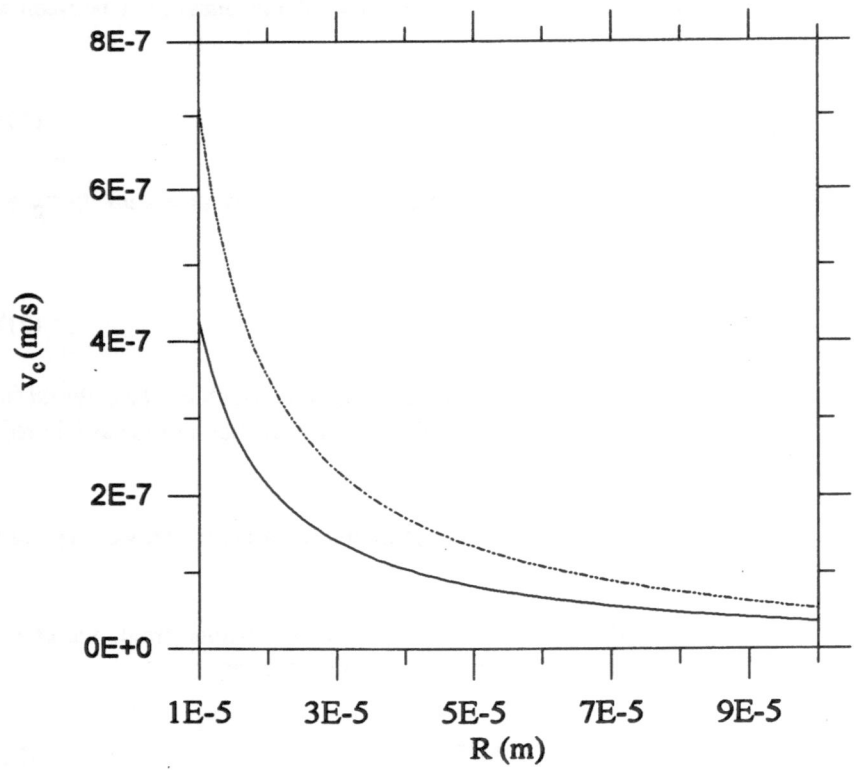

Figure 5
Comparison of front critical velocities for deformed ( dash-dot line )
and undeformed ( solid line ) interfaces

The expressions of $v_c$ for a planar or a curved solidification front [equations (22) and (24)] are represented versus the particles size for an ice-water-nylon particle system in figure 5. We can see that the deformation favours repulsion although the values of $v_c$ for both cases are not very different.

Now, let us represent in the same plot (figure 6) the experimental curve of M.A.Azouni et al (nylon particles, water-ice system) [13] and the theoretical curve using the relation (22) with $g = 0$. We assume that the fact of suspending the particle cancels the gravity effect during the first times of the interaction between the particle and the front. This figure shows that for $h = 6 \cdot 10^{-10}$ m the theoretical critical velocity is fifty times smaller than the experimental one.

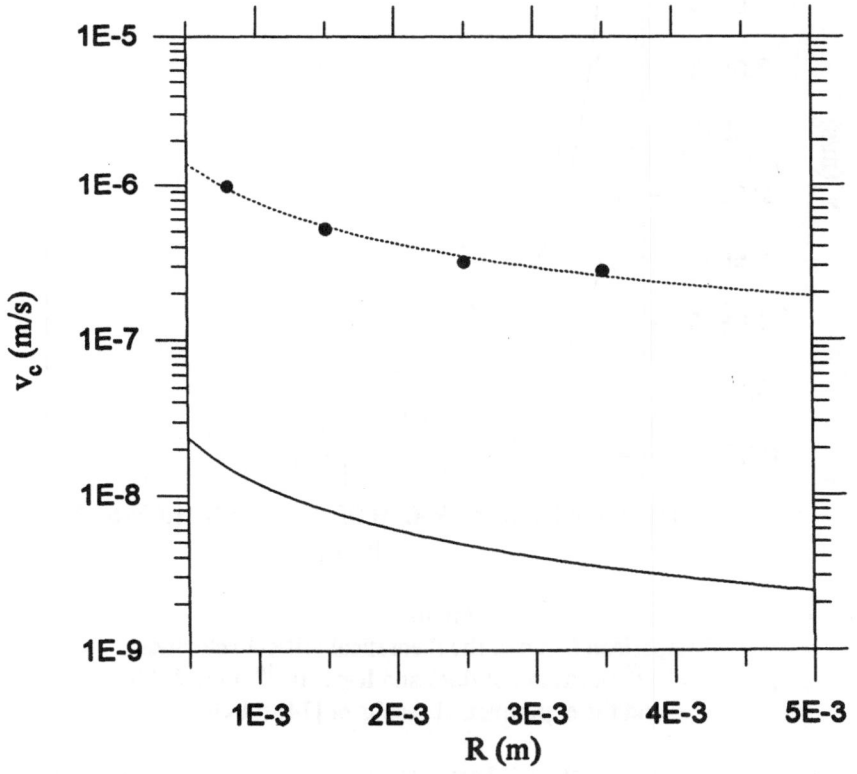

Figure 6
Comparison between the theoretical critical velocity ( solid line )
and the experimental one ( dot line and circle ) [13]

Figure 7 represents two curves plotted using equation (22). The first one corresponds to $h = 7.6 \cdot 10^{-10}$ m suggested by G.Lipp and Ch.Körber (latex particles, water-ice system) [14], the second one corresponds to $h = 3 \cdot 10^{-10}$ m. The latter value is obtained from figure 8 which represents the sum of the forces (21) in function of the distance h, for a given R and for different values of experimental velocities. In this plot, when one curve cuts the value zero (sum of the forces is null) the value of h can be deduced. For the three experimental critical velocities $v_c = 6 \ 10^{-6}$ m/s (R = 6 $10^{-6}$ m), $v_c = 8 \ 10^{-6}$ m/s (R = 4 $10^{-6}$ m), $v_c = 17 \ 10^{-6}$ m/s (R = 2 $10^{-6}$ m), the corresponding distance h is almost the same (h ≈ 3 $10^{-10}$ m).

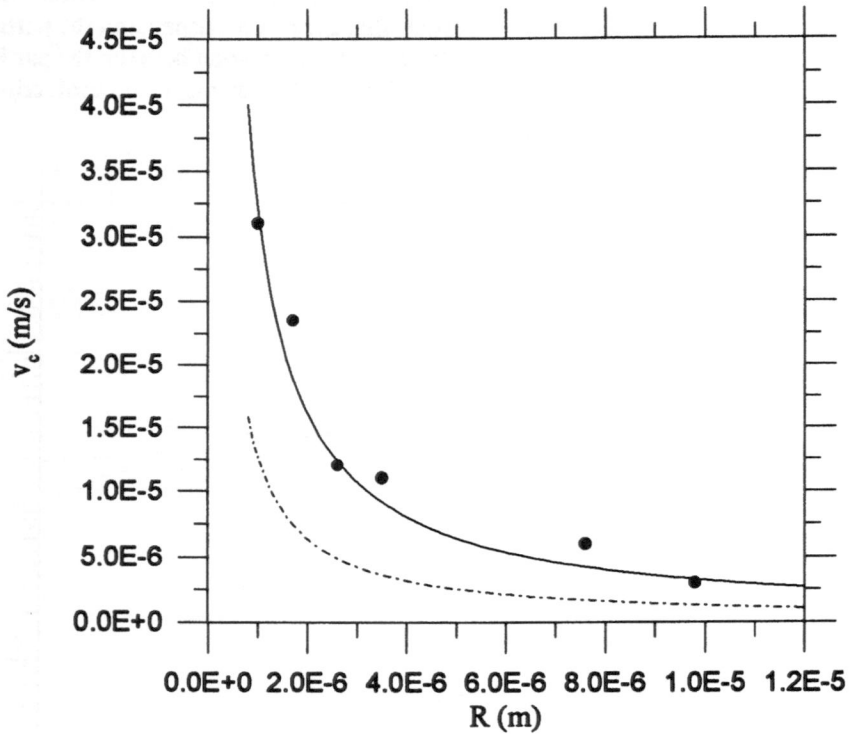

Figure 7
Comparison between the theoretical critical velocities
[ h = 7.6 $10^{-10}$ m (dash-dot line) and h = 3 $10^{-10}$ m (solid line) ]
and the experimental results of [14] (circle)

For h = 7.6 $10^{-10}$ m, we obtain theoretical values of $v_c$ twice smaller than the experimental values of G.Lipp et Ch.Körber which are represented in figure 7, but when h = 3 $10^{-10}$ m, they are in a perfect agreement.

For small sizes of particle from 1 to 100 μm, this approach [equations (22) and (24)] provides good results if we can choose the right distance h for a given range of velocities. On the other hand, these equations are not suitable for millimetric particles.

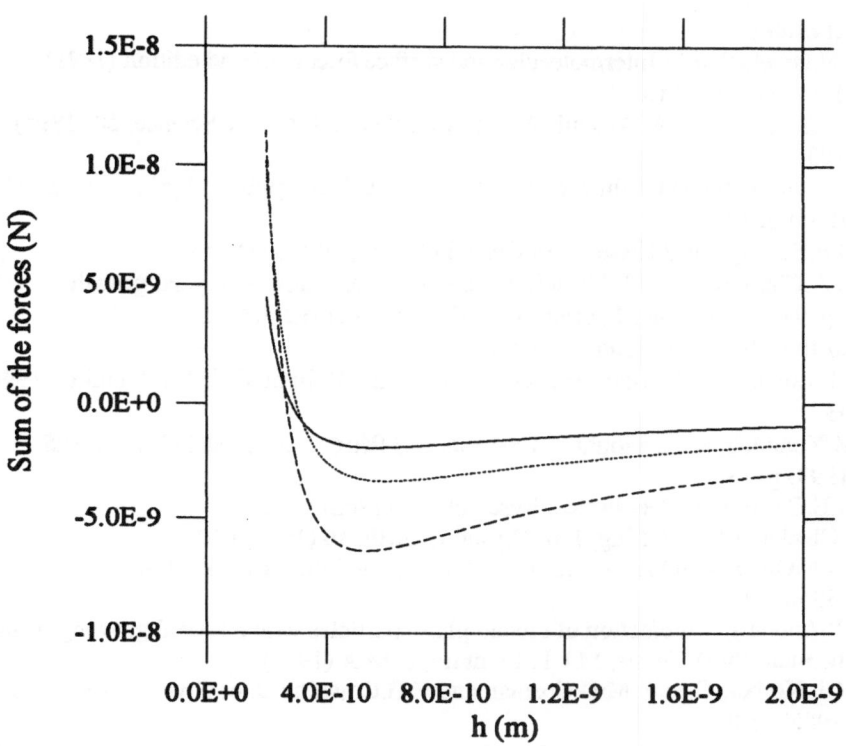

Figure 8
Sum of the forces for different radii (R = 2, 4, 6 $10^{-6}$ m)

## 5 - Conclusion

Our model has been extended to the case of particles of different surface properties (coating, absorption, roughness...) [2].

The evaluation of the different forces acting upon the particle allowed us to calculate the critical velocity for microscopic particles (see section 3) and for millimetric particles with a dynamical model [15] but also to evaluate the interface curvature due to the difference of thermal conductivities between the particle and the liquid matrix using a thermodynamics model [16].

# References

[1] J.N.Israelachvili : Intermolecular and surface forces, second edition (1991)
ed. : Academic Press

[2] P. Casses and M.A. Azouni, Adv. in Colloid and Interface Science, 50 (1994)
p103

[3] D.R.Uhlmann, B.Chalmers and K.A.Jackson, J. of Applied Physics, vol 35, n°10
(1964) p2986

[4] G.F.Bolling and J.Cissé, J. of Crystal Growth, 10 (1971) p56

[5] A.A.Chernov and D.E.Temkin : Capture of inclusions in crystal growth,
Crystal Growth and Materials, ed. : E.Kaldis and H.J.Scheel,
North Holland pub. Comp. (1977)

[6] J-L Loubet, E. Charlaix, J. Crassous, A. Tong, Bulletin de la S.F.P (Juillet 1994)
p3

[7] M.Yemmou, M.A.Azouni, P.Casses and G.Pétré, J. of Crystal Growth, 128
(1993) p1130

[8] A.E.Corte, J. of Geophysical Research, 67 (1962) p1085

[9] J.Cissé and G.F.Bolling, J. of Crystal Growth, 10 (1971) p67

[10] S.N.Omenyi and A.W.Neumann, J. of Applied Physics, vol 47, n°9 (1976)
p3956

[11] P.Aubourg : Interaction of second phase particles with a crystal growing from
the melt, Ph.D Thesis, M.I.T. Cambridge, M.A (1978)

[12] Ch.Körber, G.Rau, M.D.Cosman and E.G.Cravalho, J. of Crystal Growth, 72
(1985) p649

[13] M.A.Azouni , W.Kalita and M.Yemmou, J of Crystal Growth, 99 (1990) p201

[14] G.Lipp and Ch.Körber, J. of Crystal Growth, 130 (1993) p475

[15 ] P. Casses and M.A. Azouni, Int. Com. of Mass and Heat Transfer (sous presse)

[16] P.Casses and M.A.Azouni, J. of Crystal Growth, 130 (1993) p13

# Disintegration of Cylindrical Liquid Columns in Liquid–Fluid Systems: Direct Numerical Simulation

M. Chacha[1], R. Occelli[1], L. Tadrist[1], S. Radev[2]
[1] IUSTI – UMR 139, 13397 Marseille Cedex 20 France
[2] IMech BAS, ul Georgi Bonchev, bl 4, 1113 Sofia Bulgaria

**Abstract.** The capillary instability and breakup of a cylindrical liquid column surrounded by a bounded immiscible fluid are investigated. The process of disintegration is simulated numerically using a second order finite difference method applied to the streamfunction–vorticity formulation of the governing equations. All viscous and nonlinear terms are included. In the case of liquid jet into vacuum, the results obtained with the present numerical approach are in good agreement with those of previous authors. In the case of liquid jet into viscous liquid phase, the applicability of the numerical procedure is verified in comparison with the results of the linear stability analysis. Basic characteristics of the disintegration such as drop sizes are compared to existing experimental data obtained under the condition of zero relative velocity between the jet and the continuous phase. The influence of the physical properties is analyzed as well as that of the rigid cylinder bounding the continuous phase at some distance. The calculation code has been extended to study the disintegration of liquid bridges.

**Keywords.** Nonlinear instability, breakup, liquid cylinders

## 1. Introduction

In the recent years the capillary jets have been used in many practical applications such as printing, particle sorting, dispersing liquids, fibers spinning, etc. Most of these applications are based on jet instability, which gives rise to growing disturbances that finally break up the jet into drops. In order to optimize such applications, the fundamental mechanisms determining the behavior of the jet and the drop formation must be thoroughly understood.

Rayleigh [1] was the first to make a theoretical study of the inviscid liquid column in vacuum and to establish the possibility of droplet formation. He showed that the capillary jet is stable for all purely non-axisymmetric disturbances; however, in respect to the axisymmetric disturbances, the jet is stable or unstable depending on whether the wave length is less or greater than the circumference of the undisturbed cylinder.

Other authors, among them Weber [2], studied the linear instability of viscous liquid jets. In addition to the viscosity, Tomotika [3] took into account the density and the viscosity of the medium surrounding the jet.

Neither of the aforementioned theories predict the formation of droplets of different diameters after breakup of a liquid jet. It is assumed as a starting point that a spectrum of infinitesimal disturbances of the form of eq. (1) is initiated on the surface of stationary liquid column:

$$\delta = \delta_0 \exp(Qt + ixz) \tag{1}$$

where $x$ denotes the wavenumber and $Q$, the growth rate. $\delta$ and $\delta_0$ are, respectively, the disturbance and the initial amplitude of the disturbance. $t$ is the time and $z$ is the axial coordinate. (All those quantities are dimensionless). The aim of the linear stability analysis is to establish the dispersion relation $Q = f(x)$.

Weakly nonlinear theories were conducted first by Yuen [4]. He developed a solution for an inviscid jet. He carried out the solution up to 3rd order for a monochromatic initial displacement of the free surface. In general weakly nonlinear theories lead to good qualitative results and predict satellite drop formation as observed in experiments ([5]).

The goal of the present work is to predict the nonlinear behavior of the jet instability and breakup. To do this, a complete (numerical) solution of the full Navier-Stokes equations with appropriate boundary conditions is proposed. This solution should lead to both qualitative and quantitative results. It is not easy to carry out numerical computation of the problem due to the free and mobile surface. In the field of numerical simulation of liquid jet disintegration, the work of Shokoohi [6] appears to be of particular importance.

This paper presents a direct simulation of the process of disintegration of a viscous liquid jet in a liquid or gaseous medium. The applicability of the numerical procedure is verified in comparison with the temporal linear stability analysis. In the case of a jet in vacuum, drop sizes are compared to those obtained both numerically by Shokoohi [6] and experimentally by Rutland & Jameson [5] and Lafrance [7]. For a jet in viscous liquid phase, we compare the computed drop sizes to those obtained experimentally by Kitamura et al [8] under the condition of zero relative velocity between the jet and the continuous phase. It is shown that due to the appearance of multiple harmonics, the initially sinusoidal form of the disturbance is transformed to a complex one, which contains a satellite drop attached to a main one. The size of the drops is strongly connected to the wavelength and physical properties except the surface tension. The amplification rate of the temporally growing disturbances is time dependent except for the initial time interval. In addition, the effect of the rigid wall (see Fig. 1) limiting the continuous phase on jet breakup is shown.

Some results relative to the disintegration of liquid bridges are presented.

## 2. Physical and Mathematical Models

### 2.1 Physical Model

The geometry considered in the present work implies what has been called "infinite jet". That is, the jet has no beginning nor end. Both the jet and the surrounding fluid are supposed to be incompressible, Newtonian, immiscible and initially at rest. The gravity effects are neglected. Figure 1 shows a periodic section of the axisymmetric jet with a single wavelength disturbance (fundamental disturbance). The continuous phase

209

is limited at a distance $R_{max}$ with a rigid cylinder. This configuration ensures a physical external boundary to the computational domain. This domain is bounded elsewhere by the axis of symmetry $Oz$ and two jet cross-sections at a distance one wavelength of the prescribed disturbance.

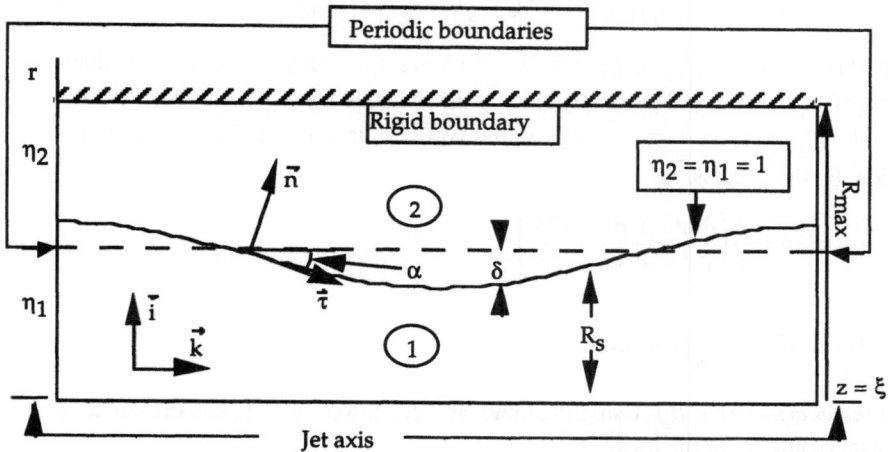

Figure 1. A wavelength-segment of the flow configuration.

Owing to the symmetry of the problem, the hydrodynamic equations are written in a cylindrical coordinate system $(O, r, z)$, using the so-called streamfunction-vorticity formulation. The stream function $\psi$ and the vorticity $\omega$ are defined in both the jet and the continuous phase as:

$$U_j = -\frac{1}{r}\frac{\partial \psi_j}{\partial r} \qquad V_j = \frac{1}{r}\frac{\partial \psi_j}{\partial z} \tag{2}$$

$$\omega_j = \frac{\partial V_j}{\partial z} - \frac{\partial U_j}{\partial r} \tag{3}$$

where $U$ and $V$ are the axial and radial components of the velocity, respectively. In equations (2) and (3), subscript $j = 1$ refers to the jet; while $j = 2$ - to the continuous phase. Hereafter, the subscript $j$ will be omitted when admissible.

## 2.2 Mathematical Model

The hydrodynamic equations are made dimensionless using the undisturbed jet radius $R_N$ and the capillary time $t_0 = \sqrt{\rho_l R_N^3/\sigma_{12}}$ as the characteristic length and characteristic time, respectively.

## 2.2.1  Governing Equations

The Navier-Stokes equations in $\psi$-$\omega$ formulation give:

$$\frac{D\omega}{Dt} - Oh_1\left(\frac{v}{v_1}\right)\left[\frac{\partial^2\omega}{\partial r^2} + \frac{1}{r}\frac{\partial\omega}{\partial r} - \frac{\omega}{r^2} + \frac{\partial^2\omega}{\partial z^2}\right] = \frac{\omega}{r^2}\frac{\partial\psi}{\partial z} \tag{4}$$

where $Oh_1$ ($Oh_1 = \mu_1/\sqrt{\rho_1 R_N \sigma_{12}}$) is the jet Ohnesorge number. $v$ is the kinematic viscosity $(v = \mu/\rho)$. $\mu$ and $\rho$ are respectively the dynamic viscosity and the density of the fluid. $\sigma_{12}$ is the surface tension. $\omega$, the vorticity, is connected to $\psi$, the stream function, by the expression:

$$\omega = \frac{1}{r}\left(\frac{\partial^2\psi}{\partial z^2} - \frac{1}{r}\frac{\partial\psi}{\partial r} + \frac{\partial^2\psi}{\partial r^2}\right) \tag{5}$$

## 2.2.2  Boundary Conditions

(a) At the interface $(r = R_s)$. Using the kinematic condition, one obtains the equation for the jet radius $R_s$ of the form:

$$\frac{\partial R_s}{\partial t} + U_s\frac{\partial R_s}{\partial z} = V_s \tag{6}$$

This equation must be used together with nonslip condition and a condition for a zero mass flux through the interface:

$$[\![U]\!] = 0 \qquad\qquad [\![V]\!] = 0 \tag{7}$$

For shortness, above and below we use the notation

$$[\![G]\!] = G_{1s} - G_{2s} \tag{8}$$

where the subscript $s$ refers to the jet surface.

Additional conditions must be derived from the balance of tangential and normal stresses at the both sides of the interface including surface tension. Thus for tangential stress condition we have

$$\left[\!\!\left[\frac{\mu}{\mu_1}\left[\left(\frac{\partial V}{\partial r} - \frac{\partial U}{\partial z}\right)\sin 2\alpha + \left(\frac{\partial V}{\partial z} + \frac{\partial U}{\partial r}\right)\cos 2\alpha\right]\right]\!\!\right] = 0 \tag{9}$$

where $\alpha$ is shown in Fig. 1. The corresponding normal stress condition is

$$\left[\!\!\left[P - Oh_1\frac{\mu}{\mu_1}\left[2\left(\frac{\partial V}{\partial r}\cos^2\alpha + \frac{\partial U}{\partial z}\sin^2\alpha\right) - \left(\frac{\partial V}{\partial z} + \frac{\partial U}{\partial r}\right)\sin 2\alpha\right]\right]\!\!\right] = \left(\frac{\cos\alpha}{R_s} - \frac{\partial^2 R_s}{\partial z^2}\cos^3\alpha\right) \tag{10}$$

Further on, the pressure difference $[\![P]\!]$ is eliminated by differentiating eq. (10) along the surface and substituting pressure derivatives by the corresponding expressions taken from Navier-Stokes "(U, V, P)" equations.

The interface boundary conditions are then expressed using the definition of the streamfunction and the vorticity.

(b) At the jet axis (r = 0). Due to the symmetry of the flow and the finiteness of the velocity, we impose the conditions:

$$\omega_1 = 0 \tag{11}$$

$$\psi_1 = 0 \tag{12}$$

(c) At the rigid wall (r = R_{max}). The nonslip and zero mass flux conditions must be satisfied:

$$\omega_2 = \frac{1}{r}\left(\frac{\partial^2 \psi_2}{\partial r^2}\right) \tag{13}$$

$$\psi_2 = 0 \tag{14}$$

(d) At both ends of one-wavelength jet segment. We have here the periodicity boundary condition

$$F^n(0, r, t) = F^n(\lambda, r, t) \tag{15}$$

where $F$ is any of the flow variables and superscript $n$, the order of differentiation. $\lambda$ is the nondimensional wavelength of the initial disturbance.

### 2.2.3 Initial Condition

At zero time, we impose upon the jet surface, an axisymmetric cosinusoidal disturbance as given by eq. (1). The initial solution is derived from the linear temporal analysis. This solution is more general than that in [3] because taking into account the presence of a rigid wall.

## 3. Numerical Model

We will focus on numerical solution of the governing equations as they appear in the preceding section.

### 3.1 Coordinate Transformation

To overcome the difficulty related to the interface location, we introduce the variables

$$\eta_j = \frac{(j-1) R_{max} - r}{(j-1) R_{max} - R_s} \tag{16}$$

which allow uniformity of grid size in the transformed coordinates, high precision on the interface by keeping the latter exactly on a mesh line ($\eta_1 = \eta_2 = 1$).

New symbols are introduced for the time and axial coordinate

$$\tau = t \qquad\qquad \xi = z \tag{17}$$

According to these transformations, the derivative operators in respect to the cylindrical coordinates $(r, z)$ are expressed in respect to the new variables $(\eta_j, \xi)$ by the following matrix relation

$$\{D^{r,z}\} = [T_j]\{D^{\eta_j,\xi}\} \tag{18}$$

where $\{D^{r,z}\}$ and $\{D^{\eta_j,\xi}\}$ are $9*1$ column matrices:

$$\{D^{r,z}\} = \left\{\frac{\partial^3}{\partial r^3}, \frac{\partial^2}{\partial r^2}, \frac{\partial}{\partial r}, \frac{\partial^3}{\partial r^2\partial z}, \frac{\partial^2}{\partial r\partial z}, \frac{\partial^3}{\partial r\partial z^2}, \frac{\partial}{\partial z}, \frac{\partial^2}{\partial z^2}, \frac{\partial^3}{\partial z^3}\right\}^T \tag{19}$$

$$\{D^{\eta_j,\xi}\} = \left\{\frac{\partial^3}{\partial \eta_j^3}, \frac{\partial^2}{\partial \eta_j^2}, \frac{\partial}{\partial \eta_j}, \frac{\partial^3}{\partial \eta_j^2\partial \xi}, \frac{\partial^2}{\partial \eta_j\partial \xi}, \frac{\partial^3}{\partial \eta_j\partial \xi^2}, \frac{\partial}{\partial \xi}, \frac{\partial^2}{\partial \xi^2}, \frac{\partial^3}{\partial \xi^3}\right\}^T \tag{20}$$

$\{X\}^T$ is the transpose matrix of $\{X\}$.

The elements of the transformation matrix $[T_j]$ which is a $9*9$ lower triangular matrix are as follows:

$T_j(1,1) = -\varepsilon_j^3$

$T_j(2,2) = \varepsilon_j^2$

$T_j(3,3) = -\varepsilon_j$

$T_j(4,1) = \eta_j\,\varepsilon_j^2\,\gamma_j$, $\quad T_j(4,2) = 2\varepsilon_j^2\,\gamma_j$, $\quad T_j(4,4) = \varepsilon_j^2$

$T_j(5,2) = -\eta_j\,\varepsilon_j\,\gamma_j$, $\quad T_j(5,3) = -\varepsilon_j\,\gamma_j$, $\quad T_j(5,5) = -\varepsilon_j$

$T_j(6,1) = -\eta_j^2\,\varepsilon_j\,\gamma_j^2$, $\quad T_j(6,2) = -\eta_j\,\varepsilon_j\left(4\gamma_j^2 + \varepsilon_j\,R_s^{"}\right)$, $\quad T_j(6,3) = -\varepsilon_j\left(2\gamma_j^2 + \varepsilon_j\,R_s^{"}\right)$

$T_j(6,4) = -2\eta_j\,\varepsilon_j\,\gamma_j$, $\quad T_j(6,5) = -2\varepsilon_j\,\gamma_j$, $\quad T_j(6,6) = -\varepsilon_j$

$T_j(7,3) = \eta_j\,\gamma_j$, $\quad T_j(7,7) = 1$

$T_j(8,2) = \eta_j^2\,\gamma_j^2$, $\quad T_j(8,3) = \eta_j\left(2\gamma_j^2 + \varepsilon_j\,R_s^{"}\right)$, $\quad T_j(8,5) = 2\eta_j\,\gamma_j$, $\quad T_j(8,8) = 1$

$T_j(9,1) = \eta_j^3\,\gamma_j^3$, $\quad T_j(9,2) = 3\eta_j^2\,\gamma_j\left(2\gamma_j^2 + \varepsilon_j\,R_s^{"}\right)$, $\quad T_j(9,3) = \eta_j\left(6\gamma_j^3 + 6\varepsilon_j\,\gamma_j R_s^{"} + \varepsilon_j\,R_s^{"'}\right)$

$T_j(9,4) = 3\eta_j^2\,\gamma_j^2$, $\quad T_j(9,5) = 3\eta_j\left(\varepsilon_j\,R_s^{"} + 2\gamma_j^2\right)$, $\quad T_j(9,6) = 3\eta_j\,\gamma_j$, $\quad T_j(9,9) = 1 \tag{21}$

All other elements are zeroes. Subscript j denotes the corresponding phase and

$$R_j = (j-1)\,R_{max}, \quad \varepsilon_j = \frac{1}{R_j - R_s}, \quad \gamma_j = \varepsilon_j\,R_s', \quad R_s' = \frac{\partial R_s}{\partial z}, \quad R_s^{"} = \frac{\partial^2 R_s}{\partial z^2} \text{ and } R_s^{"'} = \frac{\partial^3 R_s}{\partial z^3} \tag{22}$$

The expression of the $l^{th}$ element of $\{D^{r,z}\}$ in the new coordinates is given by

$$D^{r,z}(l) = \sum_{m=1}^{m=l} T_j(l,m)\,D^{\eta_j,\xi}(m) \tag{23}$$

In addition, the time derivative is transformed as

$$\left.\frac{\partial}{\partial t}\right)_{r,z} = \left.\frac{\partial}{\partial \tau}\right)_{\eta_j,\xi} + \eta_j\,\varepsilon_j\left(V_s - U_s\frac{\partial R_s}{\partial z}\right)\left.\frac{\partial}{\partial \eta_j}\right)_{\xi,\tau} \tag{24}$$

## 3.2 Formulation of the Problem in the Transformed Coordinate System

The governing equations and boundary conditions are written in $\eta_j - \xi$ coordinates using relations of the preceding section. New forms of the equations can be found in [9].

## 3.3 Numerical Procedure

The set of resulting equations belongs to the class of partial differential equations for which numerical methods have been extensively studied. But the presence of an unknown interface with no condition on the interface vorticity reveals the difficulty encountered when solving this type of problems; especially using streamfunction-vorticity formulation. In the general case $(\mu_1 \neq \mu_2)$, there are two distinct values of the vorticity at the interface (one for each side).

We use a uniform finite-difference grid of step sizes $\Delta\eta_j$ and $\Delta\xi$ in corresponding directions. The scheme is centered in space and forward in time. The solution is based on the application of the well-known alternating direction implicit method. In particular, the ADI method has been successfully used by Shokoohi [6] for the vorticity transport equation inside the jet. We have extended the method to the continuous phase vorticity equation, as well as for treating the stream function-vorticity relation in both phases. The method is necessarily iterative due to the dependance of the coefficients in the equations on the unknown jet radius and its derivatives up to the third order. The numerical procedure is as follows:

*Step 1*   Increase the time.

*Step 2*   Compute a new surface profile $R_s$ using the kinematic condition. Then differentiate the updated radius numerically using cubic splines.

*Step 3*   Solve the vorticity transport equation for the jet (the vorticity values of the preceding iteration are used at the interface).

*Step 4*   Solve the vorticity transport equation for the continuous phase and approximate the vorticity on the rigid boundary using the wall condition.

*Step 5*   Solve the $\psi-\omega$ relation for the streamfunction in the jet (a pseudo-transient approach is used by adding a fictitious time derivative of $\psi$ in $\psi-\omega$ relation. This enables us to transform the latter to a form more convenient for using ADI method).

*Step 6*   Solve the $\psi-\omega$ relation for the stream function in the continuous phase.

*Step 7*   Determine the jet vorticity at the interface using tangential stress equation. Then, approximate the corresponding $\psi-\omega$ relation at the interface to determine the continuous phase interface vorticity.

*Step 8*   Compute the velocity component values everywhere.

*Step 9*   Determine the streamfunction values at the interface using the (modified) normal stress condition.

*Sep 10*   Go to *step 2* if the convergence criterion is not satisfied (if necessary reduce the time step and correct the total time).
Go to *step 1* if convergence is obtained and the current time step is not too small (about $10^{-5}$).

*Stop*

## 4. Results

Since the jet radius $R_s$ depends on the values of $\psi$ at the jet surface, the iterations for $\psi$ and $\omega$ are coupled with the iterations for the interface position calculations. The calculation is terminated for each time step whenever $\varphi \leq \varepsilon$ with $\varphi = \max(\varphi_F)$, where $\varphi_F = \max(|F_{i,k}^{s} - F_{i,k}^{s-1}|)$ is the maximum error in the "field" of $F$ for two consecutive iterations. $F^s$ denotes $\omega$, $\psi$ or $R_s$ computed at the $s^{th}$ iteration. Tests for accuracy have been performed: a twofold variation in $\Delta\xi$ and a fivefold variation in $\Delta\tau$ make no significant difference (Figs. 2 and 3).

However, it is necessary to reduce further the time increment in the nonlinear zone (mainly near the jet breakup) to preserve the solution from very abrupt variations. The time step is divided by two whenever a given number of iterations is attained before the above time step stopping criterion is satisfied. The calculation ends off for a given case when the time step becomes too small (about $10^{-5}$). At this stage in general, the radius of the jet at minimal cross-sections is less than $0.05$: the corresponding sections and total time are considered as breakup points and time, respectively. Typically we used $\varepsilon = 10^{-6}$ and initial time step $\Delta\tau = 0.05$. Concerning the space steps, we used systematically $\Delta\eta_1 = 0.1$; $\Delta\xi = 0.5$ for $x < 0.7$ and $\Delta\xi = \lambda/20$ for $x \geq 0.7$. $\Delta\eta_2$ was chosen in function of the value of $R_{max}$.

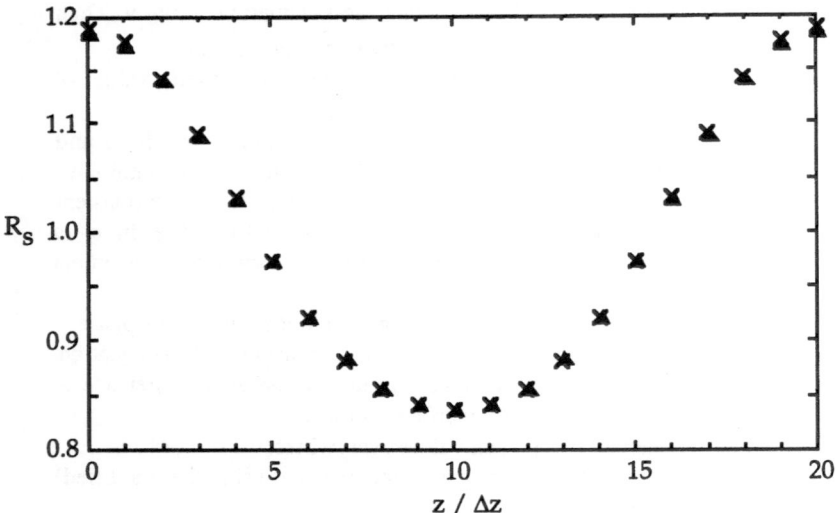

Figure 2.   Interface profile at $t = 8$   $\times$ : $\Delta z = 0.35$   $\blacktriangle$ : $\Delta z = 0.70$
            ($\lambda^* = 7 D_N$, $\delta_0 = 0.02$, $D_N = 3.5\ 10^{-5}\ m$)

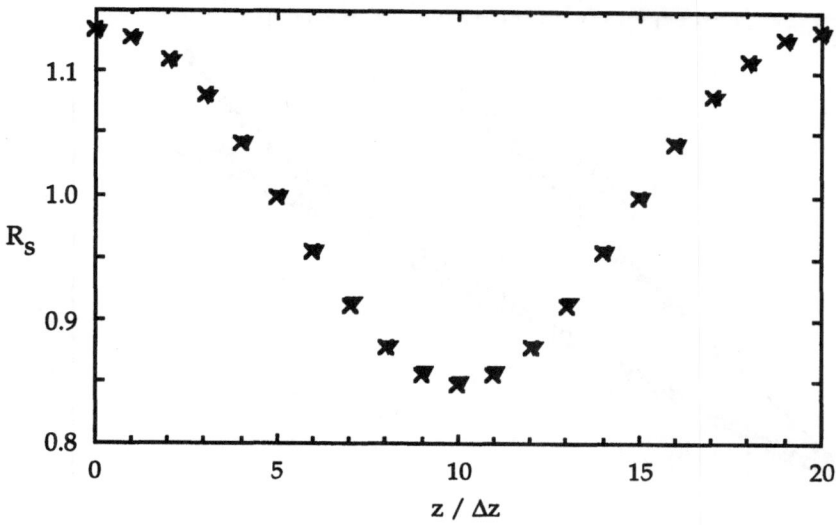

Figure 3.    Interface profile at $t = 8$     $\times$ : $\Delta \tau = 10^{-2}$   $\blacktriangle$ : $\Delta \tau = 0.05$
($\lambda^* = 3.5\, D_N$, $\delta_0 = 0.02$, $D_N = 3.5\; 10^{-5}\, m$ ).

A first group of calculations concerns the instability and breakup of infinite jets, while the second, less important, is the application of the calculation code to a particular case of finite jets; i.e. the liquid bridge.

## 4.1  Disintegration  Process

### 4.1.1  Liquid-Gas  System

*(a) Analysis of the small disturbance stage.* Figure 4 shows the amplification of the disturbance $\delta$ ($\delta = |R_s - 1|$) in time corresponding to the neck ($\xi = \lambda/2$) and the swell ($\xi = 0$) for two different nondimensional wavenumbers *(x = 0.7 and x = 0.95)* in a semi-logarithmic representation. Linear theory obtained curves (lines with constant slopes) are also drawn. The close agreement obtained in the linear range manifests the correctness of the computation. The amplitude of the disturbance grows exponentially during the early stages of the evolution. In the later stages, nonlinear effects become significant; leading to a change of the slope. According to the nonlinear simulation, the neck contracts faster than the swell grows for a short wavelength *(x = 0.95)* while they have the same amplification rate according to the linear stability analysis.

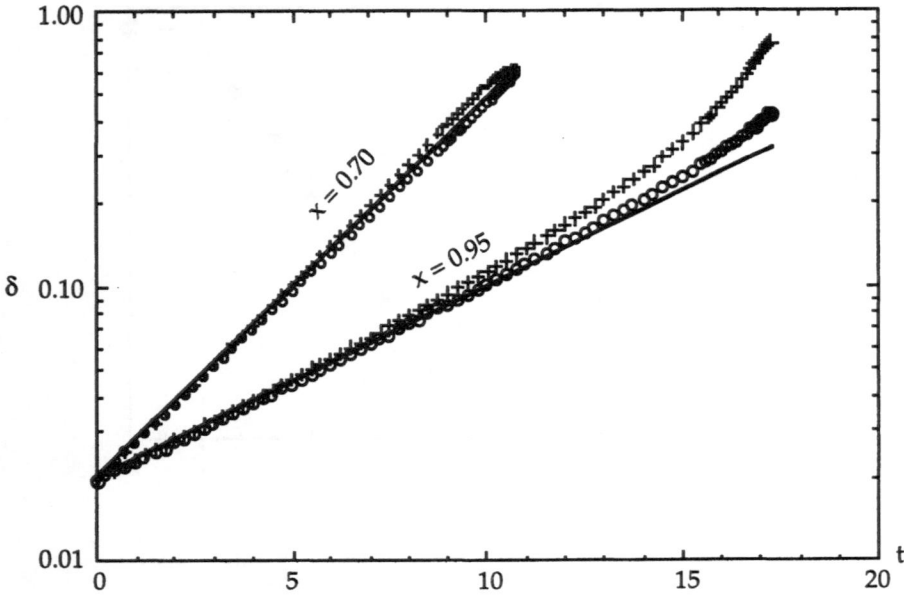

Figure 4.   Growth of the disturbance in time for a jet of water in vacuum
Linear solution: ————— ([2]); Present: + neck; O swell.

*(b) Growth rate and breakup times.* One basic characteristic of jet break-up is the duration of the interval it remains unbroken. The breakup times $T$ and the linear theory obtained growth rate $Q$ are plotted in Fig. 5, in function of the nondimensional wavenumber $x$. The growth rate curve presents a maximum at $x \approx 0.7$. There is no amplification of the perturbation for $x = 0$ (infinite wavelength) and $x = 1$ (the cut-off wavenumber) according to the linear theory analysis. However in the nonlinear simulation, we observed some growing of the perturbation at the cut-off wavelength with an "infinitely small amplification rate". According to the third order nonlinear theory of Yuen [4], the cut-off wavenumber is a little greater than *1*, depending on the initial amplitude of the perturbation. The breakup times are in good agreement with those obtained by Shokoohi [6]. The shortest breakup time is obtained when $x \approx 0.7$; which is in accordance with the linear solution. For wavenumbers included between *0.3* and *0.8*, linear theory predicts relatively the same breakup times although the breaking does not occur at the same points. Nonlinear breakup times are shorter than those of the linear theory. The breakup time becomes longer and longer when $x \longrightarrow 0$ and $x \longrightarrow 1$.

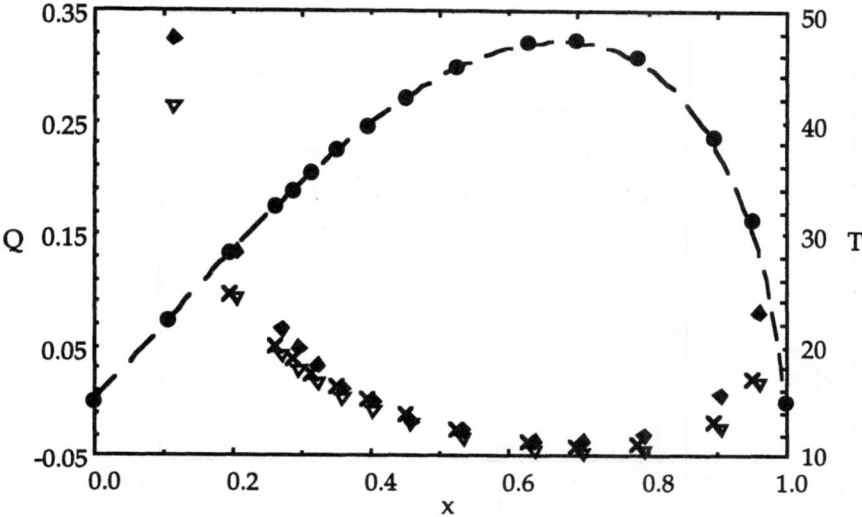

Figure 5    Nondimensional growth rate $Q$ and breakup time $T$ vs. nondimensional
            wavenumber x *(water–vacuum; $\delta_0 = 0.02$). Q:* ●ₒ • Linear theory ([2])
            *T:* ▽  Present results;  ◆ Weber's results;  ✗ Shokoohi's results


## 4.1.2  Liquid-Liquid System

*(a) Analysis of the small disturbance stage.* In Fig. 6, the growth of the swell and the
neck is shown together with that obtained from the modified linear stability analysis
of Tomotika, for $x = 0.7$ and $x = 0.95$. The slopes are constant and equal in the very
early stage of the evolution of each disturbance. Nonlinear effects occur earlier in this
system than in the previous one. And the differences between linear and nonlinear
breakup times are larger. Nonlinear effects increase the kinetics of the interface
deformation with regard to linear ones; this behavior is more pronounced than in the
liquid–vacuum case.

*(b) Typical stages in the disintegration.* The evolution of the jet surface after an
imposed disturbance of the form (1) is shown in Fig. 7, from $t = 0$ to $t = 12.51$. The
system is composed of a jet of water in dodecan continuous phase. The wavelength
corresponds to that of the fastest growing disturbance (according to the linear stability
analysis). The jet contour preserves a sinusoidal form during the early stages of the
perturbation evolution. Thus the growth rate is equal to that obtained from Tomotika
linear analysis; since $R_{max}$ is input large enough to avoid the rigid wall effect. As the
time increases, nonlinear effects become important and the initially sinusoidal shape
of the interface changes to a complex form. After the appearance of a "plateau", two
equal minima develop progressively on both sides of $\lambda/2$ and move symmetrically
toward ends, giving rise to a satellite drop.

218

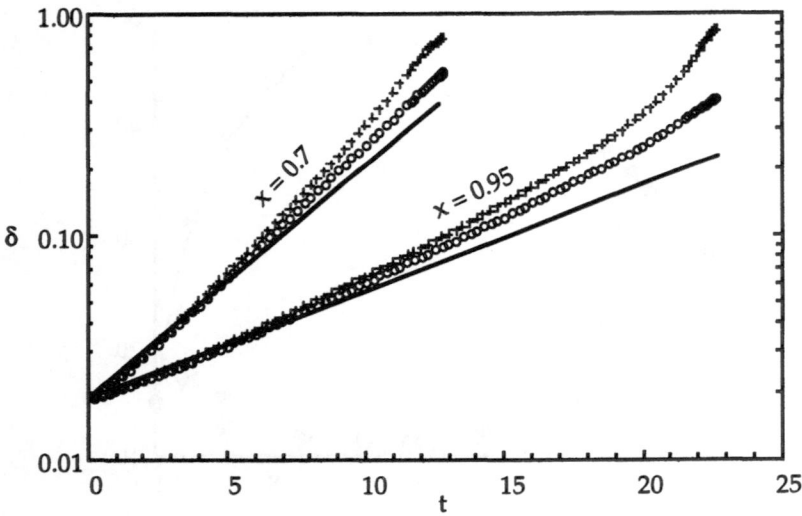

Figure 6   Growth of the disturbance in time. ———— Linear solution
Numerical solution:   + neck;   O swell. *(water–dodecan; $\delta_0 = 0.02$)*

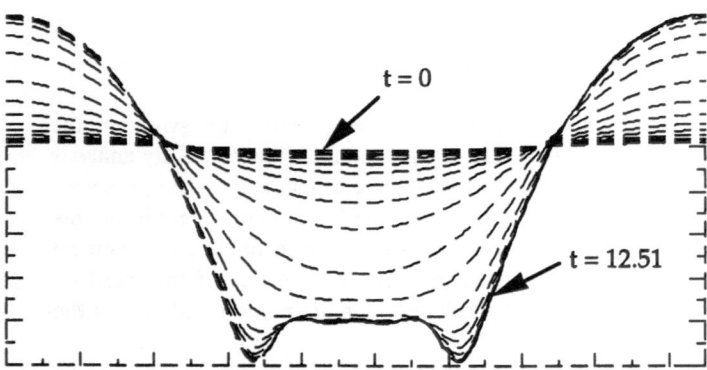

Figure 7.   Evolution of the interface profile in time for a jet of water in dodecan surrounding medium *( $\delta_0$ = 0.02, $\lambda$ = 9.52).*

*(c) Effect of the rigid wall on instability and breakup.* Figures 8 and 9 illustrate the influence of the rigid cylinder bounding the continuous phase on the evolution of perturbations of middle *($\lambda = 9.16$)* and long *($\lambda = 20$)* wavelengths.

In the first case (Fig. 8), the jet interface profiles obtained with $R_{max} = 5$ and $R_{max} = 9$ are practically identical. The corresponding breakup times do not differ (the deviation is less than *1%*). So, the radius of the rigid cylinder is large enough to avoid any wall effect. With $R_{max} = 1.5$, the shape of the interface profile at the breakup is considerably altered. The development of the swell in the radial direction is stopped by the proximity of the wall. The resulting main drop is smaller than that obtained under the condition of no wall effect. This is naturally counterbalanced by the formation of a satellite drop of more important size. The satellite drop seems not to be influenced by the presence of the wall.

In the second case (Fig. 9), there is no wall effect with $R_{max} = 5$. The swell stops to grow in the radial direction when affected by the presence of the rigid wall $(R_{max} = 2)$; but the development of the main drop continues in the axial direction. Contrarily to the case of short and middle wavelengths, the breakup gives rise to bigger main drop and smaller satellite drop. In this case, both the main and the satellite drops are affected by the presence of the rigid wall.

In both cases, the boundedness is a stabilizing factor for the jet instability.

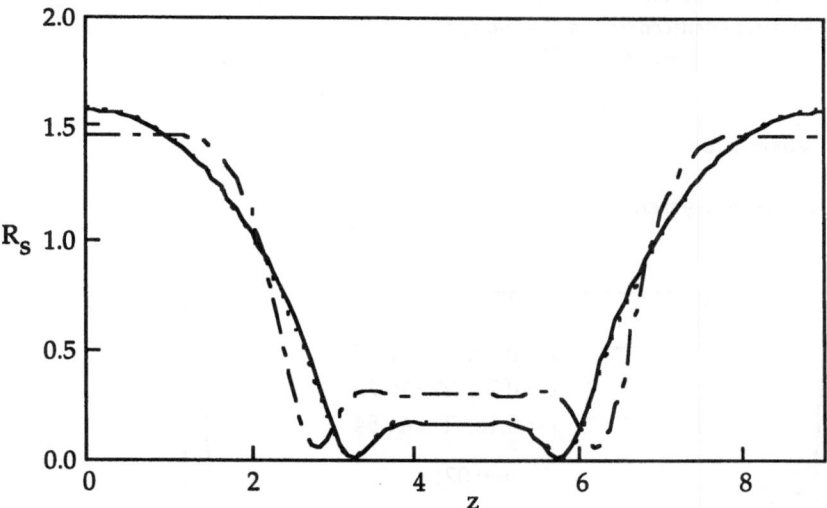

Figure 8.  Effect of the position of the rigid wall $R_{max}$ on jet instability and breakup for a water–dodecan system $(\lambda = 9.16; \delta_0 = 0.02)$.
  — – –$(R_{max} = 1.5; T = 14.8)$  . . . . .$(5; 12.68)$ ————$(9; 12.6)$

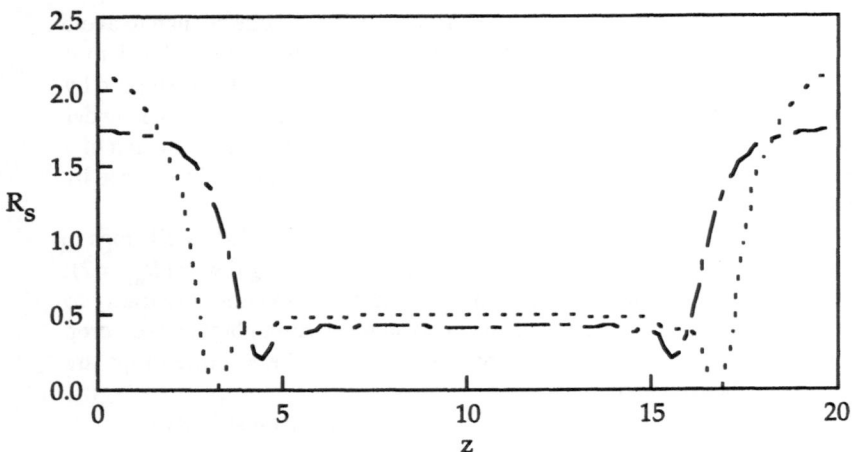

Figure 9. Effect of the position of the rigid wall $R_{max}$ on jet instability and breakup for a water–dodecan system ($\lambda = 20$; $\delta_0 = 0.02$)
——  -  - ($R_{max} = 2$, $T = 21.77$)  . . . . . ($R_{max} = 5$, $T = 18.48$)

## 4.2 Drop Formation

### 4.2.1 Liquid-Gas System

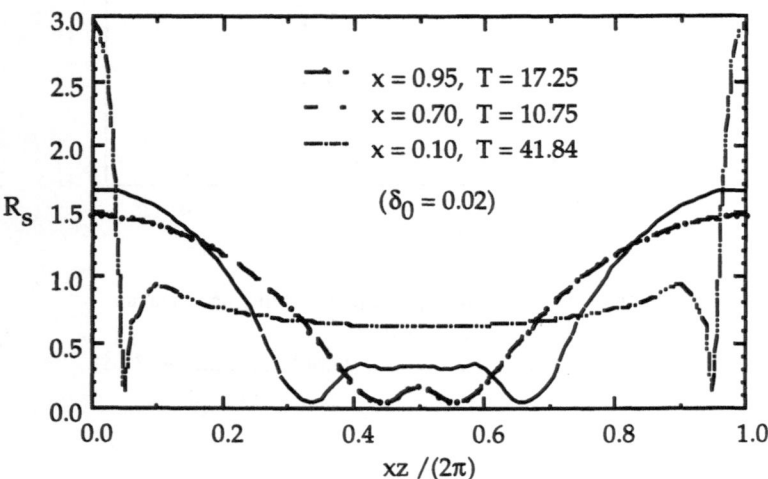

Figure 10. Interface profiles before breakup (water–). Influence of the wavenumber.

Figures 10 shows the interface profiles before breakup. The satellite drop size increases with the wavelength. At short wavelengths, the crest of the satellite and the crest of the main drop are spaced one-half wavelength apart. This indicates that the second harmonic component is mainly responsible for the formation of the satellite in the range of moderate wavelengths. With long wavelengths, the possibility of the formation of two (or more) satellites exists. Here, the satellite drop is a long ligament with a "neck".

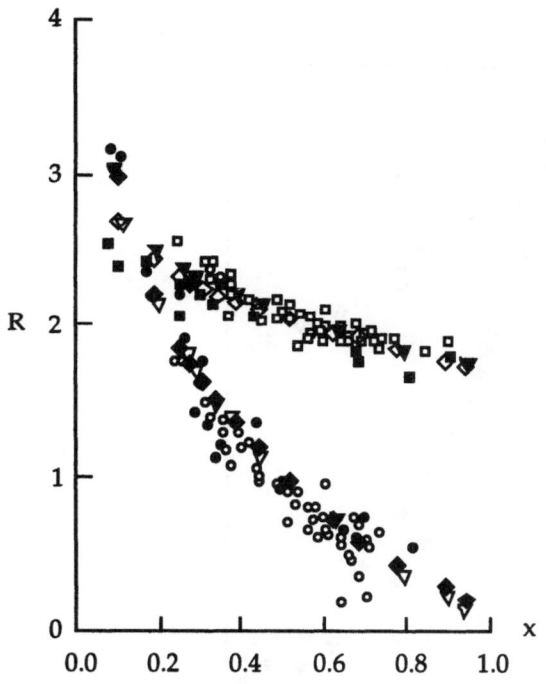

**Figure 11. Drop radii $R$ vs. wavenumber x for a jet of water in vacuum.**
Comparison between the present and the previous works

| | | | | |
|---|---|---|---|---|
| Experiment: | [5] | ■ main drop | ● satellite drop |
| | [7] | ◻ main drop | ○ satellite drop |
| Numerical: | [6] | ▼ main drop | ▽ satellite drop |
| | Present | ◊ main drop | ◆ satellite drop |

Another basic characteristic of the jet break-up is the size of the drops that are formed. After the two breakup points have been identified, the volume of liquid contained in each drop is calculated. The drop volumes are converted to the drop radii assuming that the resulting drops will be spherical. The distribution of predicted and measured drop sizes is shown in Fig. 11. The accordance is very satisfying. According to numerical results, satellite drops exist for all unstable wavenumbers.

222

## 4.2.2 Liquid-Liquid System

Predicted drop radii are represented in Fig. 12 together with those of the water–vacuum system. Satellites are found to exist for all unstable wavelengths in both systems; even if they are too small at shorter wavelengths $(x \longrightarrow 1)$, compared to the main drops. Main drops (respectively satellite drops) in the water–dodecan system are bigger (smaller) than those in the water–vacuum system. Satellite and main drops are comparatively big for long wavelengths.

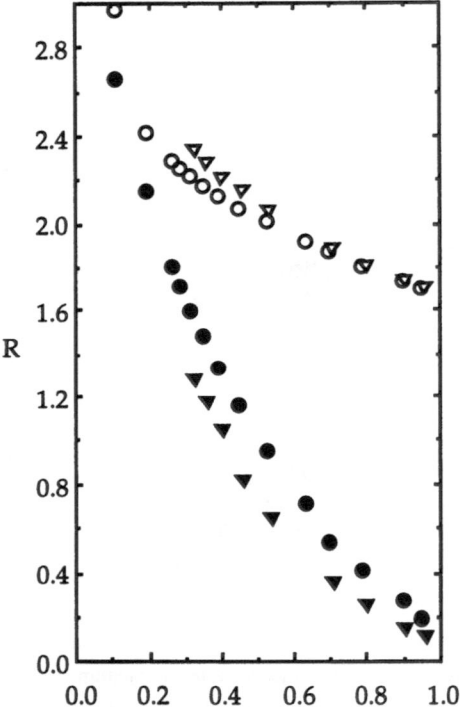

Figure 12. Drop radii $R$ vs. wavenumber $x$: Influence of the continuous phase on drop size
Water–vacuum system:    O    main drop    ● satellite drop.
Water–dodecan system:    ▽    main drop    ▼ satellite drop.

Some theoretical and experimental drop diameters $(\phi)$ are given in Table 1, together with numerical main and satellite ones. The experimental data are the mean diameter of drops formed from laminar jets under the condition of zero relative velocity between the jet and the continuous phase. Theoretical and experimental data are borrowed from Kitamura et al [8]. Experimental data agree with the numerical main diameter within *10%*.

223

Table 1. Comparison between theoretical, numerical and measured (main) drop diameter.

| System N° | Theoretical drop Ø (cm) | Experimental drop Ø (cm) | Numerical drop Ø (cm) main | satellite | Main drop Ø error (%) |
|---|---|---|---|---|---|
| 1–1 | .234 | .237 | .234 | .045 | 1.266 |
| 3–2 | .240 | .250 | .238 | .064 | 4.80 |
| 4–1 | .165 | .183 | .165 | .025 | 9.836 |
| 6–1 | .236 | .255 | .235 | .052 | 7.843 |

## 4.3 Time of Formation of Satellite Drops

Although nonlinear effects are present since the beginning of the destabilization of the jet, they really become noticeable very late and finally show themselves physically, through satellite drop formation. In what follows, we attempt to estimate the time fraction devoted to satellite drop formation. To do this, we follow the $z$-abscissa, $z_{min}$, corresponding to a minimal radius in time (Fig. 13). We count the requested time of formation of the satellite, $T_{fsat}$, from the end of the "plateau" -see § 4.1.2.(b)-, as shown in Fig. 13 for a jet of water in vacuum with $x = 0.7$.

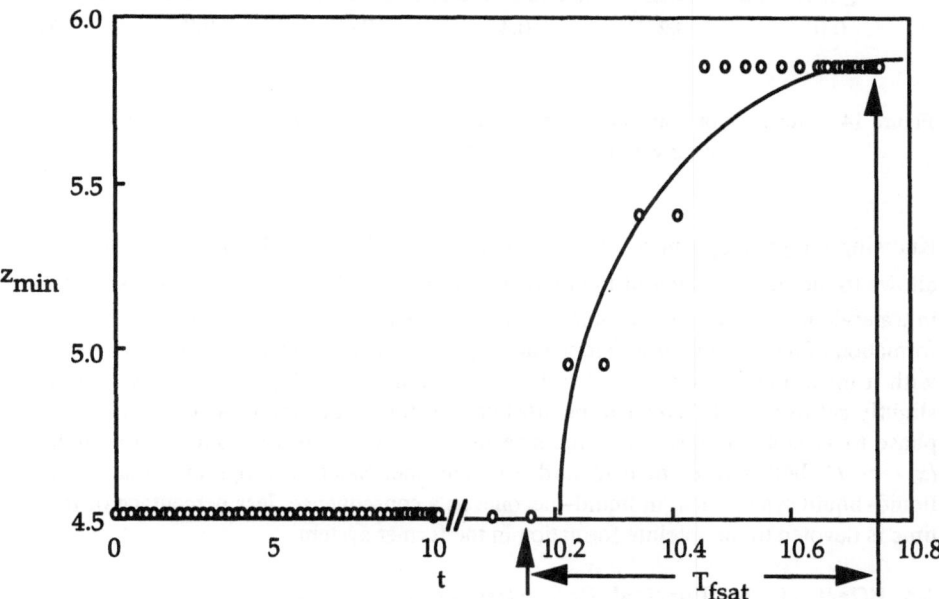

Figure 13. Variation of the $z$-abscissa corresponding to a minimal cross-section in time: determination of the time of formation of satellite drop $T_{fsat}$
(water-vacuum; $x = 0.7$, $d_0 = 0.02$, $T = 10.75$, $T_{fsat} = 0.6$).

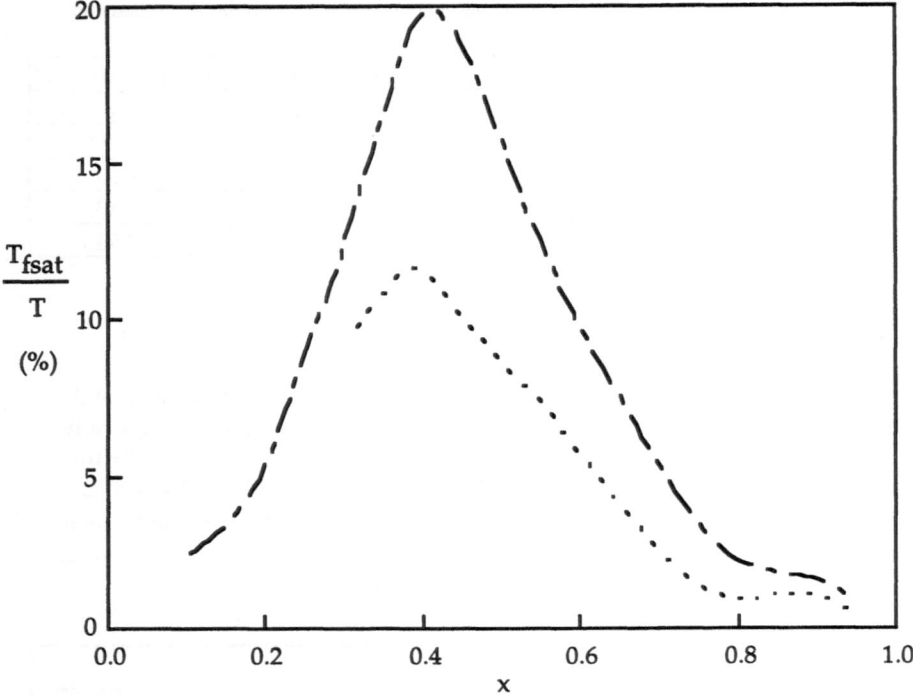

Figure 14. Percentage of time devoted to satellite drop formation $T_{fsat}/T$ vs. wavenumber x:
▬ ▬ ▬ (water-vacuum); . . . . . (water-dodecan)

Knowing the breakup times $T$ for different wavenumbers, the determination of $T_{fsat}$ allows to plot the variation of the ratio $T_{fsat}/T$ (Fig. 14). Figure 14 shows that, both in water-dodecan and water-vacuum systems, less than 20% of the time is spent for the formation of the satellite drop. These curves present the same variation characteristics, with a maximum around $x = 0.4$. The maximum for the liquid–liquid system is slightly set to the left. This can be attributed to the contribution of the continuous phase to viscous effects. When the size of the satellite drop becomes very small $(x \longrightarrow 1)$, both curves tend toward a single one. Satellite drops are smaller in liquid–liquid system than in liquid–gas one; as a consequence, less percentage of the time is devoted to the satellite formation in the former system.

## 4.4 Effect of the Physical Properties on Drop Size

The results of the parametric study dealing with the influence of the physical properties on jet structuration at the breakup are summarized in Table 2. An increase of the interfacial tension leads to a rapid breakup of the jet, but does not alter the drop size.

Increasing the dynamic viscosity and/or the density of the continuous phase reduces (respectively increases) the size of the satellite drop (respectively the main drop). The satellite drop size increases when making the jet denser or viscous in the presence of a viscous continuous phase. For liquid–vacuum systems, it has been observed (in numerical experiments) that making the jet viscous and viscous leads to smaller and smaller satellite drop.

Table 2. Variation of the drop size with the increase of physical properties in liquid-liquid systems.

| | $\mu_1$ ↗ | $\mu_2$ ↗ | $\rho_1$ ↗ | $\rho_2$ ↗ | $\sigma_{12}$ ↗ |
|---|---|---|---|---|---|
| satellite drop size | ↗ | ↘ | ↗ | ↘ | ——— |

## 4.5 Application to Liquid Bridges in Vacuum

Instead of the periodicity boundary conditions, we need to satisfy the nonslip and the zero mass flux at both ends of the jet segment; i.e.:

$$U(0, r) = 0 \qquad V(0, r) = 0 \tag{25}$$
$$U(\lambda, r) = 0 \qquad V(\lambda, r) = 0 \tag{26}$$

where $0 \le r \le 1$. One can easily obtain the corresponding conditions for $\psi$ and $\omega$.

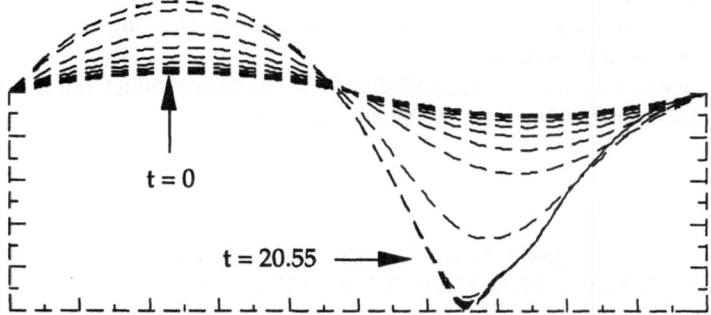

Figure 15. Variation of the interface profile in time for a bridge of water in vacuum (the initial perturbation is asymmetrical with respect to the middle plane of the bridge; $\varepsilon_0 = 0.1$, $\lambda = 6.4$).

Calculations were performed for a jet of water in vacuum, for both asymmetric and symmetric initial disturbances with respect to the middle plane of the bridge. The shapes of the jet profile at $t = 0$ are those given in [10]. For a initially unsymmetrical disturbance, the breaking of the bridge gives rise to two drops of different volumes (Fig. 15). If the initial profile is symmetrical with respect to $\lambda/2$, the breaking of the bridge will be symmetrical only if the initial amplitude of the disturbance is sufficiently big. In this latter case, a satellite drop is eventually formed between the two main drops of equal size (each of them attached to each end disk) (Fig. 16).

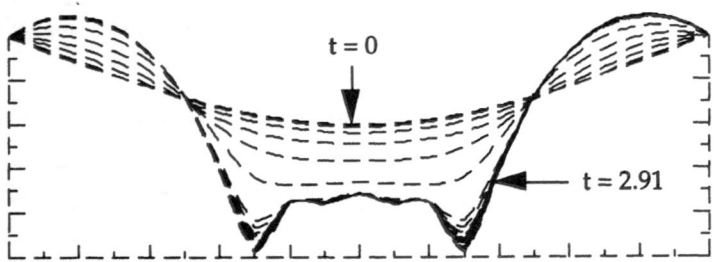

Figure 16. Variation of the interface profile in time for a symmetrically perturbed bridge of water in vacuum ($\varepsilon_0 = 0.4$, $\lambda = 6.4$).

## 5. Conclusion

Nonlinear effects are the cause of appearance of the satellite drops. Satellite drops exist for all unstable wavenumbers in the flow systems that we studied. Nonlinear breakup times are found to be shorter than linear ones, assuming the same initial conditions ($x$ and $\delta_0$). Hydrodynamic effects of the continuous phase reduce the size of the satellite drops. The satellite drop size increases when making the jet denser or viscous for liquid-liquid systems. The evolution of the instability is really affected by the boundedness; but, above a distance of about $5 R_N$ from the jet axis, the jet behaves as in an infinite surrounding medium.

The present numerical code can be applied to finite jet, compound jet, liquid bridge flows provided a modification of some boundary conditions.

## References.

[1] L. RAYLEIGH, Proc. Lond. Math. Soc., **1 0**, pp. 4-13 (1878).
[2] C. WEBER, Z. Angew, Math. Mech., **1 1**, pp. 136-154 (1931).
[3] S. TOMOTIKA, Proc. R. Soc. Lond., Cl A, pp. 322-337 (1935).
[4] M.C. YUEN, J. Fluid Mech., **3 3**, pp. 151-163 (1968).
[5] D.F. RUTLAND and G.J. JAMESON, Chem. Eng. Sci., **2 5**, pp. 1689-1698 (1970).
[6] F. SHOKOOHI, "Numerical Investigation of the Disintegration of Liquid Jets", Ph.D. Thesis, Colombia University, New York, (1976).
[7] P. LAFRANCE, Phys. Fluids, **1 8**, 4, pp. 428-432 (1975).
[8] Y. KITAMURA, H. MISHIMA and T. TAKAHASHI, The Canadian J. Chem. Eng., **60**, pp. 723-731 (1982).
[9] M. CHACHA, "Instabilités Hydrodynamiques des Jets en Systèmes Liquide–Fluide: Analyse Numérique", Thèse de doctorat,Université de Provence, Marseille, (1995).
[10] J. MESEGUER, J. Fluid Mech., **1 3 0**, pp. 123-151 (1983).

# Microwave heating as a tool for coupling Marangoni and Hickman instabilities

D. Stuerga[1], A. Steinchen-Sanfeld[2], and M. Lallemant[1].

[1] Laboratoire de Recherches sur la Réactivité des Solides URA23 CNRS
Université de Bourgogne, Fac. Mirande, BP 138 21004 Dijon Cédex, France.
[2] Laboratoire de Thermodynamique, Boulevard de l'escadrille Normandie
Niemen, Université d'Aix Marseille III, Marseille, France.

**Abstract.** The aim of this paper is the linear stability analysis of an evaporating interface in order to justify hydrodynamic behaviors observed during evaporation of water and ethanol simultaneously under microwave irradiation and reduced pressure. In relation to the local character of the thermal conversion of electromagnetic energy, and of the thermal dependency of the dielectric loss, it seems possible under microwave heating to induce a specific hydrodynamic compared to those induced by conventional heating.

**Keywords.** Microwave Heating, Rayleigh-Benard, Marangoni, Hickman

## 1 Introduction

When an evaporating system is removed from equilibrium conditions, by reducing the pressure for example, successive instabilities can emerge as described by Figure 1. Four types of convective patterns may be distinguished: convective cells (steady and unsteady convection), thermals, ribs and surface distortions (Marangoni instability). Another little known mechanism was first noted and correctly interpreted by Hickman [1952, 1972, 1976]. Since then it has been no longer been studied, despite its dramatic effects on liquid evaporation at low pressure. The Hickman instability is first characterized by located hot spots at the evaporating surface, and later by an explosive increase in the evaporation rate. At the same time, the mechanical stability of the surface is broken and violent distortions are observed. In confined medium the Hickman instability induces a spontaneous and quite violent vaporization resulting from strong distortions of evaporating surface.

The Marangoni number Ma has been defined as follow :

$$Ma = \left(\frac{\partial \sigma}{\partial T}\right)\frac{\beta \delta^2}{K\mu} \tag{1}$$

with the thermal variation of surface tension, $\beta$ the temperature gradient, $\delta$ the thickness of the layer, K the thermal diffusivity and $\mu$ the viscosity of the fluid.

particularly for a rapidly evaporating liquid. The effect of the vapour recoil instability or the Hickman instability is always neglected. Thus, it seems indispensable to take into account effects of the discontinuities in both linear momentum and kinetic energy at the evaporating interface. Palmer [1976] has proposed a theoretical interpretation of the Hickman instability and its coupling effect with Marangoni instability or surface tension effects. Following the Palmer's work, we propose now, the linear stability analysis of the boundary layer located just below the evaporating interface, with taking into account the dual effects of surface tension and vapour recoil. The microwave heating effect resulting from the shape of the thermal spatial profile is obtained experimentally.

## 2 The destabilizing mechanisms

According to Palmer [1976] many destabilizing mechanisms can be attributed that are related to wavelength of the disturbance. Five mechanisms have been defined by the analysis: differential recoil vapour, surface tension gradient, fluid inertia, viscous dissipation, and decreasing boundary layer. A local fluctuation of surface temperature increases the evaporation rate and decreases the surface tension. The increase in evaporation rate induces an increase of the normal force on the evaporating interface (recoil force) which induces a local depression, or a crater in the evaporating surface. Moreover, the surface tension gradient also induces a flow of hot liquid, and these two mechanisms produce auto-amplification of the eventual thermal fluctuations. Flexibility of the evaporating interface can also induce the instability of the evaporating system. Fluid streamlines converge in the liquid phase resulting from fluid inertia. This phenomenon produce a decrease of the pressure within the liquid, and an increase in the vapour pressure resulting in a normal force on the evaporating interface. For viscous dissipation, in the vicinity of the evaporating interface the fluid accelerations are high and the viscous friction heats the liquid, increasing the initial disturbance. This mechanism is preponderant for very short wavelength (i.e. less than one-tenth of the thermal boundary layer thickness).Decreasing of local boundary layer can enhance the heat transport to the evaporating surface, and induce amplification of the initial fluctuation. This mechanism is preponderant for very long wavelengths (i.e. more than 1000 times the thermal boundary layer thickness).

The differential vapour recoil, fluid inertia, and viscous dissipation mechanisms will be amplified by a decrease of the gas-phase pressure in relation to an increase in crater formation.

## 3 The problem formulation

The linear stability analysis is a normal mode analysis. The disturbance must verify the mass, energy, and momentum balances. The z-coordinate (normal to the evaporating interface and increasing in the vapour-phase ) used in this model moves with the unperturbed interface.

### 3.1 The unperturbed system

The balance mass at the evaporative interface is given by Equation (3):

The Hickman number Hi has been defined as :

$$Hi = \left(\frac{\partial \theta}{\partial T}\right) \frac{\theta \beta \delta^2 \mu_{Vap.}}{\rho_{Liq.} K_{Liq.} \sigma} \left(\frac{1}{\rho_{Vap.}} - \frac{1}{\rho_{Liq.}}\right) \qquad (2)$$

with $\theta$ the evaporation rate, with $\partial\theta/\partial T$ the thermal variation of the evaporation rate, $\delta$ the thickness of the layer, $\beta$ the temperature gradient, $\mu$ the viscosity, K the thermal diffusivity, $\sigma$ the surface tension, $\rho$ the specific mass. The subscripts Liq. and Vap. refer to the liquid-phase and to the gas-phase respectively.

This non-dimensional number is the ratio between destabilizing forces of the differential vapour recoil and the vapour viscosity, and the stabilizing action of the surface tension and the thermal diffusivity. In the absence of any variations in surface tension, the threshold criteria are expressed in terms of the Hickman number. Likewise in the absence of variation in the evaporation rate, the stability criteria for the system are expressed in terms of the Marangoni number. Courville [1987], Courville et al. [1991] in his study of ethanol, has used an estimation of boundary layer thickness and temperature gradient. Now, we will used our experimental results for ethanol and water Stuerga [1989], Stuerga and Lallemant [1993], and it is possible to have a true estimation of these parameters. In this study we will consider temperature difference as linear across the thickness $\delta$ of the boundary layer .

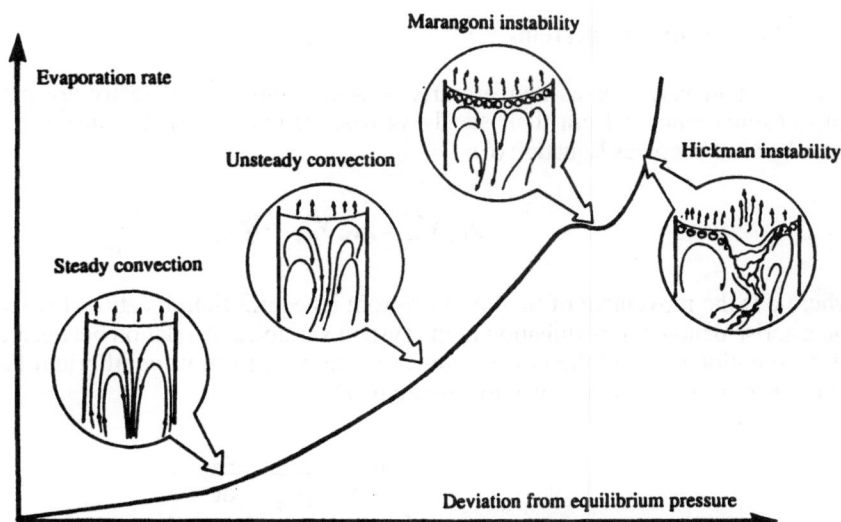

Figure 1: Convection patterns observed during evaporation in relation to deviation from equilibrium pressure.

The body of a deep layer of water in convection is rather well mixed, however near the evaporating surface, motions are constricted, and a boundary layer exists where conduction transport is preponderant. Thus, the major part of the temperature gradient occurs within this boundary layer. If the linear stability analysis of the Rayleigh-Bénard instability is well known (Chandrasekar, S. [1961]), it is different for the boundary layer,

$$\rho_{Liq.} V_{Liq.} = \rho_{vap.} V_{vap.} = \theta \qquad (3)$$

where $\rho$ is the specific mass, V the vertical component of the liquid velocity, and $\theta$ the mass rate of evaporation.

The discontinuity in fluid velocity resulting from the phase change, induces a discontinuity in the rate of momentum transport. This difference is equal to a dynamic pressure ($P_{Liq.}$-$P_{Vap.}$) which gives the vapour recoil force as described by Equation (4), and which is the balance rate transport momentum:

$$P_{Liq.} - P_{vap.} = \theta^2 \left( \frac{1}{\rho_{vap.}} - \frac{1}{\rho_{Liq.}} \right) \qquad (4)$$

If we assume that the heat conduction in the gas-phase is negligible then the energy balance is given by Equation (5):

$$\theta \Delta H_{vaporiz.} + \frac{1}{2} \theta^3 \left( \frac{1}{\rho_{vap.}^2} - \frac{1}{\rho_{Liq.}^2} \right) + k_{Liq.} \frac{dT_{Liq.}}{dz} = 0 \qquad (5)$$

where $\Delta H_{vaporiz}$ is the vaporization enthalpy of the liquid, $k_{Liquid.}$is the liquid thermal conductivity and $T_{Liquid}$ is the liquid temperature.

### 3.2 The perturbed system

If the system variables as temperature, pressure, and fluid velocity are perturbed an infinitesimal amount from their steady values, we obtain new balance equations. The mass balance becomes Equation (6):

$$\rho_{Liq.} V_{Liq.}^* - \rho_{vap.} V_{vap.}^* = \left( \rho_{Liq.} - \rho_{vap.} \right) \frac{d\varsigma}{dt} \qquad (6)$$

where $\varsigma$ is the movement of the interface from its equilibrium location, t is the time and the asterisc denote the pertubation in the system variables. A relation between the change of evaporation rate and the movement of the interface from its equilibrium location can be defined; this relation is given by Equation (7):

$$V_{vap.}^* = \frac{\theta^*}{\rho_{vap.}} + \frac{d\varsigma}{dt} \qquad (7)$$

The continuity of the tangential component of the velocity at the interface results in the following boundary condition:

$$\theta^* \left( \frac{1}{\rho_{Liq.}} - \frac{1}{\rho_{vap.}} \right) \left( \frac{\partial^2 \varsigma}{\partial x^2} + \frac{\partial^2 \varsigma}{\partial y^2} \right) = \frac{\partial V_{Liq.}^*}{\partial z} - \frac{\partial V_{vap.}^*}{\partial z} \qquad (8)$$

The normal momentum component balance gives the Equation (9):

$$\left(P_{V_{ap.}}^{\bullet} - P_{Liq.}^{\bullet}\right) + 2\theta^{\bullet}\theta\left(\frac{1}{\rho_{V_{ap.}}} - \frac{1}{\rho_{Liq.}}\right) +$$

$$2\left(\mu_{Liq.}\frac{\partial V_{Liq.}^{\bullet}}{\partial z} - \mu_{V_{ap.}}\frac{\partial V_{V_{ap.}}^{\bullet}}{\partial z}\right) - \sigma\left(\frac{\partial^2\varsigma}{\partial x^2} + \frac{\partial^2\varsigma}{\partial y^2}\right) = 0 \qquad (9)$$

where $\sigma$ is the surface tension and $\mu$ the fluid viscosity. The tangential momentum component balance gives the Equation (10):

$$\frac{\partial^2\sigma^{\bullet}}{\partial x^2} + \frac{\partial^2\sigma^{\bullet}}{\partial y^2} = \mu_{Liq.}\left[\left(\frac{\partial^2 V_{Liq.}^{\bullet}}{\partial x^2} + \frac{\partial^2 V_{Liq.}^{\bullet}}{\partial y^2}\right) - \frac{\partial^2 V_{Liq.}^{\bullet}}{\partial z^2}\right] -$$

$$\mu_{V_{ap.}}\left[\left(\frac{\partial^2 V_{V_{ap.}}^{\bullet}}{\partial x^2} + \frac{\partial^2 V_{V_{ap.}}^{\bullet}}{\partial y^2}\right) - \frac{\partial^2 V_{V_{ap.}}^{\bullet}}{\partial z^2}\right] \qquad (10)$$

where $\sigma^*$ is the perturbation in surface tension. Energy balance gives the Equation (11):

$$\theta^{\bullet}\Delta H_{vaporiz.} + k_{Liq.}\frac{\partial T_{Liq.}^{\bullet}}{\partial z} + \frac{3}{2}\theta^{\bullet}\theta^2\left(\frac{1}{\rho_{V_{ap.}}^2} - \frac{1}{\rho_{Liq.}^2}\right) -$$

$$2\theta\left(\frac{\mu_{V_{ap.}}}{\rho_{V_{ap.}}}\frac{\partial V_{V_{ap.}}^{\bullet}}{\partial z} - \frac{\mu_{Liq.}}{\rho_{Liq.}}\frac{\partial V_{Liq.}^{\bullet}}{\partial z}\right) = 0 \qquad (11)$$

All the perturbed equations are valid at the interface, i.e. $z=\varsigma$. The Equations (6) to (8) remain unchanged at $z=0$. Equation (9) can be modified with the use of a Taylor series expansion of the perturbed pressure. Equation (12) gives the required transformation:

$$\left[P_{Liq.}^{\bullet} - P_{V_{ap.}}^{\bullet}\right]_{z=\varsigma} = \left[P_{Liq.}^{\bullet} - P_{V_{ap.}}^{\bullet}\right]_{z=0} - g\varsigma\left(\rho_{Liq.} - \rho_{V_{ap.}}\right) \qquad (12)$$

where g is the acceleration due to gravity. Equations (10) and (11) can be modified by use of transformations given by the Equation (13) and (14):

$$\left[T_{Liq.}^{\bullet}\right]_{z=\varsigma} = \left[T_{Liq.}^{\bullet}\right]_{z=0} - \beta\varsigma \qquad (13)$$

$$\frac{\partial^2\sigma^{\bullet}}{\partial x^2} + \frac{\partial^2\sigma^{\bullet}}{\partial y^2} = \frac{\partial\sigma}{\partial T}\left[\frac{\partial^2 T_{Liq.}^{\bullet}}{\partial x^2} + \frac{\partial^2 T_{V_{ap.}}^{\bullet}}{\partial y^2}\right]_{z=\varsigma} \qquad (14)$$

The normal force balance applied at $z=0$ gives different destabilizing mechanisms for rapidly evaporating liquids:

$$\left(P_{Liq.}^{\bullet} - P_{Vap.}^{\bullet}\right) + 2\theta^{\bullet}\theta\left(\frac{1}{\rho_{Liq.}} - \frac{1}{\rho_{Vap.}}\right)$$

$$+2\left(\mu_{Liq.}\frac{\partial V_{Liq.}^{\bullet}}{\partial z} - \mu_{Vap.}\frac{\partial V_{Vap.}^{\bullet}}{\partial z}\right) \tag{15}$$

$$+g\varsigma\left(\rho_{Liq.} - \rho_{Vap.}\right) - \sigma\left(\frac{\partial^2\varsigma}{\partial x^2} + \frac{\partial^2\varsigma}{\partial y^2}\right) = 0$$

Far from the evaporating interface any perturbation must be equal to zero, as described by Equations (16) and (17).

$$\lim_{z\to-\infty}\left(V_{Liq.}^{\bullet}\right) = \lim_{z\to-\infty}\left(\frac{\partial V_{Liq.}^{\bullet}}{\partial z}\right) = \lim_{z\to-\infty}\left(T_{Liq.}^{\bullet}\right) = 0 \tag{16}$$

$$\lim_{z\to\infty}\left(V_{Vap.}^{\bullet}\right) = \lim_{z\to\infty}\left(\frac{\partial V_{Vap.}^{\bullet}}{\partial z}\right) = 0 \tag{17}$$

## 3.3 The marginal stability state

The linearized equations of mass, energy, and momentum balance define entirely the responses of the evaporating liquids. The z-dependent part of the perturbations must satisfy the two-dimensional wave equation:

$$\left(\frac{\partial^2 A(x,y)}{\partial x^2} + \frac{\partial^2 A(x,y)}{\partial y^2}\right) + k^2 A(x,y) = 0 \tag{18}$$

After non-dimensionalization, the equations of momentum, energy, and mass balances give the Equations (33) to (48). Previously the non-dimensional variables of the system are defined by Equations (19) to (25):

$$z \to \zeta = \frac{z}{\delta} \tag{19}$$

$$V_{Vap.}^{\bullet} \to \tilde{V}_{Vap.} = \frac{V_{Vap.}^{\bullet}\mu_{Vap.}\delta}{k_{Liq.}\mu_{Liq.}} \tag{20}$$

$$V_{Liq.}^{\bullet} \to \tilde{V}_{Liq.} = \frac{V_{Liq.}^{\bullet}\delta}{k_{Liq.}} \tag{21}$$

$$P_{Vap.}^{\bullet} \to \tilde{P}_{Vap.} = \frac{P_{Vap.}^{\bullet}\delta^2}{k_{Liq.}\mu_{Liq.}} \tag{22}$$

$$P_{Liq.}^{\bullet} \to \tilde{P}_{Liq.} = \frac{P_{Liq.}^{\bullet} \delta^2}{k_{Liq.} \mu_{Liq.}} \tag{23}$$

$$T_{Liq.}^{\bullet} \to \tilde{T}_{Liq.} = \frac{T_{Liq.}^{\bullet}}{\beta \delta} \tag{24}$$

$$\varsigma \to \tilde{\varsigma} = \frac{\varsigma}{\delta} \tag{25}$$

The non-dimension numbers which will appear in calculus are defined by Equations (26) to (32):

The crispation number:

$$Cr = \frac{\mu_{Liq.} k_{Liq.}}{\sigma \delta} \tag{26}$$

The Reynolds number:

$$Re = \frac{\theta \delta}{\mu_{Liq.}} \tag{27}$$

The Prandtl number:

$$Pr = \frac{\nu_{Liq.}}{k_{Liq.}} \tag{28}$$

The Bond number:

$$Bo = \frac{\delta^2 g \left( \rho_{Liq.} - \rho_{v\mathfrak{p}.} \right)}{\sigma} \tag{29}$$

The Brinkman number:

$$Br = \frac{\theta \nu_{Liq.}^2}{\beta k_{Liq.} \delta^2} \tag{30}$$

The viscosity and the density ratios:

$$N\mu = \frac{\mu_{Liq.}}{\mu_{v\mathfrak{p}.}} \tag{31}$$

$$N\rho = \frac{\rho_{Liq.}}{\rho_{Vap.}} \tag{32}$$

The Equations (33) and (34) are the momentum balance for each phase:

$$\left(\frac{\partial^2}{\partial\zeta^2} - k^2\right)\tilde{V}_{Vap.} - \text{Re}\,N_\mu \frac{\partial \tilde{V}_{Vap.}}{\partial \zeta} - \frac{\partial \tilde{P}_{Vap.}}{\partial \zeta} = 0 \tag{33}$$

$$\left(\frac{\partial^2}{\partial\zeta^2} - k^2\right)\tilde{V}_{Liq.} - \text{Re}\frac{\partial \tilde{V}_{Liq.}}{\partial \zeta} - \frac{\partial \tilde{P}_{Liq.}}{\partial \zeta} = 0 \tag{34}$$

The Equations (35) and (36) are obtained from the curl of the previous equations (33) and (34):

$$\left(\frac{\partial^2}{\partial\zeta^2} - k^2\right)\left(\frac{\partial^2}{\partial\zeta^2} - k^2 - \text{Re}\,N_\mu \frac{\partial}{\partial\zeta}\right)\tilde{V}_{Vap.} = 0 \tag{35}$$

$$\left(\frac{\partial^2}{\partial\zeta^2} - k^2\right)\left(\frac{\partial^2}{\partial\zeta^2} - k^2 - \text{Re}\frac{\partial}{\partial\zeta}\right)\tilde{V}_{Liq.} = 0 \tag{36}$$

The Equations (37) and (38) are the divergence of the Equations (33) and (34):

$$\left(\frac{\partial^2}{\partial\zeta^2} - k^2\right)\tilde{P}_{Vap.} = 0 \tag{37}$$

$$\left(\frac{\partial^2}{\partial\zeta^2} - k^2\right)\tilde{P}_{Liq.} = 0 \tag{38}$$

The energy balance give the Equations (39) and (40):

$$\left(\frac{\partial^2}{\partial\zeta^2} - k^2 - \text{Re}\,\text{Pr}\frac{\partial}{\partial\zeta}\right)\tilde{T}_{Liq.} = -\tilde{V}_{Liq.} \text{ for } -1 \le \zeta \le 0 \tag{39}$$

$$\left(\frac{\partial^2}{\partial\zeta^2} - k^2 - \text{Re}\,\text{Pr}\frac{\partial}{\partial\zeta}\right)\tilde{T}_{Liq.} = 0 \text{ for } -\infty \le \zeta \le -1 \tag{40}$$

Similarly, the boundary conditions in non-dimension form give the Equations (41) to (48). The Equation (7) gives Equation (41):

$$\operatorname{Re}\operatorname{Pr}\operatorname{Cr}\left(1-\frac{1}{N\rho}\right)\tilde{V}_{\text{vap.}} - \operatorname{Hi}\left(\tilde{T}_{\text{Liq.}} - \xi\right) = 0 \tag{41}$$

The Mass balance (Equation (6)) gives the Equation (42):

$$N_\rho \tilde{V}_{\text{Liq.}} - N_\mu \tilde{V}_{\text{vap.}} = 0 \tag{42}$$

The continuity of the tangential component of the velocity (Equation (8)) gives the Equation (43):

$$N\rho \frac{\partial \tilde{V}_{\text{vap.}}}{\partial \zeta} - \frac{\partial \tilde{V}_{\text{Liq.}}}{\partial \zeta} + \operatorname{Re}\operatorname{Pr}(N\rho - 1)k^2\xi = 0 \tag{43}$$

The momentum balance (10) and (15) gives the Equations (44) and (45):

$$\operatorname{Cr}\left(\tilde{P}_{\text{Liq.}} - \tilde{P}_{\text{vap.}}\right) + 2\operatorname{Cr}\left(\frac{\partial \tilde{V}_{\text{vap.}}}{\partial \zeta} - \frac{\partial \tilde{V}_{\text{Liq.}}}{\partial \zeta}\right) - 2\frac{\operatorname{HiN}\mu}{\operatorname{Pr}}\left(\tilde{T}_{\text{Liq.}} - \xi\right) - \left(k^2 + \operatorname{Bo}\right)\xi = 0 \tag{44}$$

$$k^2\operatorname{Ma}\left(\tilde{T}_{\text{Liq.}} - \xi\right) + \frac{\partial^2 \tilde{V}_{\text{Liq.}}}{\partial \zeta^2} - \frac{\partial^2 \tilde{V}_{\text{vap.}}}{\partial \zeta^2} + k^2\left(\tilde{V}_{\text{Liq.}} - \tilde{V}_{\text{vap.}}\right) = 0 \tag{45}$$

The energy balance (11) gives the Equation (46):

$$\operatorname{Pr}\frac{\partial \tilde{T}_{\text{Liq.}}}{\partial \zeta} + \frac{\tilde{V}_{\text{Liq.}}\left[1 + \operatorname{Br}\operatorname{Re}^2\left(N\rho^2 - 1\right)\right]}{\operatorname{Re}} + 2\operatorname{Br}\left(\frac{\partial \tilde{V}_{\text{Liq.}}}{\partial \zeta} - N\rho\frac{\partial \tilde{V}_{\text{vap.}}}{\partial \zeta}\right) = 0 \tag{46}$$

Equations (16, 17 and 18) give equations (47, 48):

$$\lim_{\zeta \to \infty}\left(\tilde{V}_{\text{vap.}}\right) = \lim_{\zeta \to \infty}\left(\frac{\partial \tilde{V}_{\text{vap.}}}{\partial \zeta}\right) = 0 \tag{47}$$

$$\lim_{\zeta \to -\infty}\left(\tilde{V}_{\text{Liq.}}\right) = \lim_{\zeta \to -\infty}\left(\frac{\partial \tilde{V}_{\text{Liq.}}}{\partial \zeta}\right) = \lim_{\zeta \to -\infty}\left(\tilde{T}_{\text{Liq.}}\right) = 0 \tag{48}$$

### 3.4 The general solution

The general solutions which verify Equations (35) to (40) simultaneously with the boundary conditions (Equations (41) to (48)) and with Equations (49) to (55) are :

$$\tilde{V}_{\text{vap.}} = C_{11}e^{-\alpha\zeta} + C_{12}e^{-\mathcal{N}\zeta} \tag{49}$$

$$\tilde{V}_{Liq.} = C_{21}e^{r_{Liq}\zeta} + C_{22}e^{r_{Liq.}\zeta} \tag{50}$$

$$\tilde{T}_{Vap.} = C_{31}e^{r_{Liq}\zeta} + C_{32}e^{(RePr-q)\zeta} + C_{21}e^{\frac{r_{Liq}\zeta}{a\,RePr}} + C_{22}e^{\frac{r_{Liq}\zeta}{r_{Liq.}\,Re(Pr-1)}} \quad \text{for } \zeta \geq -1 \tag{51}$$

$$\tilde{T}_{Liq.} = C_{33}e^{r_{Liq}\zeta} \quad \text{for } \zeta < -1 \tag{52}$$

with

$$r_{Vap.} = \frac{1}{2}\left(\sqrt{Re^2\,N\mu^2 + 4k^2} - Re\,N\mu\right) \tag{53}$$

$$r_{Liq.} = \frac{1}{2}\left(\sqrt{Re^2 + 4k^2} + Re\right) \tag{54}$$

$$q = \frac{1}{2}\left(\sqrt{Re^2\,Pr^2 + 4k^2} + Re\,Pr\right) \tag{55}$$

The integration constants $C_{ij}$ can be determinated from Equations (41) to (48), and with Equations (33) to (40). Following, Scriven and Sternling [1964], substitution of these results into the shear-stress balance at the interface (Equation (45)) yields the characteristic Equations (56) relating to the Hickman number, to the wave number, and to the Marangoni number.

$$Hi = \left[\frac{Re^2\,Pr^2\,Cr(N\rho - 1)}{N\mu}\right]\left[\frac{k_p q\psi_1}{N\rho\psi_2} - \frac{k_p(k_p - r_{Vap.})\psi_3 Ma}{\psi_2\,Re\,Pr}\right] \tag{56}$$

The equations defining $\psi_1$ to $\psi_5$ are very complicated and more details can be found in Palmer [1976].

## 3.5 General formulation including surface tension effects and evaporation rate variation

Because of their similarity in requiring interfacial shearing of the liquid to induce convection and amplify disturbances, the interaction between the surface tension mechanism and the differential vapour recoil mechanism can be considerable. In order to express this interaction between these two destabilizing mechanisms, the stability criteria for systems in which both mechanisms are operative are presented as a normalized critical Marangoni number versus a normalized critical Hickman number. Thus kind of coupling analysis was first made for Rayleigh and Marangoni number by Nield [1964]. These normalized critical numbers are the ratios between the critical numbers for the first mechanism in presence of the other mechanism, to the critical numbers for the first mechanism isolated from the other one. The drawing up of the coupling diagram is obtained by the Equation (56). In fact in the general solution given by the Equation (56), we consider simultaneously the Marangoni and the Hickman number. According to experimental results, the critical numbers corresponding to non-coupling conditions are known, and the critical Marangoni number with a given Hickman number can be calculated. The new critical values obtained are affected, expressing then the coupling of the two mechanisms. For perfect coupling, it was determined that the Marangoni and the Hickman normalized numbers must approximately follow the following Equation (57) at the critical threshold:

$$\left(\frac{Ma}{Ma^{\circ}}\right)_{crit.} + \left(\frac{Hi}{Hi^{\circ}}\right)_{crit.} = 1 \tag{57}$$

where $Ma^{\circ}_{crit}$ is the critical Marangoni number when the Hickman number is equal to zero (no lateral variation of evaporation rate), and $Hi^{\circ}_{crit}$ is the critical Hickman number when Marangoni number is equal to zero (no surface tension gradient). For perfect coupling the equation (57) gives a straight line in the diagram, normalized critical Marangoni number versus normalized critical Hickman number (such diagram is described by Figure 2 (a) and (b)). The slope of the straight line is fixed for a given liquid. In this diagram the values on the axis correspond to non-coupling conditions. For perfect coupling the values of the critical numbers are complementary and additive. For example for a normalized Hickman number equal to 0.75 corresponding to a stable state for non-coupling conditions, a normalized Marangoni number equal to 0.25 is sufficient to destabilize the system. The coupling diagram corresponding to our experimental results is described by Figure 2 (a) and (b).

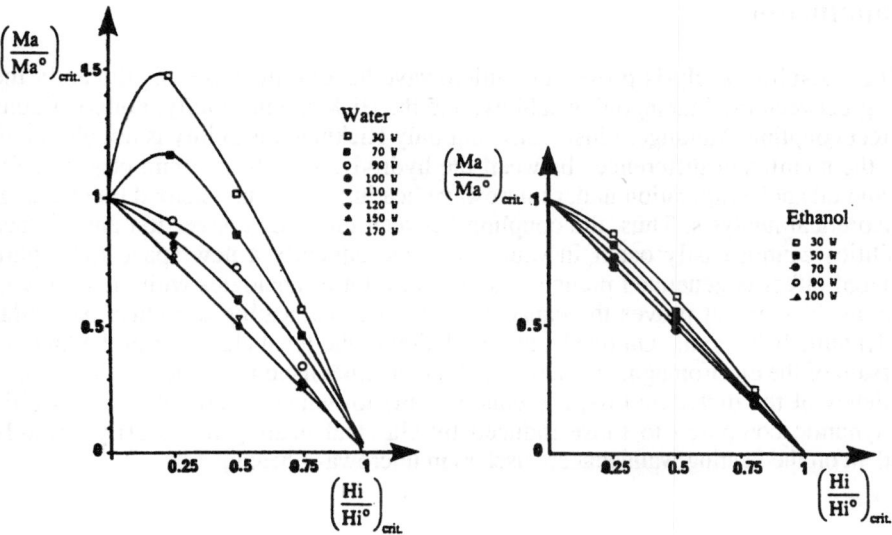

Figure 2: The coupling diagram for (a) water and (b) ethanol for different microwave power levels (normalized critical Marangoni and Hickman number).

In the two cases ((a) water and (b) ethanol) microwave heating induces a strong coupling of the two mechanisms. The intrinsic coupling of the Hickman instability and the Marangoni instability is higher for water than for ethanol. Perfect coupling appears only at high microwave power. The increase of coupling appears as convergence of wave numbers, respectively a decreasing for the Hickman instability wave number (one-dimensional phenomenon), and an increasing of the Marangoni instability wave number two-dimensional phenomenon). Thus, as the evaporation rate is increased (inducing an increase of the vapour recoil effect ) the system may first exhibit convection driven by surface tension, and then become stable undergoing a transition leading to a time periodic phenomenon, and finally become unstable owing to the effect of the differential vapour recoil. This intermediate area of stability (wider in water than in ethanol because of the coupling conditions resulting from the intrinsic properties) is due to the stabilizing effect of evaporative cooling on instability driven by surface tension.

Consequently, the discrepancy between the experimental threshold and the critical values obtained by the non-coupling approach can be explained by the intrinsic non-coupling of the two instabilities. Now, the theoretical results are in good agreement with the experimental observations. In ethanol the perfect coupling explains the absence of the Marangoni instability for high microwave power level. Because of the perfect coupling between the Hickman instability and the Marangoni effect, when the Marangoni starts, the Hickman effect immediately appears inducing destabilization of the system. For water the mechanism is quite different because of the poor coupling of the two instabilities. Thus the Marangoni instability has its own existence independent of the Hickman instability. Finally, microwave heating seems a tool which amplifies the coupling between the surface instabilities.

## 3 Conclusions

The linear stability analysis proves that microwave heating increases significantly the coupling between the Marangoni instability, and the Hickman instability. For conditions of perfect coupling Marangoni instability, but only Hickman instability is not observed.

Now, the significant differences between the hydrodynamic behaviours observed for water and ethanol evaporation under microwave heating can be understood by means of this theoretical analysis. Thus, the coupling between unsteady convection and surface instabilities cannot easily occur in water, and consequently a new space for a pure Marangoni effect is generated perhaps, as or with a lot of analogies with microgravity conditions. This result proves the capacity of microwave heating to induce particular hydrodynamic behaviours. On one hand, in relation to the local character of the thermal conversion of the electromagnetic energy, and on the other hand in relation to the thermal dependency of the dielectric loss, it seems possible to induce a particular and specific hydrodynamic compared to those induced by classical heating necessarily strongly dependent on the heating walls that are useless in microwave heating.

# 4 References

Chandrasekar, S. 1961. Hydrodynamic and hydromagnetic stability. Dover Publications Inc. (New Yok, London).

Courville, P. 1987. Modeling of bulk and interfacial instable phenomena resulting from rapidly evaporation under microwave heating and depressure. Ph. D. Dissertation Burgundy University (Dijon, France).

Courville, P., Bertrand, G., Lallemant, M., Steinchen-Sanfeld, A., and Stuerga, D. 1991.The use of microwaves to evaporate liquids. The specific role of the dielectric loss factor on hydrodynamic instabilities. J. of Microwave Power and Electromagnetic Energy 26(3):168-177.

Hickman, K. 1952. Surface behavior in the pot still. Ind. and Eng. Chem. 44(8):1892-1902.

Hickman, K. 1972. Torpid phenomena and pump oils. J. Vac. Sci. Technol. 9(2):960-976.

Hickman, K. 1976. Recoil phenomena on water evaporating rapidly. J. Vac. Sci. Technol. 13(2):585-590.

Nield, D.E 1964. Surface tension and buoyancy effects in cellular convection. J.Fluid Mech. 19:341-352.

Palmer, H.J. 1976. The hydrodynamic stability of rapidly evaporating liquids at reduced pressure. J. Fluid Mech. 75(3):487-511.

Scriven, L.E., and Sternling, C.V. 1964. On cellular convection driven by surface tension gradients: effect of mean surface tension and surface viscosity. J. Fluid Mech. 19:321-340.

Stuerga, D. 1989. Microwave technology applied to heterogeneous systems. Case of evaporating liquids. Specific effect of the electric field. PH. D. Dissertation Burgundy University (Dijon, France).

Stuerga, D., and Lallemant, M. 1993. An original way to select and control hydrodynamic instabilities: Microwave heating. Part I: Hydrodynamic background and the experimental device. J. of Microwave Power and Electromagnetic Energy. 28(4):206-218.

Stuerga, D., and Lallemant, M. 1993. An original way to selet and control hydrodynamic instabilities: Microwave heating. Part II: Hydrodynamic behaviour of water and ethanol under microwave heating and reduced pressure. J. of Microwave Power and Electromagnetic Energy. 28(4):219-233.

# Pool Boiling with an Imposed Electric Field: Main Results of a Theoretical and Experimental Research

Grassi W.[1], Di Marco P.[1], Carrica P.[2], Manetti R.[1], Grassi M.[1], Mazzoni G.[1]

[1] Dipartimento di Energetica - via Diotisalvi, 2 - 56126 Pisa - Italy
[2] Centro Atomico Bariloche & Instituto Balseiro - 8400 S. Carlos de Bariloche - Argentina

Keywords: Pool Boiling, Electric Field, Microgravity Heat Transfer.

## 1. Introduction

The research is aimed at studying the effect of an imposed electric field on the pool boiling process, in view of a set of experiments to be performed under reduced gravity. Within this frame it has the twofold goal of:

- obtaining a complete set of experimental data at normal gravity with and without electric field, to be compared with the microgravity results;
- assessing the effect of some experimental parameters (e.g. the rate of change of the heat flux) in order to optimize the flight experimental procedure.

This research was recommended for Columbus Precursor Flights and accepted for parabolic flight and sounding rocket experiments. It has followed the steps reported below.

a) Review of the theoretical aspects of the influence of gravity and electric field on pool boiling [1] and a more detailed analysis of the ground-based existing literature on the effect of an electric field [2].
b) Theoretical analysis of the stability of a vapour-liquid interface in presence of gravity and electric fields, [3]. In particular the interface behaviour in absence of gravity has been examined.
c) Pool boiling experiments in single-phase natural convection, nucleate boiling and at the CHF (Critical Heat Flux Condition) have been performed, [4].
d) An experimental campaign was dedicated to film boiling due to the peculiarities of this boiling regime. A new phenomenon was evidenced, [5], and a semi-empirical model able to explain this phenomenon has been developed later on [6].
e) Several parametric effects are being studied at present. In particular they concern the effect of pressure, subcooling and of the rate of increasing the power supplied to the test section.

The main results obtained will be shortly outlined in paragraph 3.

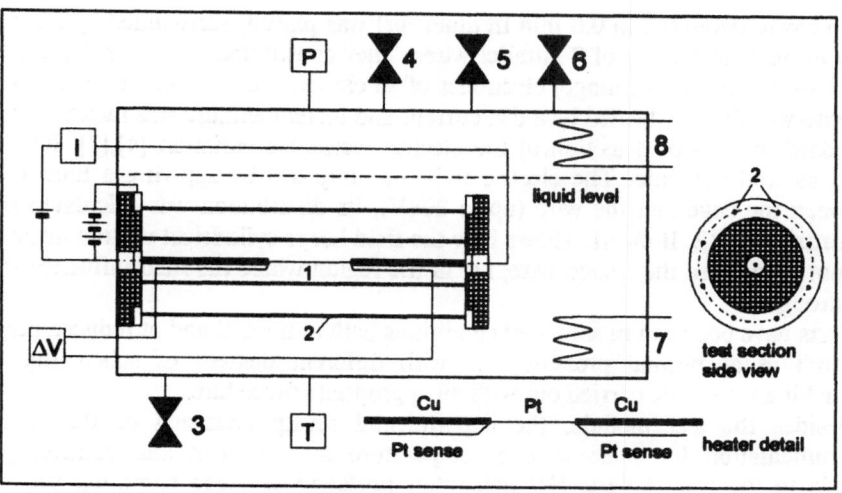

**Figure 1**: Sketch of the experimental apparatus: 1) heater; 2) high voltage cage wires; 3) drain valve; 4) fill valve; 5) to reflux condenser; 6) to vacuum line; 7) preheater; 8) condenser.

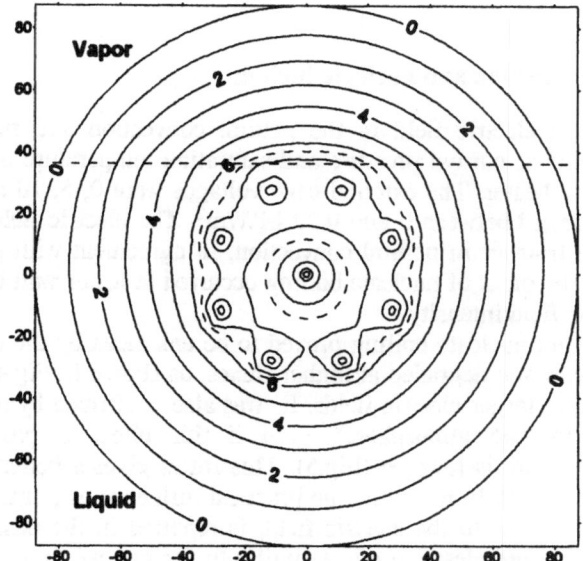

**Figure 2**: Equipotential lines around the wire (side view: the wire is in the center), as calculated by a finite element technique, for an applied potential of 10 kV, from [5].

## 2. Experimental apparatus

The experimental setup is shown in Fig.1. It consisted of a stainless steel cylindrical vessel with lateral windows for visual observation. Inside the vessel the test Pt wire (from 0.2 to 0.6 mm in diameter) was placed, surrounded by a cage (60 mm in diameter) made of 8 parallel wires. They constituted the internal (grounded) and external (at high voltage) electrodes of an electrical condenser, respectively. The Pt wire was directly heated by a d.c. current and its temperature was measured by the standard method of detection of the electric resistance variation [4],[7]. R113 was used as working fluid. The electric field was imposed by applying a high voltage between the cage and the wire (up to 20kV). Its distribution was calculated and is reported in Fig.2. It clearly shows how the field keeps cylindrical up to a distance of 10mm away from the heated wire, i.e. in the region where the field influence should be stronger.

Tests have been run in saturated conditions both at normal and at reduced pressure as well as at normal pressure, but with different degrees of subcooling. Also quenching tests were carried out with an appropriate procedure.

Besides the test module, the experimental set-up consisted of the complete instrumentation for pressure and temperature measurement and control, power supply to the heated wire, HV generator, professional video recording set, safety devices (pressure relief valve, CHF detection and shut off etc.) and a proper data acquisition system.

## 3. Experimental results

### 3-1. Natural convection and nucleate boiling

The effect of an electric field in the natural convection and nucleate boiling regions was studied at normal pressure and saturation temperature using a 0.2 mm diameter wire as a heater. The external cage voltages were 0, 5, 10 and 15 kV and the heat fluxes ranged between 0 and 0.20 MW/m$^2$. The electric field was found to enhance the heat transfer in natural convection, in agreement with previous works (see Fig.3). Also the onset of nucleate boiling occurred at lower wall overheating for increasing electric field intensity.

On the other hand nucleate boiling proved to be enhanced by the electric field at low wall heat fluxes and degraded at higher fluxes, as shown in Fig.4. This trend is more pronounced at larger electric fields. It was also confirmed by tests at reduced pressure (down to 0.45 atmosphere), even if this effect is progressively less pronounced as pressure is lowered (Fig.5). This result gives a better insight into a still controversial aspect. In fact either no practical influence or a little improvement in the heat transfer, due to the electric field, is reported in the past literature, for nucleate boiling. Nevertheless, an effect similar to the one we found is incidentally described in an old paper [8] and more recently observed for a flat geometry by Uemura et al. [9].

Visual observation and consequent video recording have been performed. In agreement with previous authors, it was observed that the presence of the electric field heavily changes the two-phase flow pattern around the wire. The size of the bubbles appeared drastically reduced at increasing electric field and they tended to organize in columns leaving the main nucleation sites. For the highest voltages, the columns springing from the lower part of the heater were directed downwards,

243

clearly indicating a dominance of the electric force on the gravity one. These features are clearly shown in the photographs of Fig.6.

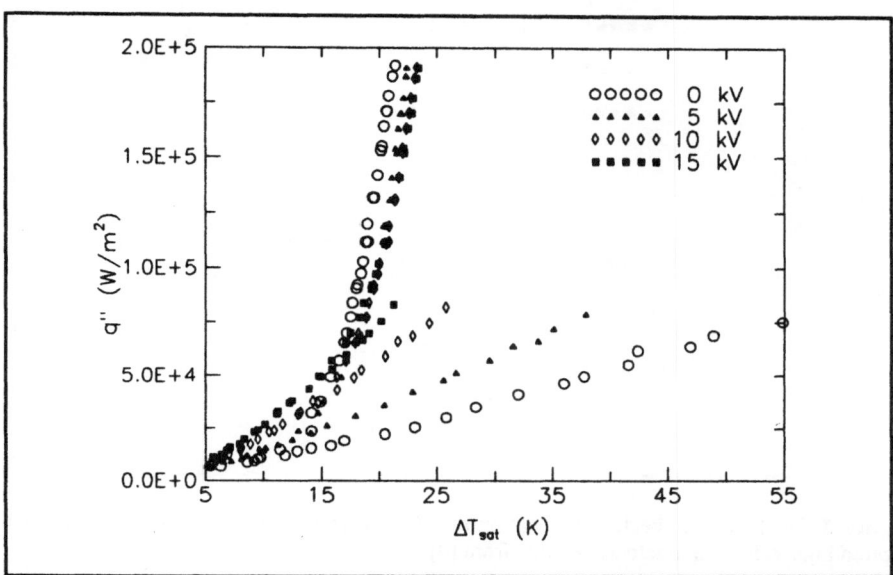

**Figure 3.** Heat flux vs. heater overheating ($T_w$-$T_{sat}$) in nucleate boiling at normal pressure, from [4].

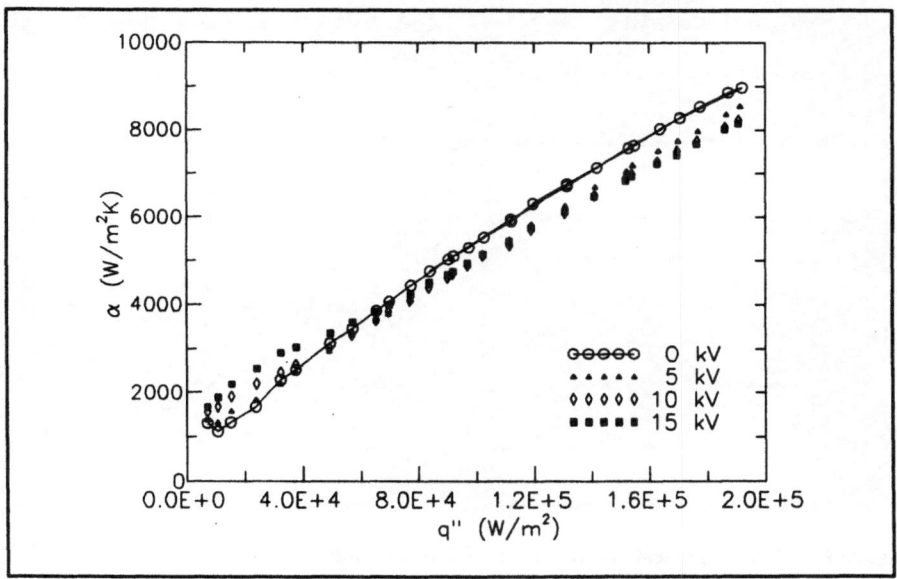

**Figure 4.** Heat transfer coefficient vs. heater applied high voltage in natural convection and nucleate boiling, from [4].

244

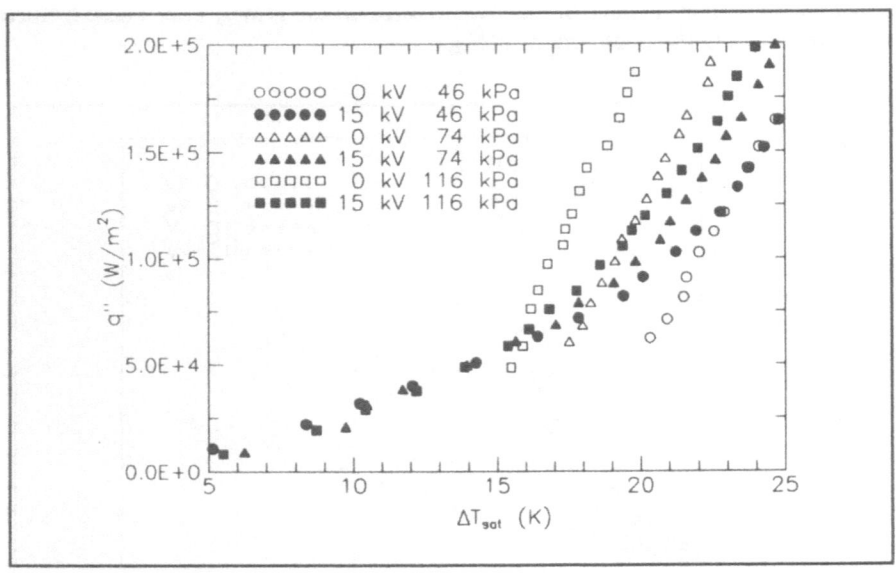

**Figure 5.** Heat flux vs. heater overheating at different pressures, for zero field and 15 kV applied high voltage in nucleate boiling, from [4].

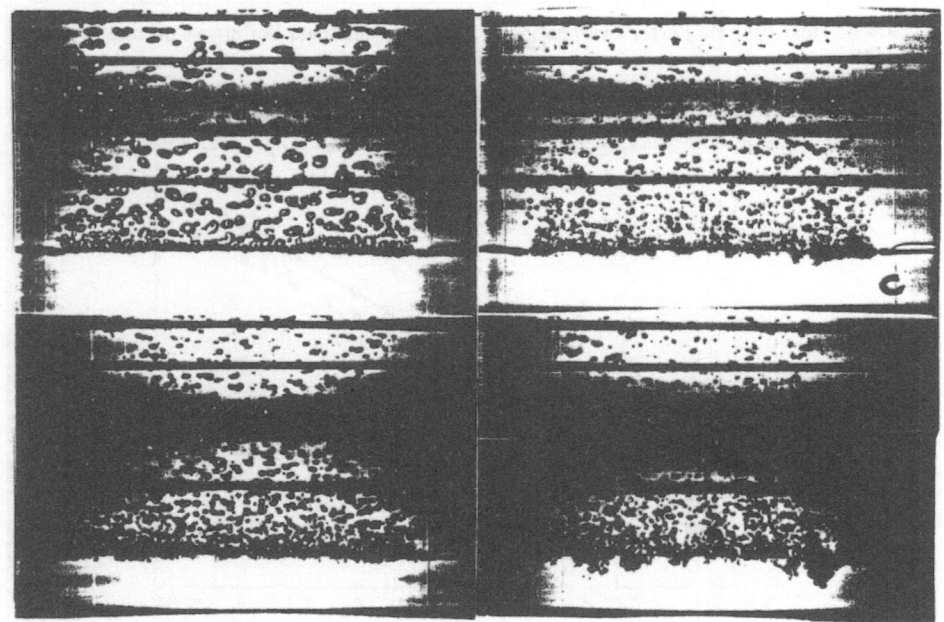

**Figure 6.** Saturated nucleate boiling over a 0.2 mm diameter wire, $q''= 252$ kW/m$^2$: a) $V = 0$ kV; b) $V = 5$ kV; c) $V = 10$ kV; d) $V = 15$ kV.

245

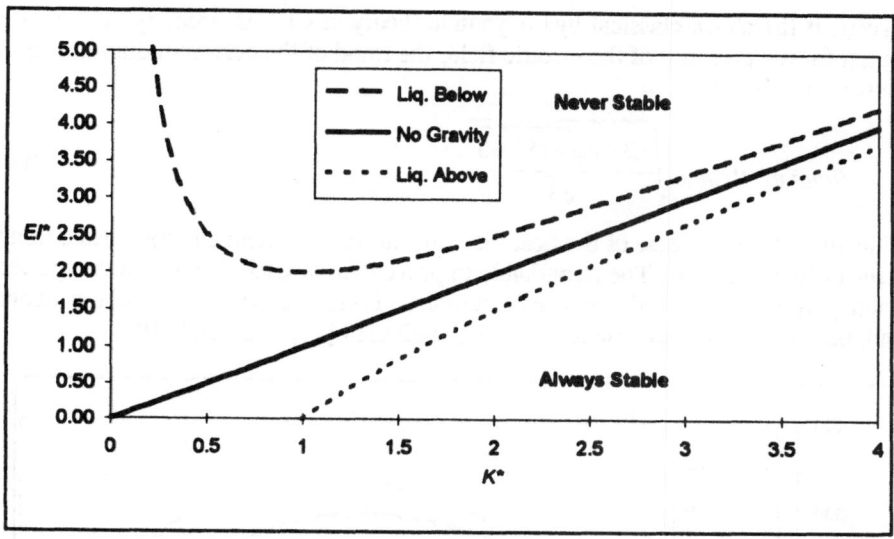

**Figure 7.** Stability plane, with liquid above, liquid below, zero-g, from [3].

### 3-2. Critical Heat Flux

A great debate exists, since several years, on the critical heat flux condition with particular regard to whether this condition is hydrodynamically controlled or not. Various models have been proposed, but none of them is yet completely assessed. In addition the influence of the electric field on CHF has only been analyzed on the basis of the hydrodynamic approach. The theory of liquid-gas interfaces stability, [3], predicts a destabilizing effect exerted by an electric field due to the reduction of the most unstable (or most dangerous) Taylor oscillation wavelength of the interface due to an imposed disturbance. For a flat liquid-vapour interface the stability plane reported in Fig.7 can be obtained, where $El^* = \varepsilon_{eq} E_{0g}^2 L / \sigma$ and $K^* = kL$. The situation for pool boiling is reproduced by the curve drawn for liquid above. It clearly shows how the oscillation wavelength, $\lambda$, of the interface decreases at increasing field. The effect of the electric field on the disturbance growth rate is presented in Fig.8. Of course the most unstable wavelength, $\lambda_u$, is the one corresponding to the highest growth rate, thus diminishing at increasing electric field. Strictly speaking all this should be only related to stable film boiling and to the minimum film boiling point (MFB), beyond which stable film boiling switches to transition boiling. In fact a definite vapour-liquid interface can be identified only in this boiling regime (see paragraph 3.3).

Nevertheless the CHF hydrodynamic approach starts from the hypothesis that, close to CHF, a pattern of vapor jets forms on the heater, spaced a distance $\lambda_u$ apart, thus using the same results obtained for film boiling. In addition the jet breakup takes place trough a Kelvin-Helmoltz instability mechanism once disturbances of wavelength $\lambda_H$ are present on its interface. For a cylindrical heater, Lienhard and Dhir [10] assumed $\lambda_u = \lambda_H$. As already said, all this matter is somehow questionable.

Anyway, if the above classical hydrodynamic theory of CHF is properly modified to account for the presence of the electric field, the trend of the corresponding heat flux is expected to be:

$$q_{CHF,E} = q_{CHF,0} \sqrt{\frac{El^* + \sqrt{El^{*2} + 3}}{\sqrt{3}}} \tag{1}$$

This implies an increase of the heat flux with the electric field and provides a trend for the critical heat flux. The remarkable improvement of CHF due to $E$ was reported by many researchers while only two authors, [11-12], attempted to fit the above trend, but with a few experimental points (3 points in [11] and 7 in [12]).

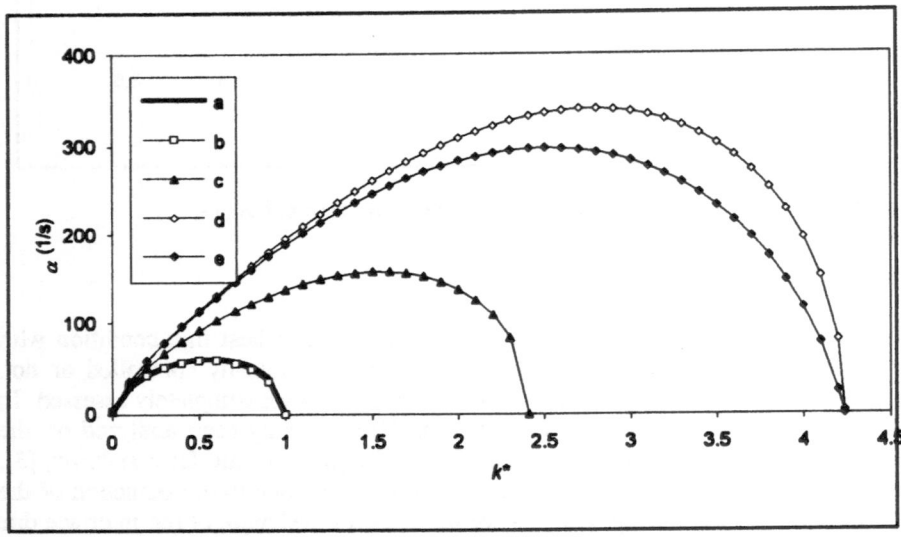

**Figure 8.** Perturbation growth rate (liquid above, R113 at 20°C) in various situations: a) $El^*=0$, $v_{eq}=0$; b) $El^*=0$, $v_{eq}=4\cdot10^{-6}$; c) $El^*=2$, $v_{eq}=0$; d) $El^*=4$, $v_{eq}=0$; e) $El^*=4$, $v_{eq}=4\cdot10^{-6}$, from [3].

The results we obtained are shown in Fig.9, from [4], in the form $q_{CHF,E}/q_{CHF,0}$ versus $V$. The fitting curve is obtained by setting:

$$El^* = cV^2 \tag{2}$$

in Eq.(1) with $c=0.038$ kV$^{-2}$. This stresses the reliability of Eq.(1) in interpreting the CHF data.

Far from being a validation of the hydrodynamic theory, these results seem to indicate that the Taylor wavelength plays a major role in the phenomenon.

One of the conclusion that can be drawn at this point is that applying an electric field could help in analysing a complex feature like CHF, by stressing some of the dominant physical aspects.

Tests were also performed with the high voltage applied only to:

247

a) the upper half cage (upper 4 cage wires),
b) the lower half cage (lower 4 cage wires),
so that two almost "semi-cylindrical" field geometries are generated.

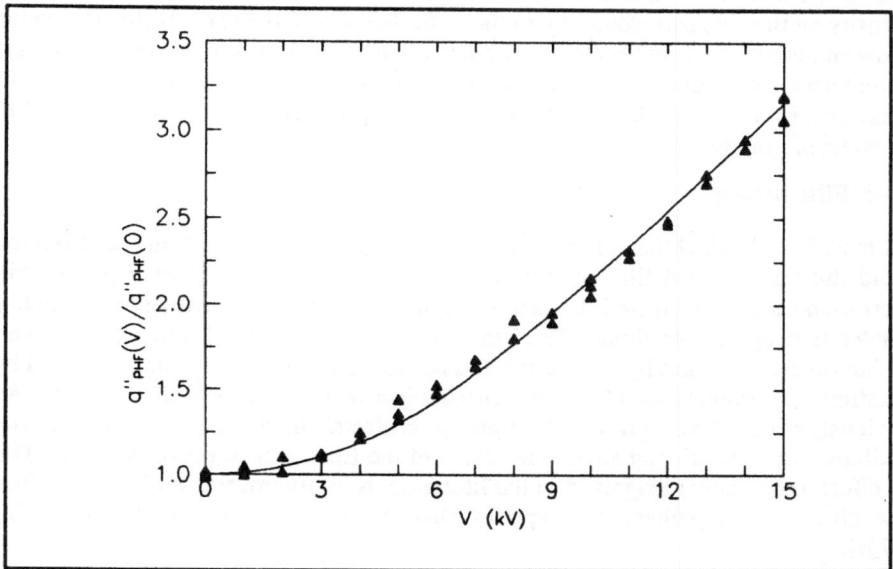

**Figure 9**. Peak heat flux (normalized to its zero field value) vs. applied high voltage for cylindrical field and semiempirical interpolation of data, from [4].

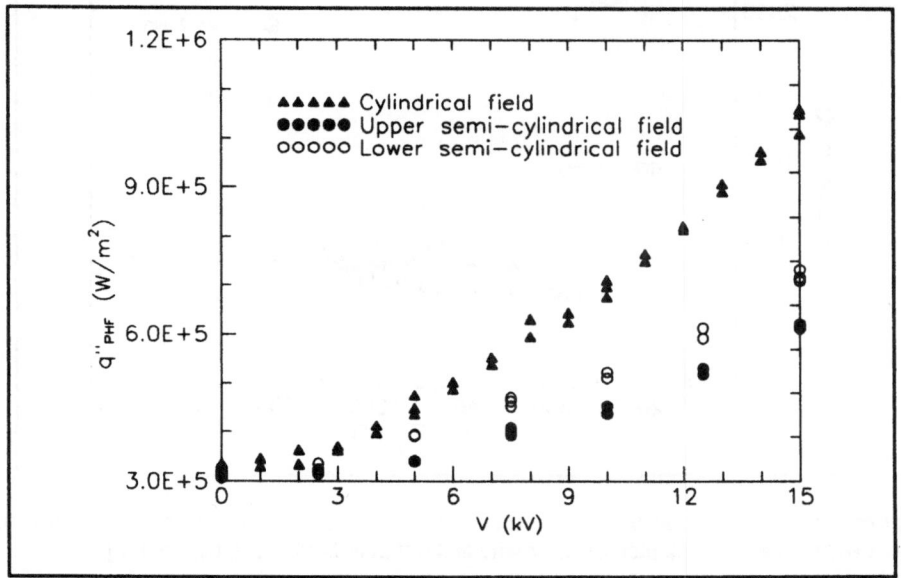

**Figure 10**. Peak heat flux vs. applied high voltage for different field configuration, from [4].

The experimental data are given in Fig.10, from [4] together with, c), those for the cylindrical geometry as in Fig.9.

Case a) and b) exhibit a lower heat flux than case c). This can be explained as follows: an electric field is anyway present all around the wire, but with a smaller intensity on the side corresponding to the unfeeded part of the cage. If the field value is low enough its contribution to the augmentation of the heat flux is very small. Part of the heater works under such a condition so that the overall heat flux enhancement is reduced. Moreover the higher heat flux in case b) than in case a) can be ascribed to the effect of gravity.

### 3-3. Film boiling

It is well established that a stable vapour layer separates the heater surface from the liquid during saturated film boiling over a horizontal cylinder. The vapour-liquid interface oscillates, with well defined wavelength, originating equally spaced vapour bubbles raising into the liquid. The heat exchange between the heating surface and the interface takes place by conduction and convection across the vapour layer and by radiation. The importance of this last mechanism increases with wall superheat. As previously outlined (paragraph 3-2) applying an electric field causes a decrease in the oscillation wavelength and an augmentation of the frequency of the oscillations. The net effect, commonly recognized in the literature, is an improvement of the heat flux at a given wall superheat, that applies also to the minimum film boiling point (MFB).

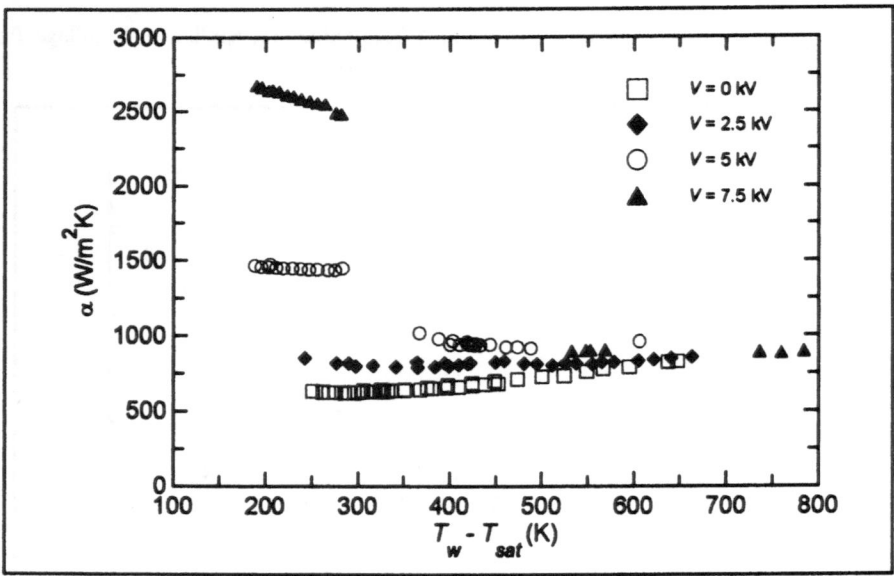

**Figure 11**: Values of heat transfer coefficient (corrected to eliminate side-end effects) vs. wire overheating, for a 0.2 mm diameter wire and different field strengths, from [5].

249

Our experimental data confirm this trend within a given set of values of heat flux, $q$, and wall superheat. In addition a new phenomenon appeared never reported by previous authors. This phenomenon consisted in a sort of second transition according to which all the curves at different high voltages tend to collapse into a single one almost coincident with the one at 0 kV. All this can be easily seen from Figs.11-12, from [5], where the film boiling heat transfer coefficient and two complete boiling curves are shown. This effect is quite clear and repeatable for different wire sizes. Quenching tests were also run to gain a clearer evidence of the occurrence of such a transition. Some typical film boiling curves are reported in Fig.13, from [5] and the above mentioned trend is quite apparent.

The visual observation indicates the existence of a related change in the vapour pattern as described by the pictures of Fig.14, from [5]. Pictures a), b), and c) clearly show the decrease of bubble size for increasing field (V from 0 to 5kV). Picture d), at the same voltage as picture c), shows how the bubble size beyond the transition is roughly equal to that at 0 kV.

A theoretical explanation of this phenomenon is not yet available even if an attempt of realizing an at least semi-empirical model is undertaken at present.

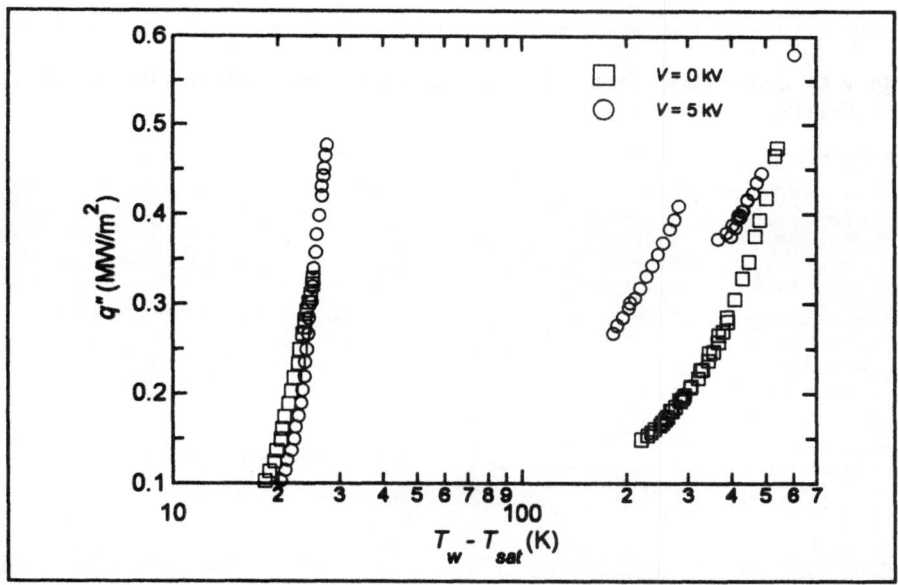

**Figure 12**: Complete pool boiling curves (including nucleate boiling section) for a 0.2 mm diameter wire for no electric field and for 5 kV applied potential, from [5].

## 4. Conclusions

A short outline of the main features of the research carried out in Pisa, about the process of boiling with an electric field, has been reported in the previous paragraphs. Several peculiar topics (related to the different regimes), deserving an *ad*

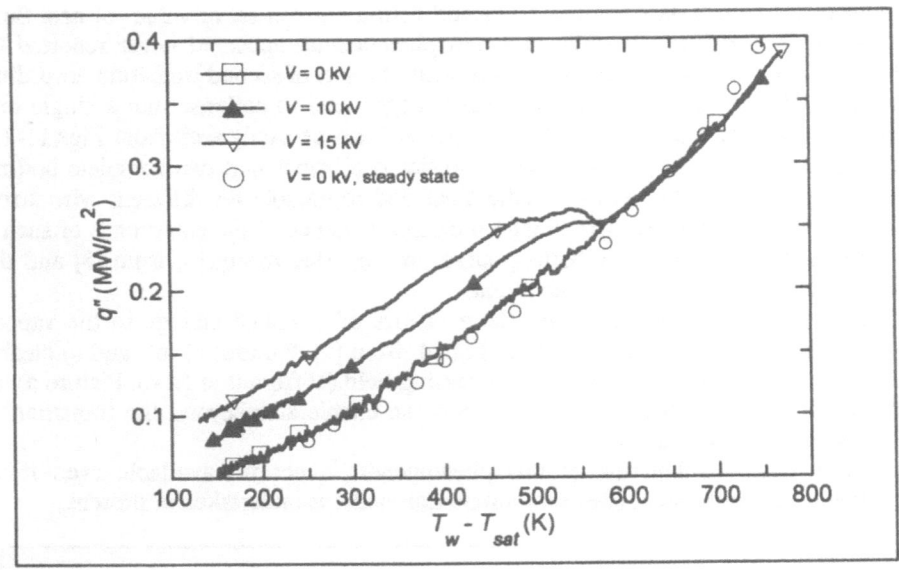

**Figure 13**: Boiling curves for transient and steady state experiments on a 0.6 mm diameter wire, from [5].

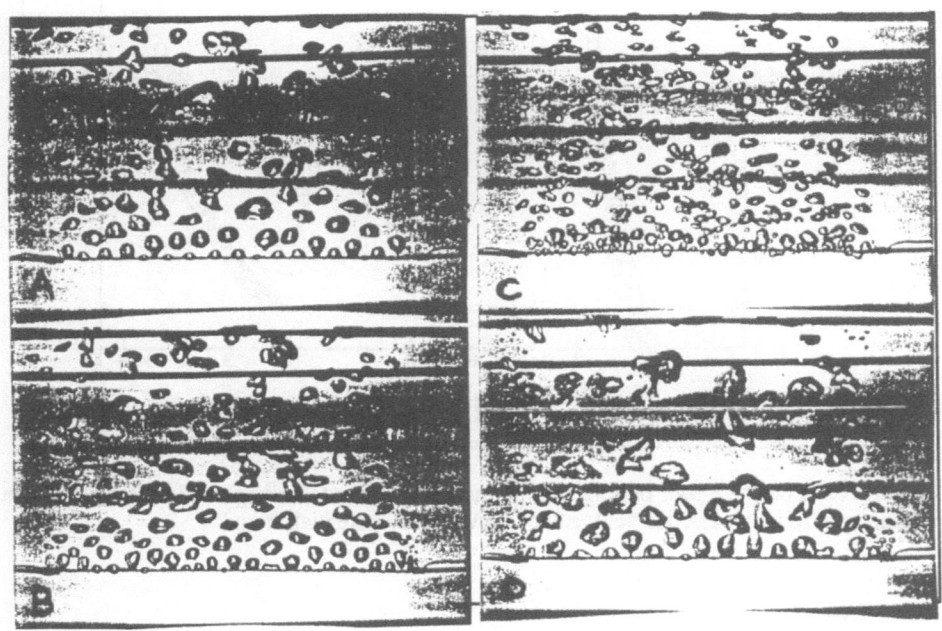

**Figure 14**: Film boiling over a 0.2 mm diameter wire: a) $V = 0$ kV, $q'' = 325$ W/m$^2$, $\Delta T = 420$ K; b) $V = 2.5$ kV, $q'' = 325$ kW/m$^2$, $\Delta T = 332$ K; c) $V = 5$ kV, $q'' = 344$ kW/m$^2$, $\Delta T = 220$ K; d) $V = 5$ kV, $q'' = 529$ kW/m$^2$, $\Delta T = 590$ K (second film boiling mode).

*hoc* analysis in reduced gravity, have been addressed. They can be quickly summarized as follows.

– *Nucleate boiling* - Most of the phenomena in on-earth conditions are gravity dependent (buoyancy). By performing μg tests under saturated conditions (to reduce Marangoni effects) would enhance the understanding of the actual role played by the electric field. This will be the scope of the already scheduled parabolic flight and sounding rocket experiment.

– *Interface dynamics*- As already stressed the behaviour of the vapour-liquid interface is fundamental in stable film boiling and MFB, while its role has to be clarified for CHF and, probably, for the film boiling transition we evidenced. It is very well known that such an interface dynamics is essentially dominated by gravity, surface tension and electric field. Moreover it is possible to model this phenomenon with a reasonable reliability, at least for a first order approach. Therefore the interface behaviour can be predicted accounting for the separate effects exerted by the above mentioned quantities. As a consequence, microgravity tests for each of the above topics (film boiling, CHF etc.) would greatly help gaining a better understanding of the physics involved. All this will be the subject of next proposals for microgravity tests.

Up to this point we have being considering microgravity as a tool to enhance the physical knowledge about the electric field behaviour. Conversely, it is possible to use the electric field as a tool to improve the performances of thermal equipment working under microgravity conditions. It is expected to occur thanks to the fact that electric forces could replace the action of gravity. This would allow for operating a phase separation in two-phase flow, due to the different electric permittivities of the two phases, and for inducing convection, in single-phase flow, through the coupling of an electrical field with non uniform thermal fields.

## Nomenclature

| | |
|---|---|
| $E$ | electric field |
| $El^* = \varepsilon_{eq} E_{0g}^2 L / \sigma$ | electrical influence number |
| $g$ | acceleration of gravity |
| $k = 2\pi/\lambda$ | wave number |
| $L = \sqrt{\sigma/g(\rho_l - \rho_g)}$ | Laplace length |
| $q$ | heat flux |
| $V$ | high voltage |
| $T$ | temperaure |
| $\lambda$ | wavelength |
| $\nu$ | kinematic viscosity |
| $\rho$ | density |
| $\sigma$ | surface tension |
| *Subscripts* | |
| $0$ | with no electric field |
| $E$ | with electric field |
| $eq$ | equivalent |

| | |
|---|---|
| *g* | gas |
| *l* | liquid |
| *o* | interface |
| w | wall |
| *Abbreviations* | |
| CHF | Critical Heat Flux |

# References

[1]    Grassi W., Di Marco P., Gravity and electric field effects on pool boiling heat transfer, VIII European Symposium on Materials and Fluid Sciences in Microgravity, Bruxelles, B, April 12-16, pp. 783-788, 1992.

[2]    Di Marco P., Grassi W., Saturated pool boiling enhancement by means of an electric field, *J. Enhanced Heat Transfer*, Vol.1, No. 1, pp. 99-114, 1993.

[3]    Di Marco P., Grassi W., Gas-liquid interface stability in presence of an imposed electric field, *Proc. 12th UIT Nat. Heat Transfer Conference*, L'Aquila, I, pp. 299-310, 1994.

[4]    Carrica P., Di Marco P., Grassi W., Effect of an electric field on nucleate pool boiling and critical heat flux: overview of the results of an experimental study, *Proc. 4th ASME-JSME Thermal Engineering Joint Conf.*, Lahaina, Maui, HI, USA, March 19-24, pp.201-208, 1995.

[5]    Carrica P., Di Marco P., Grassi W., Electric field effect on film boiling on a wire, to appear in *J. Experimental Heat Transfer*, 1996.

[6]    Carrica P., Clausse A., Masson V., Di Marco P., Grassi W., A model for film boiling in the presence of electric fields, *Proc. of NURETH-7*, Saratoga Springs, NY, USA, Sept. 10-15, pp.3218-3231, 1995.

[7]    Di Marco P., Grassi W., Influence of local phenomena in pool boiling experiments of R113 on a wire, *Proc. XIII UIT Nat. Heat Transfer Conference*, Bologna, pp.307-318, 1995.

[8]    Baboi N.F., Bologa M.K., and Klyukanov A.A, Some Features of Ebullition in an Electric Field, *Appl. Elec. Phenomena (USSR)*, Vol.20, pp.57-70, 1968.

[9]    Uemura M., Nishio S. and Tanasawa I., Enhancement of Pool Boiling Heat Transfer by Static Electric Field, *9th Int. Heat Transfer Conf.*, Jerusalem, Vol.4, pp.75-80, 1990.

[10]   Lienhard J.H., *A heat transfer textbook*, 2nd Ed., Chap. X, Prentice-Hall, 1987.

[11]   Berghmans J., Electrostatic fields and the maximum heat flux, *Int. J. Heat Mass Transfer*, Vol.19, pp.791-797, 1976.

[12]   Johnson R.L., Effect of an electric field on boiling heat transfer, *AIAA J.*, Vol.14, pp.1456-1460, 1968.

# Toward a non-equilibrium non-linear thermodynamics

Manuel G. VELARDE

Instituto Pluridisciplinar
Universidad Complutense de Madrid
Paseo Juan XXIII, n. 1
28040-Madrid, Spain

**Abstract.** Illustration is given with a few examples of the methodology recently proposed by John Ross to handle in a thermodynamically sound approach systems and (nonlinear) processes far away from equilibrium. Ross' theory builds on *excess* work and *excess* dissipation concepts very much in the spirit of Carnot and Gibbs.

**Keywords.** Irreversible thermodynamics, non-linearity, instability

## 1. Introduction

Constructing the thermodynamics of non-equilibrium phenomena poses a challenging and formidable task. Efforts have been conducted by a number of scientists since the pioneering work of Lord Rayleigh [1], Onsager [2,3] and Prigogine [4-6]. The fact that the latter two received the Nobel Prize in Chemistry for developments of non-equilibrium thermodynamics shows how much the scientific community is happy every time that some progress appears. Take, for instance, (oscillating or other) chemical reactions far from equilibirum, nonlinear transport phenomena, (nonlinear) hydrodynamics, viscoelasticity and other non-Newtonian flows, instabilities, self-organization far from equilibrium, and chaos [7-10]. What is the purpose of a thermodynamic theory for such phenomena? The thermodynamic potential functions for irreversible processes approaching equilibrium are known: for example, the Gibbs free energy change for the processes at constant temperature and pressure. Changes in this free energy yield the maximum work available from the changes. Then, by analogy, the goals of thermodynamics for all non-equilibrium processes are the establishment of macroscopic evolution criteria with physical significance such as the connection with *excess* work; the work (and power) available from a transient decay to a stationary state; necessary and sufficient criteria of stability; criteria for bifurcations and relative stability in systems with more than one (stable) attractor; and a connection of the thermodynamic formulation to fluctuations.

Linear irreversible thermodynamics [3-6] was based on the validity of Gibbs local equilibrium assumption [11,12]. Recently, J. Ross has proposed a methodology to go beyond linear theory yet retaining to some extent Gibbs assumption. The formulation offered by J. Ross [e.g 13-23] is based on the macroscopic *excess* work necessary to displace a system from a stationary state. Ross started with chemical systems. For a chemical system to be in a stationary state there must be intermediates; their concentrations vary in time but are constant at a stationary state. Hence Ross' theory for chemical kinetics in based on the concept of a species-specific affinity, which is *excess* work. Ross' approach parallels equilibrium thermodynamics, where the Gibbs free energy, at constant temperature and pressure, is the work, other than $P$-$V$ work, necessary to displace a system from equilibrium. For reaction, as well as the similar case of diffusion, applied to a single variable system with a single stationary state, the *excess* work $\Phi$, on expansion linearly away from the stationary state, yields a Gaussian distribution.

For the linear transport of mass, the *excess* work is

$$\Phi_X = \kappa T \int \ln \frac{X}{X_S}\, dX \tag{1}$$

where $X$ is the number of $X$ molecules distributed homogeneously in a volume $V$, which, say, is in a thin wedge placed between two reservoirs of different concentrations of $X$; $X_S$ is the stationary value of $X$. The assumption of local (thermal) equilibrium ensures the existence of a temperature. The time scale of establishing a Maxwell-Boltzmann distribution of velocity is taken to be much shorter than the time scale of hydrodynamic relaxation.

For linear thermal conduction, as a single transport, the *excess* work is

$$\Phi_T = C_V \int \left(1 - \frac{T_S}{T}\right) dT, \tag{2}$$

where, in a similar geometry as for difussion, $C_v$ is the heat capacity of the system at constant volume $V$, and $T$ is the temperature within a thin slice of volume $V$ wedged between two heat reservoirs of temperature $T_1$ and $T_2$; the stationary temperature $T_S$ equals $(T_1 + T_2)/2$.

For linear (Couette or Poiseuille) viscous flow disregarding heat, the *excess* work, again in a similarly idealized geometry, is

$$\Phi_V = \rho \int (v - v_S)\, dv, \tag{3}$$

with "$v$" denoting velocity. In Eqs. (1)-(3), the path of the integration needs to be specified for a multivariable system.

For nonlinear transport processes, one at a time, Ross uses the method of an *equivalent* linear system at each value of the variable ($X$ or $T$ or $v$). For example, for nonlinear thermal conduction, say of the form

$$\frac{dT}{dt} = \kappa_1 (T)(T_1 - T) - \kappa_3 (T)(T - T_3) , \tag{4}$$

where $\kappa_1$ and $\kappa_3$ are thermal conductive coefficients that are temperature dependent, the *excess* work is

$$\Phi = \int^T C_V \left(1 - \frac{T^*}{T}\right) dT, \tag{5}$$

with $T^*$ given by

$$T^* \equiv \frac{\kappa_1 (T) T_1 + \kappa_3 (T) T_3}{\kappa_1 (T) + \kappa_3 (T)} , \tag{6}$$

At each value of $T$ there is a $T^*$ (one-to-one mapping) that equals $T_S$ in the *equivalent* linear system at $T$.

Thus, in Ross' theory, the *excess* work, $\Phi$, for chemical reactions and for the individual transport processes of diffusion, thermal conduction, viscous flow, and electrical conduction, provides the basis for a thermodynamic and stochastic theory, as shown by the following properties of $\Phi$ [16]:

(1) $\Phi$ is an extremum at all stationary states.

(2) $\Phi \geq 0$ for a displacement from stable stationary states, and $d\Phi/dt \leq 0$ for the relaxation to a stable stationary state; since $\Phi$ is bounded, $\Phi$ is a Lyapunov function with a physical significance analogous to the Gibbs free energy (maximum work at constant temperature and pressure) for relaxation o equilibrium states. Here $\Phi$ is the thermodynamic potential ("driving force"), an *excess* work, for the motion toward a stable stationary state.

(3) With $\Phi$ comes a *necessary and sufficient* condition for stable stationary states: if for every conceivable variation from a stable stationary state $\delta\Phi > 0$, then the stationary state is stable.

(4) For a stable stationary state (for one- variable systems), for example for diffusion, it is $d^2\Phi/dX^2 > 0$; for an unstable stationary state, $d^2\Phi/dX^2 < 0$; for a marginally stable state, $d^2\Phi/dX^2 = 0$, and for a critical stationary state, $d^3\Phi/dX^3 = 0$, as well.

(5) $d\Phi/dt$ is a *component* of the *total* dissipation related to the relaxation toward a stable stationary state.

(6) The differential *excess* work is related to differences in thermodynamic functions for variations at $X$ minus that for variations at $X_S$ (or $X^*$): for diffusion this function is a combination of Gibbs and Helmholtz free energies; for thermal conduction it is the entropy; and for viscous flow it is the energy.

(7) For homogeneous systems with multiple stable stationary states the *excess* work provides a criterion of *relative* stability of such states.

(8) Finally, fluctuations in the concentration of $X$, in the temperature, and in the mass velocity are given, respectively, by a master equation, a Fokker-Planck equation with probability diffusion coefficient proportional to temperature, and a Fokker-Planck equation with a constant probability diffusion coefficient. In all these cases, then, the corresponding $\Phi$ yields the stationary probability density distribution. For multivariable problems the differential *excess* work is inexact; integration requires specification of a path. The choice of the deterministic trajectory (or its reverse) has been suggested to give an integrated *excess* work, which yields the stationary probability distribution of the appropriate stochastic equation. With the specification of the deterministic path of relaxation toward a stationary state $\Phi$ is a state and Lyapunov function.

For hydrodynamics is interesting to recall past failures and successes. Helmholtz, Korteweg, and Rayleigh (for a comprehensive account see Lamb's classical treatise [24]) introduced the *dissipation* function,

$$
D = \iiint dx \, dy \, dz \eta \left[ 2 \left(\frac{\partial v_x}{\partial x}\right)^2 + 2\left(\frac{\partial v_y}{\partial y}\right)^2 + 2\left(\frac{\partial v_z}{\partial z}\right)^2 \right.
$$
$$
\left. + \left(\frac{\partial v_z}{\partial y} + \frac{\partial v_y}{\partial z}\right)^2 + \left(\frac{\partial v_x}{\partial z} + \frac{\partial v_z}{\partial x}\right)^2 + \left(\frac{\partial v_x}{\partial y} + \frac{\partial v_y}{\partial x}\right)^2 \right] , \qquad (7)
$$

associated with viscous flow in the Fourier thermohydrodynamic equation, where $\eta$ is the dynamic viscosity. Their common aim was to provide a variational principle, and so an evolution criterion with a thermodynamic potential for the Navier-Stokes equations and their steady states. They where able to treat only their linearized approximation. Indeed, a variational principle and thus a potential or Lyapunov function embracing both (nonlinear) inertial terms and dissipation cannot be obtained for the full Navier-Stokes equations [25]. Helmholtz, Korteweg, and Rayleigh did provide for this case a minimum *total* dissipation evolution criterion. Ross $d\Phi/dt$ is part of this dissipation. Note that in the absence of dissipation the nonlinear Euler equations do derive from a potential with a pertinent variational principle.

Realizing that Rayleigh's dissipation, which is related to the entropy production in the system, cannot provide a thermodynamic evolution criterion far from equilibrium, Glansdorff and Prigogine [5] considered the second variation of the entropy as a criterion for evolution and stability. Their approach, however, was limited to small deviations from a stationary state, and in the form of a (restricted) variational principle provides only *sufficient* conditions for instability. Keizer [26] formulated a stochastic approach for the relaxation to stationary states and fluctuations around a (single) stationary state by assuming Gaussian fluctuations. His work was limited to small fluctuations as related to linearized kinetics (for chemical reactions). Generally, fluctuating hydrodynamics adds Gaussian fluctuations to the *linearized* Navier-Stokes equations, and the connection to the thermodynamics may be sufficient for systems approaching equilibrium, but not for evolution to a stationary state far from equilibrium. In none of the earlier studies are connections made to work, or power, nor the connection of work to a Lyapunov function for the system, and there is no discussion of relative stability when several steady states are available. Such connections are essential in the formulation of thermodynamics whether far from, near to or, at equilibrium.

## 2. Non-linear thermal transport

An example of nonlinear thermal transport is radiation. The heat transport between two objects (black body) by radiation is the temperature difference to the fourth power. In the simplest possible geometry of a system (denoted by 2) between two reservoirs (1 and 3) the process is described by the equation

$$dT/dt = a_1 (T_1 - T)^4 - a_3 (T - T_3)^4 \qquad (2.1)$$

In general, the total thermal transport can be a combination of different processes, for which

$$dT/dt = \kappa_1 (T) (T_1 - T) - \kappa_3 (T) (T - T_3) \qquad (2.2)$$

When there is only linear thermal conduction, then in the simplest case $\kappa_1$ and $\kappa_3$ are constants. Unlike the case of mass transport, where each individual (forward and reverse) flux can be identified at each instant, there is not in general that information about individual thermal fluxes, even in the linear problem.

Ross uses thermodynamic and kinetic linear *equivalence* and writes

$$dT/dt = \{\kappa_1 (T) + \kappa_3 (T)\}\{T^* (T) - T\} \qquad (2.3)$$

where the reference temperature at each instant is defined as $T^* \equiv (\kappa_1(T)T_1 + \kappa_3(T)T_3)/(\kappa_1(T)+\kappa_3(T))$. The reference temperature $T^*$ is the temperature in the stationary state of the *instantaneously equivalent* linear system. For the nonlinear system, $T^*$ approaches the temperature of the stationary state when the system approaches any stationary state. The system $dT/dt = \{\kappa_1(T)+\kappa_3(T)\}\{T^*(T)-T\}$ is mapped to a linear one $dT'/dt = \{\kappa'_1 + \kappa'_3\}\{T'_s - T\}$ at each value of $T$, or at each instant. The definition of $T^*$ ensures that the driving force $(1-(T^*/T))$ is zero when the system is at the stationary state. The function

$$\Phi = \int^T C_V \left(1 - \frac{T^*}{T}\right) dT \qquad (2.4)$$

is interpretable at each instant as *excess* work. Clearly, the driving force is the efficiency of a *Carnot engine* working between two thermal reservoirs $T^*$ and $T$, and $C_V (1- (T^*/T))dT = (1-(T^*/T))dQ = dW$ is the differential work input from the external world (or output to the external world). The integral of the differential work, $\Phi$, is the minimum *excess* work necessary to move the system from the steady state through a finite temperature difference. Then

$$\frac{d\Phi}{dT} = -\frac{C_V}{T(\kappa_1(T) + \kappa_3(T))}\frac{dT}{dt} = 0 \quad \text{at stationary states} \quad (2.5)$$

$$\frac{d^2\Phi}{dT^2} \begin{cases} > 0 \text{ at a stable stationary state} \\ < 0 \text{ at an unstable stationary state} \end{cases} \quad (2.6)$$

Thus we know that $\Phi$ is a minimum at a stable stationary state, a maximum at an unstable stationary state, and $\Phi(T) - \Phi(T_S) \geq 0$ is valid near a stable stationary state. As

$$\frac{d\Phi}{dT} = \frac{d\Phi}{dT}\frac{dT}{dt} = -\frac{C_V}{T(\kappa_1(T) + \kappa_3(T))}\left(\frac{dT}{dt}\right)^2 \leq 0 \quad (2.7)$$

in a spontaneous process, the system goes away from an unstable stationary state (maximum $\Phi$) and approaches a stable stationay state (minimum $\Phi$). If there are multiple stationary states in the system, then $\Phi$ is a Lyapunov function of global validity.

An equation for the probabity distribution of fluctuations of the temperature, for which $f(T) \exp(-\Phi/RT_C)$ gives the stationary distribution; $T_C = (T_1 + T_3)/2$ is a chosen constant which yields the correct stationary and equilibrium (where $T_C = T_{eq}$) distribution. This is a Fokker-Planck equation

$$\frac{\partial P(T,t)}{\partial t} =$$

$$-\frac{\partial}{\partial T}\{(\kappa_1(T) + \kappa_3(T))(T^* - T)P(T,t)\} + \frac{\partial^2}{\partial T^2}D(T)P(T,t) \quad (2.8)$$

with probability diffusion coefficient $D(T)$

$$D(T) = \frac{RT_C T(\kappa_1(T) + \kappa_3(T))}{C_V} \quad (2.9)$$

and the stationary solution in terms of the *excess* work $\Phi$

$$P_S(T) = P_0 \frac{C_V}{RT_cT(\kappa_1(T) + \kappa_3(T))} \exp\left(-\frac{\Phi}{RT_C}\right) \qquad (2.10)$$

Notice that the pre-exponential factor $f(T)$ in the stationary distribution $P_S(T)$ $= f(T)\exp(-\Phi(T)/RT_C)$ does not affect the maximum in a macroscopic system $(C_V\to\infty)$. The maximum distribution temperature is determined by $dP_S/dT=0$; which is equivalent to $(df(T)/dT)-(f(T)/RT_C)(d\Phi/dT)=0$; in the thermodynamic limit, the factor $C_V$ in $\Phi$ makes the second term be the dominant term, so that $d\Phi/dT=0$ is equivalent to $dP_S/dT=0$.

The relative stability of stable stationary states can be determined from the stationary distribution. The requirement of equistability of two stable stationary states at temperatures $T_A$ and $T_B$ is

$$\frac{P_S(T_A)}{P_S(T_B)} = \exp\left(\frac{1}{RT_C}\int_{T_A}^{T_B} d\Phi\right) = 1 \qquad (2.11)$$

If the unstable stationary state at temperature $T_M$ separates the two stable homogeneous stationary states at temperatures $T_A$ and $T_B$, then equistability means

$$\int_{T_A}^{T_B} d\Phi = 0 \quad \text{or} \quad \int_{T_A}^{T_M} d\Phi = \int_{T_B}^{T_M} d\Phi \qquad (2.12)$$

that is at equistability the *excess* work from $T_A$ to $T_M$ equals the *excess* work from $T_B$ to $T_M$. The concept of relative stability is given a thermodynamic interpretation in terms of *excess* work.

## 3. Non-linear Flow

The linear relationship between the strain $e_{ij}$ and the stress $\sigma_{ij}$

$$\sigma_{ij} = \eta e_{ij} = \frac{1}{2}\eta\left(\frac{\partial V_i}{\partial x_j} + \frac{\partial V_j}{\partial x_j}\right) \qquad (3.1)$$

with the dynamical viscosity coefficient $\eta$ being a constant, is a good approximation for many simple fluids. However, flows in polymer solutions and powders (non-Newtonian fluids).follow complex nonlinear relations. The transition from a branch of more viscous states to another branch of less viscous states has been found on increasing the shear rate (strain); on decreasing the shear rate the reverse transition occurs, sometimes whith hysteresis. For one-dimensional Poiseuille flow (and similarly for Couette flow) of a non-Newtonian fluid, there may be two stationary states at given values of external constraints, such as at a constant pressure difference (or constant force).

Nonlinear relations between strain and stress can be described phenomenologically with a strain-dependent viscosity coefficient $\eta$, rather than constant $\eta$ for a Newtonian fluid, $\eta=\eta(\text{eij})$ ; we write the macroscopic equation of mass velocity in the middle section of a three-section model like in Sect. 2,

$$\rho \frac{dv}{dt} = f - 2\eta (v) v \qquad (3.2)$$

Since $\eta$ is the function of strain, the equation is nonlinear. The strain reduces to just v. As in other cases Ross uses a linear system which is at any instant *equivalent* to the nonlinear system. Then

$$\rho \frac{dv}{dt} = f - 2\eta (v) v = 2\eta(v) (v^*-v) \qquad (3.3)$$

where $v^*=f/2\eta(v)$ is a reference state, which is the stationary state for the *instantaneously* thermodynamically and kinetically *equivalent* linear system at each instant of time. The function

$$\Phi = V \int\limits^{v} \rho(v - v^*)\, dv \qquad (3.4)$$

is such that its diferential , $\rho(v-v^*)$ dv, is the relative change of kinetic energy of a system of momentum $\rho(v-v^*)$ due to a small velocity change dv, and is the differential *excess* work for viscous flow. The integral of the differential excess work, $\Phi$, is the total *excess* work necessary to remove a system from a stationary state to a given nonstationary velocity. It is

$$\frac{d\Phi}{dv} = \rho\, V (v - v^*) = -\frac{\rho^2 V}{2\eta(v)} \frac{dv}{dt} = 0 \qquad (3.5)$$

at all stable and unstable stationary states. The second derivative of $\Phi$ at stationary states is

$$\frac{d^2\Phi}{dT^2} \quad \left\{ \begin{array}{l} > 0 \text{ at a stable stationary state} \\[2mm] < 0 \text{ at an unstable stationary state} \end{array} \right. \qquad (3.6)$$

Hence $\Phi$ is a minimum at a stable stationary state and a maximum at an unstable one. Further it is

$$\frac{d\Phi}{dt} = -\frac{\rho^2 V}{2\eta(V)} \left(\frac{dv}{dt}\right)^2 \leq 0 \qquad (3.7)$$

In a spontaneous process, the system goes away from an unstable stationary state (maximum $\Phi$) and approaches a stable stationary state (minimum $\Phi$); therefore, $\Phi$ is a globally valid Lyapunov function.

Is there an equation for the probability distribution of fluctuations of the macroscopic mass velocity for which the form $F(v) \exp(-\Phi/RT)$ gives the stationary distribution? It is a Fokker-Planck equation

$$\frac{\partial P(v,t)}{\partial t} = -\frac{\partial}{\partial v}\left(\frac{f-2\eta(v)v}{\rho} P(v,t)\right) + \frac{\partial^2}{\partial v^2}D(v) P(v,t) \quad (3.8)$$

with probability diffusion coefficient

$$D(v) = \frac{2RT\eta(v)}{\rho^2 V} \qquad (3.9)$$

The stationary solution in terms of excess work $\Phi$ is

$$P_S(v) = \frac{\rho^2}{2RT\eta(v)} \exp\left(-\frac{\Phi}{RT}\right) \qquad (3.10)$$

The solution has maxima at stable homogeneous stationary states and minima at unstable ones. The relative stability of two stable stationary states, corresponding to two different values of the strain $v_{up}$ and $v_{low}$, is determined by the relative heights of two peaks

$$\frac{P(v_{up})}{P(v_{low})} = \exp\left(-\frac{1}{RT}\int_{v_{low}}^{v_{up}} d\Phi\right) = \exp\left(-\frac{V}{RT}\int_{v_{low}}^{v_{up}}\rho(v-v^*)dv\right) \quad (3.11)$$

The equistability condition is $P(v_{up})/P(v_{low}) = 1$, which leads to

$$V\int_{v_{low}}^{v_{up}}\rho(v-v^*)\,dv = \int_{v_{low}}^{v_{up}} d\Phi \;, \;\text{ or }\; \int_{v_{mid}}^{v_{up}}d\Phi = \int_{v_{mid}}^{v_{low}}d\Phi \qquad (3.12)$$

This last equation gives a physical interpretation for the equistability condition; the *excess* work to remove the system from each of stable stationary states, $v_{up}$ and $v_{low}$, to a reference state, vmid, say the unstable stationary state in between the two stable stationary states, is the same.

## 4. Instability and bifurcation

Let us see how Ross theory provides a Lyapunov function with thermodynamic significance for a combination of transport processes and hydrodynamics with flow due to body forces. In view of the remarks made in the Introduction we can only refer to a truncation of the Navier-Stokes equations. Let us take the simplified Lorenz [27] version of Rayleigh-Bénard convective instability, i.e. the onset of convection past a critical temperature gradient, defined by a dimensionless Rayleigh number, R [17,28]. The function applies not only in the motionless, conductive stationary case, but also in the convective state close to the bifurcation. Although the Lorenz equations were obtained by Saltzmann [29] from a first-order Fourier decomposition approach to the Navier-Stokes and Fourier equations, hence valid only in a small enough neighborhood of the instability threshold, it has been shown that they are a valid model for not just Rayleigh-Bénard convection but also for laser dynamics and other fields of science [7,8,10].

Let us consider a liquid layer of depth $h$ confined between two horizontal parallel flat plates; the vertical coordinate is labeled z and the two horizontal are x,y. There is translational symmetry along one horizontal direction, say the y axis and thus we need consider only one horizontal coordinate, x. In the motionless state the properties of the liquid have arbitrary translational symmetry in the x direction. At the bifurcation point $R=R_c$, this symmetry is broken; there appears a stationary cellular convective pattern with wave number q. It suffices to consider only one convective cell of 'volume' $V = \{(0 \leq x \leq 2\pi / q); (-h / 2 \leq z \leq h/2)\}$.

The total differential *excess* work of the system in $V$ is the sum of the *excess* work contributed by the viscous flow and that by the thermal conduction. Then the total *excess* work is

$$\Phi (X,Y,Z) = \int d\Phi$$

$$= \int \frac{64 \, \rho \kappa^2 \pi^5}{b^3 q^3} (X-X^*) \, dX + \frac{C_v \beta^2 h^4}{r^2 \pi q T_{00}}$$

$$\times [(Y - Y^*) \, dY + (Z-Z^*) \, dZ] \qquad (4.1)$$

The quantities $(X^*,Y^*,Z^*)$ are the values of these variables at the stationary state of the *equivalent* linear system since to the first-order terms in Saltzmann's expansion [29], the system described by the thermodynamic function $\Phi$ is indeed equivalent to that described by Lorenz equations [17,27]. $T_{00}$ is a reference value. The other quantities are given parameters of the problem and their meaning can be found in Ref. 17. Using $(X^*,Y^*,Z^*)$ from the *equivalent* linear Lorenz equations, it appears that $d\Phi$ is not an exact differential; $\Phi$ is not an state function; its integral is *path dependent*. The choice of the deterministic path as the path of the integration connecting the stationary state $(X^S, Y^S, Z^S)$ and $(X, Y, Z)$ has been shown by examples to yield a state function $\Phi$, which is an *excess* work and also yields the stationary solution of appropriate stochastic equations.

For Rayleigh-Bénard convection it appears that

(1) At all stationary states of the Lorenz model considered, (0,0,0) for $(0 < r < 1 + \delta)$, and

$$[\pm \sqrt{b(r-1)} \ , \pm \sqrt{b(r-1)} \ , (r-1)] \text{ for } (1 \leq r < 1 + \delta),$$

the first derivatives of $\Phi$ with respect to $X, Y$ and $Z$ are zero. Hence $\Phi$ is an extremum at any stationary state, stable or unstable.

(2) When $\Phi$ has minimum at the stationary state $(X^S, Y^S, Z^S)$, then due to the fact that its time derivate is negative semidefinite, the system has a tendency to approach this stationary state from an off-stationary point in the neighborhood, and thus this stationary state is stable. Otherwise when $\Phi$ has a maximum at that stationary state, the system has the tendency to go away from this stationary state, thus this stationary state is unstable. Hence $\Phi$ is a Lyapunov function of global validity, not restricted to one stationary state at a time.

(3) To move the system away from the stable stationary state to its neighborhood causes an increase in $\Phi$, and to move the system away from an unstable stationary state to its neighborhood causes a decrease in $\Phi$. Hence, if for every variation from a stationary state $\delta\Phi \leq 0$, then the stationary state is stable.

(4) The *excess* work is

$$\Phi(X, Y, Z) = \int\limits_{X,Y,Z} d\Phi_V + d\Phi_T$$

$$= \int\limits_{S.S.} \left\{ \int dV \left[ \rho(v-v^*) \cdot dv \right. \right.$$

$$\left. \left. + C_V \left( 1 - \frac{T^*}{T} \right) dT \right] \right\} \tag{4.2}$$

The first term is a difference of kinetic energies, $dE - dE^* = \rho(v-v^*) \, dv$, and the second term a difference of entropies, $-T^*(dS - dS^*) = C_V(1 - T^*/T) dT$. Hence, $\Phi$ has the dimension of energy. The path of the integral in Eq. (4.2) is from a stationary state; thus $\Phi$ is the difference in energy necessary to displace the system from an arbitrary state compared to that in a stationary state, that is, the *excess* work.

(5) The *total* dissipation in the volume of one convective cell, in terms of the rate of *entropy production* is

$$\langle \sigma \rangle = \frac{\lambda \beta^2}{T_{00}^2} \langle 1 \rangle + \frac{\lambda \beta^2}{T_{00}^2} \langle \theta v_z \rangle + \frac{R\eta \kappa^2}{T_{00} h^4} \langle \theta v_z \rangle$$

$$- \frac{\lambda \beta^2}{T_{00}^2} \left\langle \theta \frac{\partial \theta}{\partial t} \right\rangle - \frac{\rho \kappa^3}{T_{00} h^4} \left\langle v_j \frac{\partial v_j}{\partial t} \right\rangle \tag{4.3}$$

where the brackets denote integration over a cell. The first three terms have straightforward explanations: the first one is the contribution to the rate of entropy production due to the thermal conduction in the average temperature gradient; the second one is the correction to the first one due to the vertical component of the velocity (which distorts the linear gradient); and the third term is the contribution due to the buoyancy force. We denote $\langle \sigma \rangle = \langle \sigma_1 \rangle + \langle \sigma_2 \rangle$ with

$$\left\langle \sigma_1 \right\rangle = \frac{\lambda \beta^2}{T_{00}^2} \langle 1 \rangle + \frac{\lambda \beta^2}{T_{00}^2} \langle \theta v_z \rangle + \frac{R\eta \kappa^2}{T_{00} h^4} \langle \theta v_z \rangle \tag{4.4}$$

as the known part of the rate of *entropy production* and

$$\left\langle \sigma_2 \right\rangle = - \frac{\lambda \beta^2}{T_{00}^2} \left\langle \theta \frac{\partial \theta}{\partial t} \right\rangle - \frac{\rho \kappa^3}{T_{00} h^4} \left\langle v_j \frac{\partial v_j}{\partial t} \right\rangle \tag{4.5}$$

Each term in $\langle \sigma_2 \rangle$ consists of the product of two parts; the variable and the rate of change of that variable. Introducing the reference state, $\theta^*$ and $v_j^*$, which is the stationary state of the linear *equivalent* system yields

$$\left\langle \sigma_2 \right\rangle = - \frac{\lambda \beta^2}{T_{00}^2} \left\langle (\theta - \theta^* + \theta^*) \frac{\partial \theta}{\partial t} \right\rangle - \frac{\rho \kappa^3}{T_{00} h^4} \times \left\langle (v_j - v_j^* + v_j^*) \frac{\partial v_j}{\partial t} \right\rangle$$

$$= - \frac{\kappa}{T_{00} h^2} \frac{d\Phi}{dt} + \left\langle \sigma_2^* \right\rangle \tag{4.6}$$

$\langle \sigma_2^* \rangle$ is the *entropy production* at the stationary state of the linear *equivalent* system, but with the same rate of change of all the variables as at the state $(\theta, v_j)$. Hence $d\Phi/dt$ is a well-defined part of the total dissipation.

(6) The *excess* work $\Phi$ is used for assessing stability of steady states in the problem and show the correspondence to the predictions of stability from Lorenz

equations. To determine the stability of a stationary state $(X^S, Y^S, Z^S)$, it is necessary to calculate the change of $\Phi$ with respect to a small displacement from that stationary state. It has been shown that $\Phi$ is a minimum at the only stable stationary state $(X^S, Y^S, Z^S) = (0,0,0)$ when $r<1$ ($r=R/Rc$). For $r>1$, the $(0,0,0)$ is still an extremum of $\Phi$; however, the sign of $\delta\Phi$ changes, $\delta\Phi<0$, so that the stationary state is unstable; $r=1$ is the bifurcation point.

Another pair of stationary states appear when $r \geq 1$. These stationary states correspond to cellular convective states. Again the *excess* work $\Phi$ determines their stability. Now the change of $\Phi$ subject to a small and arbitrary displacement of the system from these stationary states

$$[\pm\sqrt{b(r-1)}, \pm\sqrt{b(r-1)}, (r-1)]$$

is

$$\delta\Phi = -\left(\frac{64\rho\,\kappa^2\pi^5}{b^3q^3}P(Y-Y^*)^2 + \frac{C_V\beta^2h^4}{r^2\pi q T_{00}}[(Y-Y^*)^2 + (Z-Z^*)^2]\right)\frac{\delta Y}{-YZ-Y+rY} \quad (4.7)$$

The first factor on the rhs of the equation is positive definite, thus the sign of the second part is the only item needed. It has been shown that for all possible variations from the convective state, $\delta\Phi>0$; $\Phi$ always increases for an arbitrary displacement from the stationary states and as it is a minimum at these stationary states they are stable. These conclusions about the stability of the Lorenz system obtained from the *excess* work $\Phi$ agree with the results of (linear) stability analysis of the Lorenz equations [28]. Further details about this convective instability and, in particular, the relation between theory and experiment can be found in Ref. 17.


# 5. Conclusion

I have sketched a few developments of Ross' thermodynamics for systems and (nonlinear) processes far away from equilibrium. Ross' theory for chemical kinetics and individual transport process is connected to stochastic (but not molecular) formulations: for chemical reaction and diffusion his *excess* work yields the stationary probability distributions of birth-death master equations corresponding to linear and nonlinear macroscopic equations. For flow and (linear and nonlinear) heat transport, stochastic equations of the Fokker-Planck type with stationary distributions satisfy the same connection to the respective *excess* work. In all cases these stationary probability distributions go to the appropiate limit at equilibrium. The fluctuations from stationary states may be of arbitrary magnitude and are not restricted to Gaussian fluctuations close to a stationary state.

Dissipation is essential for a thermodynamic formulation of hydrodynamic processes. Ross' $(d\Phi/dt)$ is *a component of the total dissipation* and $\Phi = \int(d\Phi/dt)dt$ is the *excess* work. Application of these concepts to the Lorenz truncation of the Navier-Stokes equations has proven useful in the descripion of (steady convective) instability. For Rayleigh-Bénard convection this approximation yields the homogeneous stationary state for reduced Rayleigh number $r<1$, and the transition to an inhomogeneous convective stationary state for $r>1$. The work of displacement of the system from an arbitrary state minus the work of the same displacement from a reference state is the *excess* work. The reference state is the stationary state in the *equivalent* linear system, defined uniquely. The differential *excess* work is related to thermodynamic functions of the arbitrary and the reference state; the reference state is not the equilibrium state unless the system approaches equilibrium ($r = 0$). With the thermodynamic concept of *excess* work, Ross provides *necessary and sufficient* criteria of stability; a Lyapunov function for relaxation to stable stationary states with physical significance parallel to the Gibbs free energy changes (maximum work at constant temperature and pressure) or entropy changes (in an isolated system) as Lyapunov functions for relaxation toward an equilibrium state; and a thermodynamic description of the bifurcation from homogeneous to inhomogeneous stationary states in the Rayleigh-Bénard problem. If the constraints on the system are those of equilibrium, then $\Phi$ is the minimum work necessary for displacement of the system from equilibrium and the maximun work available on the return from the displacement back to equilibrium; it is the extension of the availability defined at equilibrium to the nonequilibrium systems.

As the thermodynamic theory here sketched is still being developped (interfacial phenomena belongs to a territory yet to be explored) my only pretension has been to indicate how Ross' methodology is thermodynamically sound. For further refined developments, particularly concerning fluctuations far from equilibrium, the reader is advised to proceed to Refs. 19-23.

## Acknowledgment

My knowledge of thermodynamics has benefited from the many times, many hours of enjoyable discussions with John Ross, and with Xiao-Lin Chu and Marcel O. Vlad. John taught me how to appreciate the classics and comply with their heuristic yet meaningful rules. This contribution was delivered at a mid-term meeting of an EU Network held in Meudon. It was written while the author was visiting the Laboratoire de Mécanique, Université Aix-Marseille III. I gratefully acknowledge the hospitality offered by Prof. Henri Gouin as well as enlightning discussions with him and with Professors Annie Steinchen-Sanfeld and Albert Sanfeld. Research supported by European Grant Network 106 and by Fundación "Ramón Areces" (Spain).

## References

[1] Lord Rayleigh, Philos. Mag. 26, 1913, 776
[2] L. Onsager, Phys. Rev. 37, 1931, 405; ibidem 38, 1931, 2265
[3] S. R. de Groot and P. Mazur, Non-Equilibrium Thermodynamics, North-Holland, Amsterdam, 1962
[4] I. Prigogine, Etude Thermodynamique des Phénomènes irreversibles, Désoer, Liège, 1947

267

[5] P. Glansdorff and I. Prigogine, Thermodynamic Theory of Structures, Stability, and Fluctutations, Wiley, New York, 1971

[6] G. Nicolis and I. Prigogine, Self-organization in Non-equilibrium Systems, Wiley, New York, 1977

[7] see e.g. H. Haken, Synergetics (3rd Ed.), Springer-Verlag, Berlin, 1983

[8] see e.g. H. Haken, Advanced Synergetics, Springer-Verlag, Berlin, 1983

[9] see e.g. M. G. Velarde, in Stability of Thermodynamic Systems, edited by J. Casas and G. Lebon, Springer-Verlag, Berlin, 1982, pp. 248-278

[10] see e.g. M. G. Velarde, in Evolution of Order and Chaos in Physics, Chemistry, and Biology, edited by H. Haken, Springer-Verlag, Berlin, 1982, pp. 132-145

[11] M. G. Velarde and J. Wallenborn, Phys. Lett. A 26, 1968, 584

[12] G. Nicolis, J. Wallenborn and M. G. Velarde, Physica 43, 1969, 263

[13] J. Ross, K.L.C. Hunt and P. M. Hunt, J. Chem. Phys. 88, 1988, 2719

[14] P. M. Hunt, K. L. C. Hunt and J. Ross, J. Chem. Phys. 92, 1990, 2572

[15] J. Ross, K. L. C. Hunt and P. M. Hunt, J. Chem. Phys. 96, 1992, 618

[16] J. Ross, X.-L. Chu, A. Hjelmfelt and M. G. Velarde, J. Phys. Chem. 96, 1992, 11065

[17] M. G. Velarde, X.-L. Chu and J. Ross, Phys. Fluids 6, 1994, 550

[18] B. Peng, K. L. C. Hunt, P.M. Hunt, A. Suárez and J. Ross, J. Chem. Phys. 102, 1995, 4548

[19] M. O. Vlad and J. Ross, J. Chem. Phys. 100, 1994, 7268

[20] M. O. Vlad and J. Ross, J. Chem. Phys. 100, 1994, 7279

[21] M. O. Vlad and J. Ross, J. Chem. Phys. 100, 1994, 7295

[22] A. Suárez, J. Ross, B. Peng, K.L.C. Hunt and P.M. Hunt, J. Chem. Phys. 102, 1995, 4563

[23] A. Suárez and J. Ross, Thermodynamic and stochastic theory of coupled transport processes: Rayleigh scattering in a fluid in a temperature gradient, preprint

[24] H. Lamb, Hydrodynamics, Dover Ed., New York, 1945

[25] see e.g. B.A. Finalyson, The Method of Weighted Residuals and Variational Principles, Academic Press, New York, 1972

[26] J. Keizer, Satistical Thermodyamics of Nonequilibrium Processes, Springer-Verlag, New York, 1987

[27] E. N. Lorenz, J. Atmos. Sci. 20, 1963, 130

[28] C. Normand, Y. Pomeau and M. G. Velarde, Rev. Mod. Phys. 49, 1977, 581

[29] B. Saltzmann, J. Atmos. Sci. 19, 1962, 329